Jose I. Diaz MD, PhD
Dept of Pathology
EVMS

Capillary Electrophoresis of Nucleic Acids
Volume II

METHODS IN MOLECULAR BIOLOGY™

John M. Walker, SERIES EDITOR

METHODS IN MOLECULAR BIOLOGY™

Capillary Electrophoresis of Nucleic Acids

Volume II: Practical Applications of Capillary Electrophoresis

Edited by

Keith R. Mitchelson

Australian Genome Research Facility, University of Queensland, Brisbane; and Walter & Eliza Hall Institute, Parkville, Australia

and

Jing Cheng

Biochip Research and Development Center, State Key Laboratory for Biomembrane and Membrane Biotechnology, Tsinghua University, Beijing, China; and Aviva Biosciences Corporation, San Diego, CA

Humana Press ✳ Totowa, New Jersey

Dedication

To Nanne and Moon
Each is unique.

Cover design by Patricia F. Cleary.

For additional copies, pricing for bulk purchases, and/or information about other Humana titles, contact Humana at the above address or at any of the following numbers: Tel.: 973-256-1699; Fax: 973-256-8341; E-mail: humana@humanapr.com; Website: http://humanapress.com

Photocopy Authorization Policy:
Authorization to photocopy items for internal or personal use, or the internal or personal use of specific clients, is granted by Humana Press Inc., provided that the base fee of US $10.00 per copy, plus US $00.25 per page, is paid directly to the Copyright Clearance Center at 222 Rosewood Drive, Danvers, MA 01923. For those organizations that have been granted a photocopy license from the CCC, a separate system of payment has been arranged and is acceptable to Humana Press Inc. The fee code for users of the Transactional Reporting Service is [0-89603-765-7-7/01 $10.00 + $00.25].

Printed in the United States of America. 10 9 8 7 6 5 4 3 2 1

Library of Congress Cataloging in Publication Data

Main entry under title:

Methods in molecular biology™.

Capillary electrophoresis of nucleic acids/edited by Keith R. Mitchelson and Jing Cheng.
    p. cm. -- (Methods in molecular biology; 162-163)
  Includes bibliographical references and index.
  Contents: v. 1. Introduction to the capillary electrophoresis of nucleic acids-- v.2. Practical applications of capillary electrophoresis
  ISBN 0-89603-779-7 (v. 1: alk. paper) -- ISBN 0-89603-765-7 (v.2: alk. paper)
  1. Nucleic acids--Separation. 2. Capillary electrophoresis. I. Mitchelson, Keith R. II. Cheng, Jing. III. Series.
QP620.C36 2000
572.8--dc21
                                                    00-038911
                                                    CIP

Preface

The development of PCR, which enables extremely small amounts of DNA to be amplified, led to the rapid development of a multiplicity of analytical procedures to utilize this new resource for analysis of genetic variation and for the detection of disease causing mutations. The advent of capillary electrophoresis (CE), with its power to separate and analyze very small amounts of DNA, has also stimulated researchers to develop analytical procedures for the CE format. The advantages of CE in terms of speed and reproducibility of analysis are manifold. Further, the high sensitivity of detection, and the ability to increase sample throughput with parallel analysis, has led to the creation of a full range of analysis of DNA molecules, from modified DNA-adducts and single–strand oligonucleotides through to PCR-amplified DNA fragments and whole chromosomes. *Capillary Electrophoresis of Nucleic Acids* focuses on such analytical protocols, which can be used for detection and analysis of mutations and modification, from precise DNA loci through to entire genomes of organisms. Important practical considerations for CE, such as the choice of separation media, electrophoresis conditions, and the influence of buffer additives and dyes on DNA mobility, are discussed in several key chapters and within particular applications. The use of CE for the analysis of drug–DNA interactions and for examination of the metabolism of therapeutic oligonucleotides and modified nucleosides illustrate the burgeoning applications of CE for the analysis of a wide range of medical and molecular diagnostic problems. The study of interactions between nucleic acids and diverse ligands by CE is also a new and rapidly developing area. Microanalytical devices, which include capillary electrophoretic separations, have resulted from the fusion of semi-conductor devices and microdiagnostic analysis. Several seminal papers on the application of microchip-based capillary electrophoresis for genotyping and DNA sequence determination signal the revolution that is occurring in both capillary electrophoresis and laboratory analysis.

Capillary Electrophoresis of Nucleic Acids comprises two volumes divided into twelve parts, each part containing chapters that address particular general goals, or experimental approaches. Broadly, Volume I addresses instrumentation, signal detection, and the capillary environment, as well as the integration of mass spectrometry and CE for the analysis of small oligonucle-

v

otides and modified nucleotides. Volume II broadly addresses techniques for high-throughput analysis of DNA fragments of less than 1 kb, employing SNP detection, mutation detection, DNA sequencing methods, and DNA-ligand interactions.

Volume I

Volume I, Part I presents basic CE instrumentation and the theoretical background to the separation of DNA by CE. The choice of different modes of CE and detection systems, the characteristics of sieving matrices, sample preparation, and quantitation of measurement are emphasized. DNA sequencing by CE is discussed. Methods for the manufacture of microchip devices are illustrated.

Volume I, Part II presents important factors, such as choice of capillary coatings and separation media, which offer new possibilities for separations. The theoretical background for the selection and composition of sieving polymers for optimal separation is illustrated.

Volume I, Part III presents choices of separation media and the CE environment for size-based (fragment length) separations of DNAs. Emphasis is made on new co-polymer matrix materials and liquid agaroses.

Volume I, Part IV discusses the fast separations of large DNA molecules and chromosomes, using pulsed-field capillary electrophoresis formats. Methods of collection of separated CE fractions for the analysis of supercoiled DNAs are also presented.

Volume I, Part V presents practical choices for the analysis of small therapeutic oligonucleotides, nucleosides, ribonucleotides, and DNA metabolism products by CE. Practical applications for the control of quality of nucleotides and oligonucleotides are illustrated.

Volume I, Part VI presents choices for the sensitive CE analysis of nucleotides and DNA metabolism products arising from environmental and cellular damage through disease. Practical applications of mass spectrometry and capillary zone electrophoresis for the detection of oncogenic change are also presented.

Volume II

Volume II, Part I presents practical considerations necessary for very rapid and accurate separations of linear DNAs by CE using short capillaries.

Volume II, Part II presents details of the high sensitivity detection of single nucleotide polymorphism in DNA fragments by CE. Particular emphasis is given to the use of denaturing gradient environments. Techniques include SNuPE, heteroduplex analysis, ARMS analysis of DNA, and SSCP analysis.

Volume II, Part III details various modes of genotyping by CE. Particular emphasis is given to the use of multiplex separations. The techniques include parallel analysis of multiple simple repeat loci in human and CE analysis of gene markers following degenerate oligonucleotide amplification from single cells. Methods for the application of chemical cleavage of mismatched DNA for mutation detection are illustrated.

Volume II, Part IV presents techniques for the quantitative estimation of gene expression using CE. Both quantitative RT-PCR and differential display analyses are highlighted.

Volume II, Part V presents practical choices of separation media and the capillary electrophoresis environment for DNA sequencing by CE. Rapid sequencing at elevated temperature and sequencing on micro- and array-CE devices are discussed. Methods for sequencing with selective primers are also illustrated.

Volume II, Part VI presents the analysis of DNA protein and DNA ligand interactions using CE. Techniques discussed include mobility shift assays and capillar affinity-gel electrophoresis.

Keith Mitchelson
Jing Cheng

Contents

Contributors

ARAM ADOURIAN • *Whitehead Institute for Biomedical Research, Cambridge, MA*

PAULO C. ANDRÉ • *Center for Environmental Health Sciences, Massachusetts Institute of Technology, Cambridge, MA*

DONALD H. ATHA • *Biotechnology Division, National Institute of Standards and Technology, Gaithersburg, MD*

YOSHINOBU BABA • *CREST, Japan Science and Technology Corp., Department of Medicinal Chemistry, Faculty of Pharmaceutical Sciences, University of Tokushima, Tokushima, Japan*

IVAN BIROŠ • *Australian Equine Blood Typing Research Laboratory, University of Queensland, Brisbane, Australia*

SERENA BONIN • *Department of Clinical, Morphological, and Technological Sciences, University of Trieste; and International Centre for Genetic Engineering and Biotechnology, Trieste, Italy*

JOHN M. BUTLER • *Biotechnology Division, National Institute of Standards and Technology, Gaithersburg, MD*

PAOLA CARRERA • *I. R. C. C. S., H. San Raffaele, Laboratorio Biologia Molecolare Clinica, Milano, Italy*

KING C. CHAN • *Analytical Chemistry Laboratory, National Cancer Institute at Frederick, Frederick, MD*

JING CHENG • *Biochip Research and Development Center, and State Key Laboratory for Biomembrane and Membrane Biotechnology, School of Life Sciences and Engineering, Tsinghua University, Beijing, China; and Aviva Biosciences Corporation, San Diego, CA*

DAN CHISHOLM • *Whitehead Institute for Biomedical Research, Cambridge, MA*

HILARY A. COLLER • *Fred Hutchinson Cancer Research Center, Seattle, WA*

LAURA CREMORESI • *Istituto Scientifico H. S. Raffaele, DIBIT, Milano, Italy*

DAVID DAILEY • *Celera AgGen, Davis, CA*

JOHN DAVIS • *Australian Genome Research Facility, Brisbane Division, Gehrmann Laboratories, University of Queensland, Brisbane, Australia*

JOSEPH M. DEVANEY • *Transgenomic, Gaithersburg, MD*

DANIEL EHRLICH • *Whitehead Institute for Biomedical Research, Cambridge, MA*

FELICIA A. ETZKORN • *Department of Chemistry, Virginia Polytechnic Institute and State University, Blacksburg, VA*

JEROME P. FERRANCE • *Department of Chemistry, University of Virginia, Charlottesville, VA*

MAURIZIO FERRARI • *I. R. C. C. S., H. San Raffaele, Laboratorio Biologia Molecolare Clinica, Milano, Italy*

IAN FINDLAY • *Australian Genome Research Facility, Brisbane Division, Gehrmann Laboratories, University of Queensland, Brisbane, Australia*

FRANTIŠEK FORET • *Barnett Institute, Northeastern University, Boston, MA; and Institute of Analytical Chemistry, Academy of Sciences of Czech Republic, Brno, Czech Republic*

PAOLO FORTINA • *Department of Pediatrics, University of Pennsylvania School of Medicine and The Children's Hospital of Philadelphia, Abramson Pediatric Research Center, Philadelphia, PA*

GLENN J. FOULDS • *Department of Chemistry, University of Virginia, Charlottesville, VA*

CECILIA GELFI • *CNR, Institute of Advanced Biomedical Technologies, Milano, Italy*

KIMBERLY S. GEORGE • *SRA Life Sciences, Rockville, MD*

BRADEN GIORDANO • *Department of Chemistry, University of Virginia, Charlottesville, VA*

IMAD I. HAMDAN • *Department of Pharmaceutical Chemistry, University of Jordan, Amman, Jordan*

KENSHI HAYASHI • *Division of Genome Analysis, Institute of Genetic Information, Kyushu University, Fukuoka, Japan*

MASAKAZU INAZUKA • *Division of Genome Analysis, Institute of Genetic Information, Kyushu University, Fukuoka, Japan*

HALEEM J. ISSAQ • *Analytical Chemistry Laboratory, National Cancer Institute at Frederick, Frederick, MD*

STEPHEN C. JACOBSON • *Chemical and Analytical Sciences Division, Oak Ridge National Laboratory, Oak Ridge, TN*

MARTIN D. JOHNSON • *PE Corp. Applied Biosystems, Foster City, CA*

MASAO KAMAHORI • *Central Research Laboratory, Hitachi Ltd., Tokyo, Japan*

HIDEKI KAMBARA • *Central Research Laboratory, Hitachi Ltd., Tokyo, Japan*

BARRY L. KARGER • *Department of Chemistry, Barnett Institute, Northeastern University, Boston, MA*

TAKAO KASUGA • *Roche Molecular Systems, Alameda, CA*

KONSTANTIN KHRAPKO • *Beth Israel Deaconess Medical Center and Harvard Medical School, Boston, MA*

YURIKO KIBA • *CREST, Japan Science and Technology Corp., Department of Medicinal Chemistry, Faculty of Pharmaceutical Sciences, University of Tokushima, Tokushima, Japan*

ANDREA S. KIM • *Division of Bioengineering and Environmental Health, Center for Environmental Health Sciences, Massachusetts Institute of Technology, Cambridge, MA*

KAREL KLEPÁRNÍK • *Institute of Analytical Chemistry of the Czech Academy of Sciences, Brno, Cczech Republic*

LANCE KOUTNY • *Whitehead Institute for Biomedical Research, Cambridge, MA*

LARRY J. KRICKA • *Department of Pathology and Laboratory Medicine, University of Pennsylvania Medical Center, Philadelphia, PA*

JAMES P. LANDERS • *Departments of Chemistry and Pathology, University of Virginia, Charlottesville, VA*

TAO LI • *Central Research Laboratory, Hitachi Ltd., Tokyo, Japan*

PETER LINDBERG • *Department of Analytical Chemistry, Royal Institute of Technology, Stockholm, Sweden*

XIAO-CHENG LI-SUCHOLEIKI • *Division of Bioengineering and Environmental Health, Center for Environmental Health Sciences, Massachusetts Institute of Technology, Cambridge, MA*

MAURIZIO LUGLI • *Beckman Analytical s. p. a., Milano, Italy*

RAMAKRISHNA S. MADABHUSHI • *Biology and Biotechnology Research Program, Lawrence Livermore National Laboratory, Livermore, CA*

ELAINE S. MANSFIELD • *ACLARA BioSciences, Mountain View, CA*

MICHAEL A. MARINO • *Transgenomic, Gaithersburg, MD*

PAUL MATSUDAIRA • *Whitehead Institute for Biomedical Research, Cambridge, MA*

KEITH R. MITCHELSON • *Australian Genome Research Facility, University of Queensland, Brisbane; Walter and Eliza Hall Institute, Parkville, Australia*

ODILO M. MUELLER • *Agilent Technologies, Hewlett-Packard Str., Waldbronn, Germany*

KAZUNORI OKANO • *Central Research Laboratory, Hitachi Ltd., Tokyo, Japan*

CHRISTINE A. PIGGEE • *Laboratory of Neurotoxicology, National Institute of Mental Health (NIMH), National Institutes of Health (NIH), Bethesda, MD*

J. MICHAEL RAMSEY • *Chemical and Analytical Sciences Division, Oak Ridge National Laboratory, Oak Ridge, TN*

DENNIS J. REEDER • *Biotechnology Division, National Institute of Standards and Technology, Gaithersburg, MD*

JICUN REN • Department of Chemistry, Hunan Normal University, Changsha, China

PIER GIORGIO RIGHETTI • CNR, Institute of Advanced Biomedical Technologies, Milano, Italy

JOHAN ROERAADE • Department of Analytical Chemistry, Royal Institute of Technology, Stockholm, Sweden

TOMONARI SASAKI • Division of Genome Analysis, Institute of Genetic Information, Kyushu University, Fukuoka, Japan

DIETER SCHMALZING • Whitehead Institute for Biomedical Research, Cambridge, MA

GRAHAM G. SKELLERN • Department of Pharmaceutical Sciences, Strathclyde Institute for Biomedical Sciences, Glasgow, UK

GIORGIO STANTA • Department of Clinical, Morphological, and Technological Sciences, University of Trieste; and International Centre for Genetic Engineering and Biotechnology, Trieste, Italy

TOMOKO TAHIRA • Division of Genome Analysis, Institute of Genetic Information, Kyushu University, Fukuoka, Japan

WILLIAM G. THILLY • Center for Environmental Health Sciences, Massachusetts Institute of Technology, Cambridge, MA

ROGER D. WAIGH • Department of Pharmaceutical Sciences, Strathclyde Institute for Biomedical Sciences, Glasgow, UK

LARRY C. WATERS • Chemical and Analytical Sciences Division, Oak Ridge National Laboratory, Oak Ridge, TN

H. MICHAEL WENZ • PE Corp. Applied Biosystems, Foster City, CA

PETER WILDING • Department of Pathology and Laboratory Medicine, University of Pennsylvania School of Medicine, Philadelphia, PA

P. MICKEY WILLIAMS • Cell Biology and Technology, Genentech, South San Francisco, CA

STEPHEN J. WILLIAMS • ACLARA BioSciences, Mountain View, CA

ROBERT B. WILSON • Department of Pathology and Laboratory Medicine, University of Pennsylvania Medical School, Philadelphia, PA

JUN XIAN • Genome Therapeutics Corporation, Waltham, MA

XILIN ZHAO • National Institutes of Health/National Institute of Diabetes and Digestive and Kidney Diseases, Bethesda, MD

I

PRACTICAL APPLICATIONS FOR RAPID DNA FRAGMENT SIZING AND ANALYSIS

1

Development of a High-Throughput Capillary Electrophoresis Protocol for DNA Fragment Analysis

H. Michael Wenz, David Dailey, and Martin D. Johnson

1. Introduction

Since the first descriptions of electrophoresis in small diameter tubes in the 1970s and 1980s *(1,2)*, capillary electrophoresis (CE) has been recognized for its potential to replace slab-gel electrophoresis for the analysis of nucleic acids *(3,4)*. In particular, the availability of commercial instrumentation for CE over the last several years has made both the size determination and quantitation of DNA restriction fragments or polymerase chain reaction (PCR) products amenable to automation. Due to the same charge-to-mass ratio, the electrophoretic mobility of nucleic acid molecules in free solution is largely independent of their molecular size *(5)*. Therefore, a sieving medium is required for the electrophoretic analysis of DNA fragments based on their size. Typically, two different principal types of separation matrix are used. The first type of matrix is of high viscosity polymer (e.g., polyacrylamide) with a well-defined crosslinked gel in regard to the structure and size of its pores. The second type of matrix is a noncrosslinked linear polymer network of materials such as, linear polyacrylamide, agarose, cellulose, dextran, poly(ethylene oxide), with lower viscosity than the former type and with a more dynamic pore structure. Although the first type of matrix is attached covalently to the capillary wall and may provide better separation for small (sequencing) fragments, the second matrix format has the advantage of being able to be replenished after each electrophoretic cycle. This typically extends the lifetime of a capillary, prevents contamination of the system, avoids sample carryover and allows the use of temperatures well above room temperature. Most matrices used in both systems are tolerant to the addition of DNA denaturants. Many different media useful for the separation of DNA have now become commercially available *(6)*.

In summary, the application of CE for DNA related research is attractive for numerous reasons:

1. The high degree of automation avoids cumbersome gel pouring and sample loading.
2. High mass sensitivity eliminates the need to label DNA with carcinogenic stains, or with radioactive DNA precursors.

From: *Methods in Molecular Biology, Vol. 163:*
Capillary Electrophoresis of Nucleic Acids, Vol. 2: Practical Applications of Capillary Electrophoresis
Edited by: K. R. Mitchelson and J. Cheng © Humana Press Inc., Totowa, NJ

3. Very reproducible size information is achieved through the use of an internal size standard, which compensates for run-to-run variations.
4. Quantitative information is obtained after on-line detection.
5. Differences in fragment length as small as one base can be visualized by utilizing appropriate separation conditions.

1.1. Fast-Cycle CE

Typical DNA separations by CE are considered fast, ranging from 10 to 60 min. However, single capillary instruments do not achieve the same productivity as slab gels, which have longer run times, but have higher throughput owing to a multitude of simultaneously addressable lanes. Attempts have been made to substantially decrease the run times in capillaries by using very short effective lengths and high electric field strength *(7,8,9)*. These approaches considerably shorten the electrophoresis times to 3 min or less. However, none of these protocols has been implemented on a commercially available instrument.

We have developed a "fast-cycle capillary electrophoresis protocol" to address the need for high throughput and to make it amenable for commercially available instrumentation. This protocol allows the electrophoretic separation of DNA fragments up to 500 bp in length in less than 5 min with a total cycle time from one sample injection to the next of approx 7 min.

Analyses are performed on the ABI PRISM® 310 Genetic Analyzer that allows the simultaneous analysis of fragments that are tagged with different fluorophors. In order to achieve fast analysis times, several conventional electrophoresis factors are modified:

1. Both the separation polymer (2%) and electrophoresis buffer (60 mM) are at low concentration to accelerate electro-migration.
2. Electrophoresis run temperature is elevated to 60°C, with DNA molecules separated as single-strands.
3. The capillary length is shortened (effective separation length of 30 cm).

We show that typically 306 consecutive injections can be performed under these conditions without the need to change either the capillary or the electrophoresis buffer. This protocol can be used for applications that require the resolution of fragments that differ by at least 5 bp in length with a sizing precision of 0.4 bp. We present data that demonstrate the use of this protocol for the sizing and quantification of PCR fragments, the analysis of minisequencing reactions, the analysis of DNA fragments that are the product of an oligonucleotide ligation assay (OLA), and the quality control of phosphorylated short synthetic oligonucleotides.

2. Materials

2.1. Instrumentation and Electrophoresis

1. The ABI PRISM® 310 Genetic Analyzer (PE Biosystems, Foster City, CA), a laser-based CE instrument, is used for all experiments. This instrument uses a multi-line argon-ion laser, adjustable to 10 mW, which excites multiple fluorophores at 488 and 514 nm.
2. Fluorescence emission is recorded between 525 and 650 nm on a cooled CCD camera. This configuration currently allows the multiplexing and sizing of samples that overlap in

size by using three different fluorophors, plus an additional fluorophore that is attached to an internal size standard.
3. The instrument controls temperature between ambient and 60°C with an accuracy of ± 1°C.
4. Electrophoresis voltage is controlled between 100 and 15,000 V.
5. A sample tray holds 48 or 96 samples for unattended operation.
6. Data are collected and automatically analyzed, using an instrument specific collection software and GeneScan analysis software (PE Biosystems, Foster City, CA).
7. The separation medium in the capillary is automatically replaced after each sample run. Samples are introduced by electrokinetic injection, typically for 5–10 s at 7–15 kV.
8. The features that allow the use of this high throughput protocol are implemented in the PRISM 310® Collection Software, version 1.2 (*see* **Note 1**).

3. Methods
3.1. Polymer Preparation
1. GeneScan polymer (PE Biosystems, Foster City, CA) is a hydrophilic polymer that provides molecular sieving and noncovalent wall coating, when used in uncoated fused silica capillaries (PE Biosystems, Foster City, CA) (*see* **Note 2**).
2. GeneScan polymer is provided as a 7% stock solution in water that can be diluted and mixed with different additives, such as urea or glycerol. The polymer is most commonly diluted in Genetic Analyzer Buffer containing EDTA (PE Biosystems, Foster City, CA), but is also compatible with other buffers (*10*).
3. To prepare a 2% solution of GeneScan polymer, combine 14.3 mL of the polymer and 3 mL Genetic Analyzer buffer with EDTA in a 50-mL polypropylene tube, bring to 50 mL with deionized water and mix thoroughly. For the preparation of the 0.6X electrophoresis buffer, combine 3 mL of the Genetic Analyzer buffer with EDTA with 47 mL of distilled water. Both solutions are stable for at least 4 wk refrigerated at 4°C. Before use, the solutions have to be warmed up to room temperature.

3.2. Sample Preparations
3.2.1. PCR Samples
1. To evaluate the robustness of the fast protocol, five short tandem repeat (STR) markers with repeat units of 4 bp are individually amplified by PCR. Samples are labeled with 6-Fam (blue), Hex (green), and Ned (yellow). Markers are pooled in a ratio to provide comparable intensities when injected into the capillary.
2. Four μL of the pool are added to 15 μL of deionized formamide and 0.25 μL of GeneScan 500 size standard, labeled with Rox (red). Up to 16 injections are performed from each sample tube. Samples are injected for 30 s at 15 kV.
3. It is critical to dilute the oligonucleotide sample into high-quality deionized formamide for loading onto the CE instrument. To deionize formamide, mix 50 mL of formamide and 5 g of AG501 X8 mixed bed resin and stir for at least 30 min at room temperature. Check if the pH of formamide is greater than 7.0. If it is not, repeat above step. When the pH is greater than 7.0, dispense the deionized formamide into aliquots of 500 μL and store for up to 3 mo at –15 to –25°C. Usually, there is no need to purify the DNA sample before diluting it into deionized formamide. Should a signal, even with extended injection time/voltage prove to be insufficient, purifying the sample, and thereby removing salt anions that might compete with the DNA sample during electrokinetic injection, might increase the DNA signal.

3.2.2. Minisequencing Samples

1. Minisequencing reactions are generated in a single tube using 5 µL of the SNaPshot minisequencing reaction premix (PE Biosystems, Foster City, CA) along with 0.15 pmol of primer for the A, C, and T reaction and 0.75 pmol primer for the G reaction. pGEM (0.4 µg) is used as template.

2. Following primer extension, reactions are treated with 0.5 U shrimp alkaline phosphatase (SAP) (USB, Cleveland, OH) to modify the mobility of the unincorporated fluorescently labeled ddNTPs.

3. One µL of the SAP treated sample is diluted into 10 µL of deionized formamide. Samples are injected for 5 s at 15 kV.

3.2.3. Oligonucleotide Ligation Assay (OLA)

1. For the OLA reaction, a DNA sample heterozygous for locus 621+1 G/T of the *CFTR* gene is interrogated with two allele specific probes and one common probe *(11)*.

2. The allele specific oligo (ASO) detecting wild-type is 17 nt long and labeled with 6-Fam, the ASO detecting the mutation is 18 nt long and labeled with Vic (green); the common probe is 41 nt long (including a 24-nt modifier sequence).

3. OLA conditions are essentially as described in **ref. *11***, with the exception that 80 OLA cycles are used. Typically, 0.5 µL of the sample is diluted into 9 µL of deionized formamide. Samples are injected for 5 s at 15 kV.

3.2.4. Oligonucleotide Probes

1. Seven-mer oligonucleotides are synthesized in 50-nmol scale on a DNA synthesizer Model 3984 (PE Biosystems, Foster City, CA) using standard amidite chemistry. Oligonucleotides are labeled on the 3'-end with 6-Fam, followed by two random mixed base sequences. The terminal 5'-nt is chemically phosphorylated through PhosphoLink reaction (PE Biosystems, Foster City, CA). Unpurified oligonucleotides are analyzed by ion-exchange high performance liquid chromatography (HPLC) and oligonucleotides with less than 70% purity are discarded. Typically, 1 µL of a sample is diluted into 9 µL of deionized formamide, and is then injected for 5 s at 15 kV.

3.3. Protocol Optimization

1. Our goal was to develop a CE protocol that provides at least 5-bp resolution between DNA fragments as well as a fast-analysis time to detect DNA fragments in the size range between 75 and 500 bp, the typical size range for PCR products. This protocol is useful to confirm the presence or absence of an expected amplification product, provide information about the quality of the amplification, and if necessary, allow the determination of the ratio in peak height or area of adjacent DNA peaks (*see* **Note 3**).

2. We started with a protocol that was previously recommended for the analysis of dsDNA in the size range between 50 and approx 5000 bp under nondenaturing electrophoresis conditions *(12)*. This protocol uses a hydrophilic polymer (GeneScan polymer) of low viscosity, that accomplishes both the separation of DNA fragments and the dynamic coating of the capillary walls when used together with uncoated fused silica glass capillaries (*see* **Table 1**, #1). To monitor the effect of the described changes from the initial protocol, we injected a DNA ladder (GeneScan 500-size standard) into the capillary. This ladder consists of DNA fragments ranging in size from 50 to 500 bp; one of the strands is labeled with the fluorophore Tamra (red). We determined the electrophoresis time for the 100-, 300-, and 500-bp fragments and calculated the resolution in the 150- and 500-bp

Table 1
Calculation of Electrophoresis Time and Resolution Relative
to Changes in Electrophoresis Conditions[a]

#	Conditions	Et (min) 100 bp	Et (min) 300 bp	Et (min) 500 bp	160 bp Rs 1/5 bp	500 bp Rs 1/5 bp
1	L=47 cm E=277 V/cm	8.30	9.45	10.49	0.33/1.65	0.20/1.00
2	L=47 cm E=319 V/cm	6.93	7.89	8.71	0.27/1.35	0.18/0.89
3	L=41 cm E=366 V/cm	4.85	6.02	6.49	0.24/1.20	0.13/0.66
4	Temperature 30°C	4.81	5.44	5.98	0.31/1.54	0.16/0.80
5	35°C	4.50	5.12	5.62	0.31/1.54	0.16/0.79
6	40°C	4.25	4.83	5.31	0.32/1.59	0.16/0.78
7	45°C	4.04	4.58	5.05	0.35/1.74	0.14/0.71
8	50°C	3.87	4.39	4.84	0.23/1.14	0.14/0.71
9	55°C	3.77	4.26	4.68	NA	NA
10	60°C	3.64	4.11	4.52	NA	NA
11	Temperature 45°C	4.46	5.72	6.73	0.46/2.28	0.10/0.5
12	50°C	4.29	5.51	6.44	0.46/2.28	0.09/0.45
13	55°C	4.14	5.33	6.19	0.46/2.28	0.08/0.39
14	60°C	4.00	4.64	5.95	0.44/2.21	0.07/0.33
15	GSP 3%	3.98	5.10	5.85	0.49/2.43	0.10/0.48
16	2.5%	3.75	4.67	5.28	0.40/1.99	0.10/0.51
17	2.0%	3.43	4.08	4.51	0.29/1.47	0.09/0.45
18	Buffer 1X	3.73	4.44	4.89	0.29/1.45	0.09/0.47
19	0.8X	3.69	4.42	4.88	0.33/1.64	0.11/0.55
20	0.6X	3.63	4.36	4.82	0.36/1.80	0.15/0.77
21	0.4X	3.47	4.20	4.63	0.35/1.77	0.12/0.60

[a]Electrophoresis times (Et) for the 100-, 300-, and 500-bp fragments of the GeneScan 500-size ladder as they relate to the applied condition are listed. Resolution (Rs) in the 160- and 500-bp size ranges were calculated. Values both for single-base and five-base resolution are displayed. As peak widths approach the peak interval, individual adjacent peaks become more difficult to distinguish. A resolution value of 0.5 represents the resolution limit, where peaks share significant area, but can still be discriminated; at values below 0.5 adjacent peaks have merged and cannot be further discriminated. Each table entry represents the average of four injections. Conditions in addition to the listed are for: #1–3: 3% GSP in 1X buffer at 30°C, sample buffer: water; #4–10: 3% GSP in 1X buffer, L = 41 cm, E = 366 V/cm, sample buffer: water; #11–14: 3% GSP in 1X buffer, L = 41 cm, E = 366 V/cm, sample buffer: distilled formamide; #15–17: 1X buffer, L = 41 cm, E = 366 V/cm at 60°C, sample buffer: distilled formamide; #18–21: in 2% GSP, L = 41 cm, E = 366 V/cm at 60°C, sample buffer: distilled formamide.

size range. We report values both for single nucleotide and 5-nt resolution, assuming that a value of 0.5 represents the limit of resolution between adjacent peaks (*see* **Note 4**).

3. We initially raised the electric field strength (E) to the maximum supported by this instrument, 15 kV (**Table 1**, #2). In the second set of experiments, we cut the capillary to the shortest length possible for use on this instrument, 41 cm, which represents an effective length *(l)* of 30 cm (**Table 1**, #3). For these experiments, the samples were diluted in distilled water before electrokinetic injection into the capillary. With these changes in the running conditions, the electrophoresis time decreased for the 100-bp fragment from 8.3 to 4.85 min and for the 500-bp fragment from 10.49 to 6.49 min.

4. Next, we examined the effect of raising the electrophoresis temperature from 30°C, in increments of 5, to 60°C (**Table 1**, #4–10). The effects on overall electrophoresis time are probably due to a decrease in polymer viscosity which reduces the capillary fill time and increases the speed by which DNA fragments migrate through the polymer mesh. The electrophoresis time for the 100-bp fragment decreased from 4.81 to 3.64 min and for the 500-bp fragment from 5.98 to 4.52 min.

5. Peaks became increasingly broad at elevated run temperatures (above 50°C), which could have been caused by a partial denaturation of the dsDNA injected from water. Therefore, we resuspended the DNA in deionized formamide and repeated the experiment (**Table 1**, #11–14). Between 30 and 40°C the peak pattern was not discernible. We speculate that the reason for this observation is that the denatured single-stranded fragments partially reannealed or formed single-stranded conformations as they entered the neutral polymer, causing them to migrate independent from their respective size *(13)*.

6. At temperatures at and above 45°C, the expected peak pattern is observed. It should be noted that the 250-bp and 340-bp fragments did not completely run according to their size. This is similar to what has been previously noted under highly denaturing conditions *(14)*. The total electrophoresis time for same sized DNA fragments is greater for DNA injected out of formamide than out of water, indicating the single-stranded nature of the molecules in formamide, compared to DNA in water where it is thought to be (usually) double-stranded.

7. Next, as the polymer concentration is reduced from 3% to 2% (**Table 1**, #15–17), the electrophoresis time decreases from 3.98 to 3.43 min for the 100-bp fragment, and from 5.85 to 4.51 min for the 500-bp fragment.

8. The final set of experiments examines the effect of reducing the ionic strength of the electrophoresis buffer from 1X to 0.4X (**Table 1**, #18–21). Although the effect of buffer dilution on electrophoresis times is modest, the resolution of DNA fragments in the selected size ranges increased significantly between 1X and 0.6X buffer concentration.

3.4. Separation of Simple Repeat Alleles

1. In order to visualize separation between closely spaced, rapidly migrating DNA fragments, we had to change the frequency and time that the system-specific software samples the fluorescence emission of peaks passing the detector. Using the default peak integration time of 200 ms, microsatellite markers were minimally resolved. Increasing the sampling frequency by decreasing the integration time to 65 ms yielded significant improvements in resolution (**Fig. 1**).

2. The final conditions for the fast electrophoresis protocol consisted of a capillary of 41 cm total length, an electrophoresis voltage of 15 kV and 2% GeneScan polymer in 0.6X electrophoresis buffer and 60°C electrophoresis temperature. The samples were injected out of formamide (**Table 1**, #20). **Figure 2** shows the electropherogram for the GeneScan

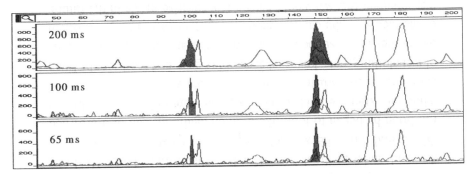

Fig. 1. Effect of change in sampling rate on resolution of DNA fragments. The CCD integration time in the system specific firmware is changed from the default value of 200 ms (*top*) to 100 ms (*middle*) to 65 ms (*bottom*). The changes in resolution for peaks of two microsatellite markers are shown. The highlighted peaks indicate the resolution that the GeneScan software can recognize with the provided data points.

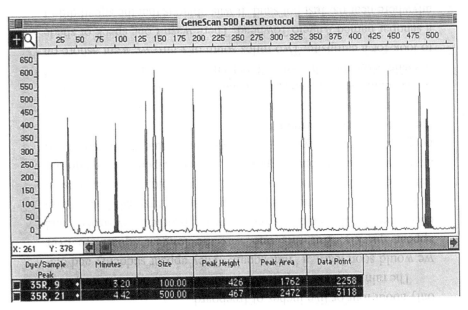

Fig. 2. Electropherogram of the GeneScan 500 size ladder run with the fast electrophoresis protocol. The size ladder is injected out of formamide and electrophoresed under the conditions described in the text (*see also* **Table 1**, #20).

500-size ladder run under these conditions. The 490- and 500-bp fragments are base line resolved, indicating that under these conditions fragments differing in size by 5 bp can be resolved. The electrophoresis time for the 500-bp fragment is 4.42 min. The run time is then 4.42 min plus an additional 150 s that are needed for filling the capillary with fresh polymer, injecting a sample into the capillary and other instrument related functions. This

A   **B**

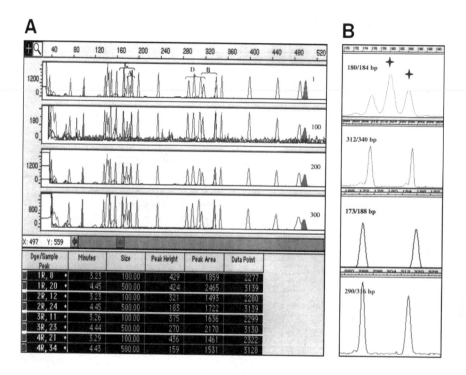

Fig. 3. Reproducibility of the fast electrophoresis protocol. A sample containing GeneScan 500-size standard and five different microsatellite markers were injected a total of 306 times into the same capillary. The microsatellite markers and size standard can be discriminated by the color. Electropherograms for injections 1, 100, 200, and 300 are shown: (**A**), The 100- and 500-bp peaks of the size standard are highlighted and their respective electrophoresis time displayed in the table. In (**B**), the four microsatellites used for the determination of sizing and quantitation precision are shown magnified. Their respective position within the sample is indicated in (**A**).

results in a total cycle time from one sample injection to the next of approx 7 min for the detection for fragments up to 500 bp in length. With this protocol, a full autosampler tray of 96 samples can be analyzed in less than 12 h.

3.5. PCR Product Analysis

1. To evaluate the above protocol for consistency of performance, a sample is injected repeatedly into the same capillary. The sample contains five different overlapping microsatellite markers, labeled with three different fluorophores. On each of three consecutive days, the same sample is therefore injected 102 times. Using the ABI PRISM 310®, the syringe pump needs to be refilled with polymer after each set of experiments, however, the reservoir of electrophoresis buffer was sufficient for use throughout the 306 injections.
2. **Figure 3A** shows the electropherograms for injection number: 1, 100, 200, and 300 of this sample. The listed electrophoresis times for the 100- and 500-bp fragments indicate

the high reproducibility in mobility achieved with this protocol and convey no obvious change in peak appearance over repeated use of the capillary.

3. We determined the sizing precision for four of the microsatellite markers (**Fig. 3B**) for all 306 injections. The sizing for all eight fragments across the three sets is very reproducible (**Table 2A**). The highest standard deviation encountered (for the 180-bp fragment) was 0.49 bp. This allows the accurate sizing of DNA fragments that differ in size by 4 bp. The DNA fragment sizing could be reproduced with 99.7% precision.

4. As a further measure of reproducibility using these fast run conditions, we determined the ratio of both peak height and peak area for two adjacent peaks (**Table 2B**). The ratios between peak area or peak height within one set and across the three sets consisting of 306 injections are comparable and very reproducible with standard deviations ranging from 0.01 to 0.10.

3.6. Minisequencing Product Analysis

1. Single nucleotide polymorphisms (SNPs) are used both as direct measures of mutations (e.g., sickle cell anemia) and as genetic markers in linkage analysis studies. A variety of techniques are currently employed to examine specific nucleotide compositions at defined positions in DNA *(15)*. One of these techniques, minisequencing *(16)*, or single nucleotide extension, employs template directed primer extension by a single fluorescently labeled nucleotide to interrogate individual loci.

2. The peaks highlighted in **Fig. 4** represent a set of four single nucleotide extension products. The pGEM plasmid is used as a template for extension and is interrogated by four different primers that should end in a fluorescently labeled A, G, C, or T respectively, after single nucleotide extension. The extension products are 30 nt (A reaction), 35 nt (T reaction), 39 nt (G reaction), and 46 nt (C reaction) in length. The two blue doublets flanking the highlighted peaks represent dichlororhodamine R110 labeled custom synthesized oligonucleotides that are 15, 19, 65, and 70 nt in length respectively, and are used as sizing standards for this run. Detection of the extension products could be accomplished within an electrophoresis time of less then 3 min. This protocol allows the rapid assessment of the quality and fidelity of an extension reaction during reaction optimization experiments, and should also be useful for the rapid typing of SNPs.

3.7. OLA Product Analysis

The OLA is used to detect DNA polymorphisms (SNPs) with high specificity. We use the fast electrophoresis protocol to quickly evaluate the fidelity of an OLA reaction.

1. **Figure 5** shows the individual detection of sample 621+1 G/T of the *CFTR* gene *(11)* for homozygous wild-type (top panel), mutation (middle panel), and heterozygous alleles (bottom panel). The allele specific probes are designed to detect the wild-type and mutation, and are labeled with two different fluorophors for better discrimination. The total lengths for the ligation products are 58 nt for the wild-type and 59 nt for the mutant ligation product.

2. The size difference of 1 nt between wild-type and mutation can be resolved (**Fig. 5**, bottom), probably aided by fluorophore-induced mobility differences. Electrophoresis times for both fragments are approx 3 min. This protocol allows for a fast assessment of performance of newly designed probes in an OLA reaction and for the determination of equal peak sizes in a multiplex OLA experiment *(18)*.

Table 2
Sizing and Quantitation Reproducibility for 306 Consecutive Injections with the Fast Electrophoresis Protocol[a]

A

		Fam 180 bp	Fam 184 bp	Fam 312 bp	Fam 340 bp	Hex 290 bp	Hex 316 bp	Ned 173 bp	Ned 188 bp
	Size								
Injections 1–102	Average	180.06	184.13	311.60	339.18	290.27	316.87	173.02	188.06
	SD	0.49	0.28	0.31	0.25	0.30	0.31	0.26	0.23
Injections 103–204	Average	180.20	184.19	311.67	339.24	290.17	316.89	173.19	188.03
	SD	0.30	0.29	0.37	0.30	0.28	0.42	0.32	0.27
Injections 205–306	Average	180.18	184.12	311.71	339.24	290.23	316.80	173.19	188.04
	SD	0.23	0.20	0.29	0.25	0.26	0.23	0.22	0.23

B

		Fam 180 bp/184 bp		Fam 312 bp/340 bp		Hex 290 bp/316 bp		Ned 173 bp/188 bp	
	Peak ratio	Height	Area	Height	Area	Height	Area	Height	Area
Injections 1–102	Average	1.71	1.80	1.02	0.99	1.23	1.21	1.11	1.09
	SD	0.10	0.09	0.04	0.04	0.07	0.07	0.05	0.03
Injections 103–204	Average	1.72	1.75	1.05	0.98	1.26	1.20	1.13	1.09
	SD	0.07	0.10	0.05	0.03	0.04	0.05	0.04	0.02
Injections 205–306	Average	1.73	1.79	1.01	0.99	1.24	1.18	1.14	1.09
	SD	0.06	0.05	0.02	0.02	0.03	0.05	0.04	0.01

[a]A DNA sample containing five microsatellite markers is injected 102 times each on three consecutive days. For the indicated markers (see **Fig. 3**) the average size (**A**) and peak ratio (**B**) is determined. Three out of the 306 injections did not provide sizing information for the 180/184-bp fragments due to insufficient resolution. Eleven injections could not be used for the quantification of the 170/195-bp fragments, because the peak size was below the threshold of 75 fluorescent units; overall 3 out of 306 injections could not be used due to insufficient peak resolution.

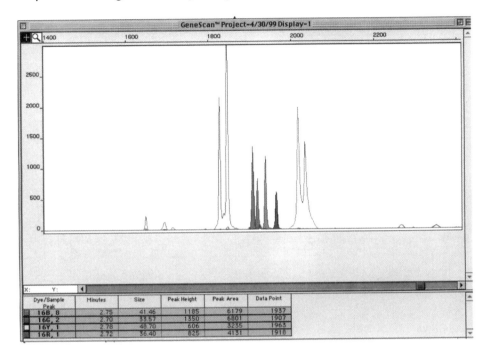

Dye/Sample Peak	Minutes	Size	Peak Height	Peak Area	Data Point
16B, 8	2.75	41.46	1185	6179	1937
16G, 2	2.70	33.57	1350	6801	1907
16Y, 1	2.78	48.70	606	3235	1963
16R, 1	2.72	36.40	825	4131	1918

Fig. 4. Separation of minisequencing reactions with the fast electrophoresis protocol. Single nucleotide extension products ending in A (green), T (red), G (blue), or C (black) are shown. Separations are performed as described in the text.

3.8. Assessment of Oligonucleotide Probe Quality

During ligation assays, the downstream probe requires a phosphate group attached to the 5'-end for the enzymatic ligation to an upstream probe to occur. The 5'-end phosphorylation can be accomplished either by polynucleotide kinase, or chemically by using a phospholink during the probe synthesis.

1. We used the fast electrophoresis protocol to routinely assess the completeness of phosphorylation of probes of 7 nt in length. The phosphate group provides additional charge to similar sized probes, which results in a higher mobility to the phosphorylated molecule, allowing the discrimination between phosphorylated and nonphosphorylated oligonucleotides (**Fig. 6**).
2. To prove the usefulness of this protocol, probes are synthesized that were not phosphorylated (**Fig. 6**, top), phosphorylated (**Fig. 6**, middle), or a mix of both (**Fig. 6**, bottom). In order to compensate for run-to-run variations that might interfere with the interpretation of results (**Fig. 6A**), we included an internal size standard consisting of a 6-Fam or Tamra labeled dinucleotide (AT) for normalization between runs (not shown). Although this fast electrophoresis protocol does not provide single nucleotide resolution, it can be used to assess the overall quality of an oligonucleotide synthesis (*see* **Fig. 6A** vs **B**). The electrophoresis times for short oligonucleotide probes are less than 3 min.

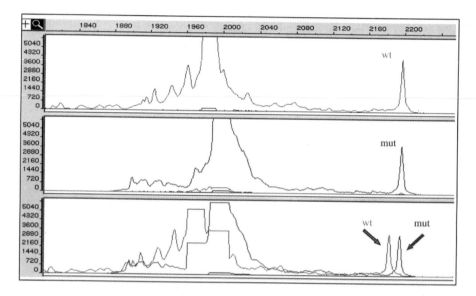

Fig. 5. Analysis of products of the OLA with the fast electrophoresis protocol. The 621+1 G/T locus of the *CFTR* gene is interrogated with two allele specific oligos designed to detect the wild-type (labeled with 6-Fam, blue) and mutant (labeled with Vic, green) genotype. Ligation assays are performed to detect only the wild type (*top*), mutant (*middle*), or heterozygous (*bottom*) genotype.

3.9. The Fast Electrophoresis Protocol

We have developed a fast electrophoresis protocol for a commercially available CE system that allows the analysis of DNA fragments ranging in length from short oligonucleotide probes to PCR products of up to 500 bp in length.

1. The cycle time for each sample is at maximum 7 min. This protocol is robust in our hands, allowing the analysis of 300 samples with a single capillary.
2. Simply diluting a commercially available polymer solution can easily produce the separation polymer.
3. This protocol is routinely used to assess the quality of PCR products during the development of assays for multiplexed genotyping assays, or gene expression profiling prior to the actual high precision analyses that require up to 30 min run times (*14*).
4. Moreover, this protocol has been invaluable to quickly optimize and fine tune OLA and minisequencing reactions during the development of new chemistries and products.
5. The effectiveness of the protocol is evident in its application for rapid analysis of more than 1000 oligonucleotides as the Quality Control (QC) assay for the completion of the chemical phosphorylation step during probe synthesis operations.

4. Notes

1. Software features that allow this fast-cycle protocol to run on the ABI PRISM 310® are implemented in the PRISM 310® Collection Software, version 1.2. This software version

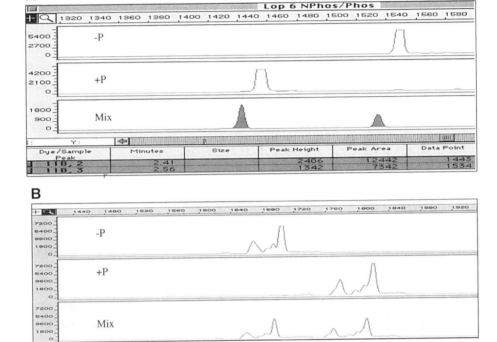

Fig. 6. Discrimination of phosphorylated from non-phosphorylated oligonucleotides with the fast electrophoresis protocol. Seven-mer oligonucleotides are synthesized with a 6-Fam label on the 3-end either not phosphorylated (*top*) or phosphorylated (*middle*) during the synthesis. The *bottom* panel represents a mixture of both. The samples in: (**A**) represent a typical "good" synthesis, whereas (**B**) represents a "bad" synthesis.

can be down loaded from the PE Biosystems web site. The protocol is described in detail in the ABI PRISM 310® Genetic Analyzer User Bulletin "A Fast Native Protocol for the Analysis of PCR Products."

2. The hydrophilic and low-viscous GeneScan polymer (GSP) is versatile. It has also been used after mixing with glycerol for the analysis of single-strand conformation polymorphisms (SSCP) (*10,20,21*). It is also the basis for the polymer preparations POP-4 and POP-6 that are used for high-precision genotyping (*14*) and for DNA-sequencing operations.

3. The fast-cycle capillary electrophoresis protocol constitutes a tool that allows for a rapid analysis of DNA fragments as they are encountered often during routine DNA-based assays. It should be noted that although this protocol is valuable for the sizing and quantification of DNA fragments that differ in size by at least 5 bp, other protocols are recommended if higher resolution and precision is required (*14*).

4. The sizing of unknown DNA fragments on the fast-cycle platform is achieved by utilizing an internal size standard that is labeled with a different fluorophore than the fluorophore label(s) used for the fragment(s) in the sample. After assigning the appropriate size values to the size standard peaks, the GeneScan Analysis software uses a sizing algorithm

(usually the Local Southern algorithm) to automatically size the unknown sample. It is good practice, once all samples are sized, to overlay the electropherograms of the internal size standards for each run. Anomalies that may occur during a run are indicated by a misalignment of the fragments of the internal size standard of this run, relative to the mobility of the internal size standard seen in other runs with the same system. The anomalous sample then has to be run again. Alternatively, the sizes for the internal standard have to be reassigned, and the sizing for this run has to be repeated.

Acknowledgments

We would like to thank D. Sherman for providing us with the OLA probes and help with the OLA reactions. B. Williams helped to test the protocol for the separation of STR loci. K. Wang tested the protocol during QC of oligonucleotide probes. D. Hershey and B. Johnson helped with the calculation of fragment resolution. Helpful suggestions and critical reading of the manuscript by S. Baumhueter, A. Karger, and M. Hane are also acknowledged.

References

1. Mikkers, F. E. P., Everaerts, F. M., and Verheggen, T. P. E. M. (1979) High performance zone electrophoresis. *J. Chromatogr.* **169**, 11–20.
2. Jorgenson, J. W. and Lukacs, K. D. (1981) Zone electrophoresis in open tubular glass capillaries. *Anal. Chem.* **53**, 1298–1302.
3. Landers, P. J., Oda, R. P., Spelsberg, T. C., Nolan, J. A., and Ulfelder, K. J. (1993) Capillary Electrophoresis: A powerful microanalytical technique for biologically active molecules. *BioTechniques* **14**, 98–111.
4. Karger, B. L., Chu, Y. H., and Foret, F. (1995) Capillary electrophoresis of proteins and nucleic acids. *Annu. Rev. Biophys. Biomol. Struct.* **24**, 579–610.
5. Olivera, B. M., Baine, P., and Davidson, N. (1964) Electrophoresis of nucleic acids. *Biopolymers* **2**, 245–257.
6. Wehr, T., Zhu, M., and Mao, D. T. (2001) Sieving matrix selection, in *Capillary Electrophoresis of Nucleic Acids*, Vol. 1 (Mitchelson, K. R., and Cheng, J., eds.), Humana Press, Totowa, NJ, pp. 167–187.
7. Mueller, O., Minarik, M., and Foret, F. (1998) Ultrafast DNA analysis by capillary electrophoresis. *Electrophoresis* **19**, 1436–1441.
8. Muth, J., Williams, P. M., Williams, S. J., Brown, M. D., Wallace, D. C., and Karger, B. L. (1996) Fast capillary electrophoresis-laser induced fluorescence analysis of ligase chain reaction products: Human mitochondrial DNA point mutations causing Leber's hereditary optic neuropathy. *Electrophoresis* **17**, 1875–1883.
9. Kléparník, K., Malà, Z., Havàč, Z., Blazkova, M., Holla, L., and Boček, P. (1998) Fast detection of a $(CA)_{18}$ microsatellite repeat in the IgE receptor gene by capillary electrophoresis with laser-induced fluorescence detection. *Electrophoresis* **19**, 249–255.
10. Atha D. A., Wenz H.-M., Morehead H., Tian J., and O'Connell C. (1998) Detection of p53 point mutations by single strand conformation polymorphism (SSCP): analysis by capillary electrophoresis. *Electrophoresis* **19**, 172–179.
11. Brinson, E. C., Adriano, T., Bloch, W., Brown, C. L., Chang, C. C., Chen, J., et al. (1997) Introduction to PCR/OLA/SCS, a multiplex DNA test, and its application to cystic fibrosis. *Genetic Testing* **1**, 61–68.
12. GeneScan Reference Guide, ABI PRISM 310® Genetic Analyzer (1997) http://www.pebio.com/ga/pdf/310GSRG.pdf

13. Orita M., Iwahana H., Kanazawa H., Hayashi K., and Sekiya, T. (1989) Detection of polymorphisms of human DNA by gel electrophoresis as a single-strand conformation polymorphism. *Proc. Natl. Acad. Sci. USA* **86**, 2766–2770.

14. Wenz, M.-H., Robertson, J. R., Menchen, S., Oaks, F., Demorest, D. M., Scheibler, D., et al. (1998) High Precision Genotyping by Denaturing Capillary Electrophoresis. *Genome Res.* **8**, 69–80.

15. Landegren, U., Nilsson, M., and Kwok, P. Y. (1998) Reading bits of genetic information: methods for single-nucleotide polymorphism analysis. *Genome Res.* **8**, 769–776.

16. Syvänen, A.-C., Aalto-Setala, K., Harju, L., Kontula, K., and Soderlund, H. (1990) A primer-guided nucleotide incorporation assay in the genotyping of apolipoprotein E. *Genomics* **8**, 684–692.

17. Landegren, U., Kaiser, R., Sanders, J., and Hood, L. (1988) A ligase-mediated gene detection technique. *Science* **241**, 1077–1080.

18. Grossman, P. D. Bloch, W., Brinson, E., Chang, C. C., Eggerding, F. A., Fung, S., et al. (1994) High-density multiplex detection of nucleic acid sequences: oligonucleotide assay and sequence-coded separation. *Nucleic Acids Res.* **22**, 4527–4534.

19. Day, D. J. Speiser, P. W., White, P. C., and Barany, F. (1995) Detection of steroid 21-hydroxylase alleles using gene-specific PCR and a multiplexed ligation detection reaction. *Genomics* **29**, 152–162.

20. Hayashi, K., Wenz, H.-M., Inazuka, M., Tahira, T., Sasaki, T., and Atha, D. H. (2001) SSCP analysis of point mutations by multicolor capillary electrophoresis, in *Capillary Electrophoresis of Nucleic Acids*, Vol. 2 (Mitchelson, K. R., and Cheng, J., eds.), Humana Press, Totowa, NJ, pp. 109–126.

21. Ren, J. (2001) SSCP analysis by capillary electrophoresis with laser-induced fluorescence detector, in *Capillary Electrophoresis of Nucleic Acids*, Vol. 2 (Mitchelson, K. R., and Cheng, J., eds.), Humana Press, Totowa, NJ, pp. 127–134.

2

Ultra-Fast DNA Separations Using Capillary Electrophoresis

Karel Klepárník, Odilo M. Mueller, and František Foret

1. Introduction

For decades electrophoresis, together with chromatography and centrifugation, has been one of the most important tools in biochemistry and biology. Size selective separations using synthetic polyacrylamide sieving gels (SDS-PAGE) *(1)*, development of the two dimensional electrophoresis (2D-PAGE) *(2)* together with the use of the polyacrylamide gels for DNA sequencing *(3,4)* revolutionized the field of bio-separations *(5–7)*. At present, slab-gel electrophoresis is the most frequently used technique for analysis of proteins and DNA. At the same time, it is also laborious and relatively slow since only low electric field strength can be applied without excessive Joule heating. Numerous efforts to increase the separation speed have been taken, typically applying very thin gel slabs, allowing higher electric field strength during the separation.

Alternative approach for fast electrophoresis utilized a narrow capillary as the separation column. This technique was pioneered by Hjertén *(8)* and Virtanen *(9)* and refined by Mikkers *(10)* and Jorgenson *(11)*. Although capillary electrophoresis (CE) first emerged as a free solution technique, sieving media for size selective separations have been developed soon after *(12)*. It is worth mentioning that the first papers on electrophoresis in gel-filled capillaries had been published in the early 1960s when the term "capillary gel electrophoresis" had also been used for the first time *(13–14)*. Although the potential of capillary gel electrophoresis has clearly been demonstrated in these early works and even instrumentation for multiple CE was developed, the lack of a suitable detector prevented any practical success and the works have been forgotten for the next two decades.

At present, CE is a mature technique routinely applied for inorganic, organic, environmental, pharmaceutical, and biological analyses. The potential for very high separation speed and multiplexing, together with sensitive laser-induced fluorescence (LIF) detection make the technique ideal for a number of applications in DNA analysis. The

From: *Methods in Molecular Biology, Vol. 163:*
Capillary Electrophoresis of Nucleic Acids, Vol. 2: Practical Applications of Capillary Electrophoresis
Edited by: K. R. Mitchelson and J. Cheng © Humana Press Inc., Totowa, NJ

development of replaceable sieving matrices *(12)* allows reusing the separation column for hundreds of consecutive runs. This enables the use of completely automated systems that were previously impossible with the slab gels. The separation of DNA fragments by capillary array electrophoresis (CAE) is currently having an increasing impact on the speedy completion of the Human Genome Project.

Most CE separations are performed using columns of 25–50 cm long with typical analysis time on the order of tens of minutes. However, in many cases, the analysis time can be substantially shortened simply be decreasing the separation distance, increasing the electric field strength, or both. The reduction of the separation time, resulting in increased analysis throughput, is of great importance. Much shorter analysis times can be achieved easily with miniaturized instrumentation, using either standard capillary columns, or micro fabricated chips. This chapter aims at reviewing the principles and limitations of CE for ultra-fast DNA separations using replaceable sieving matrices.

1.2. Practical Considerations for Fast DNA Separations

A series of experimental factors have to be considered for fast DNA separations, where the goal is to achieve resolution of two consecutive zones in a minimum amount of time. From the definition of migration time t (**Eq. 1**), it is evident that effective capillary length and electrical field strength are two key parameters which influence the speed of a separation.

$$t = \frac{l_{eff}}{\mu \cdot E} \tag{1}$$

The symbols in equation 1 are: l_{eff} = effective capillary length, E = electric field strength, and μ = electrophoretic mobility of the analyte.

Decreasing the separation distance (effective capillary length) and increasing the applied electrical field strength will result in a decrease in the separation time. Clearly, there are fundamental limitations as to how fast a DNA separation can be. First, the production of Joule heat is the restrictive factor that limits the electric field strength. For any given applied electrical field strength, the smaller the dimensions of the separation system, the lower the electric current generated. This dictates the application to systems with a minimal cross-sectional area of the separation channel.

Another factor limiting the speed of analysis is the zone dispersion during the separation. For practical description, it is useful to relate the zone width (in the form of a variance of the zone concentration distribution σ^2) to the length of the separation column l^2 as the separation efficiency, N. The separation efficiency scales with the capillary length as:

$$N = \frac{l^2_{eff}}{\sigma^2_{tot}} = \frac{l^2_{eff}}{\sigma^2_{ext} + \sigma^2_{diff} + \sigma^2_{therm} + \sigma^2_{other}} \tag{2}$$

where σ^2_{tot} is the total peak variance, σ^2_{ext} is the extra column dispersion, σ^2_{diff} is the diffusion dispersion, σ^2_{therm} is the thermal dispersion, and σ^2_{other} is dispersion caused by other factors. Clearly, if short separation distances are to be used, each source of dispersion has to be minimized to achieve high separation efficiencies.

To characterize the practical consequence of zone dispersion, one can follow electrophoretic separation of two DNA species with the zone resolution, R_s, defined in terms of selectivity, α, and separation efficiency, N:

$$R_s = \frac{1}{4} \cdot \frac{\Delta\mu}{\bar\mu} \cdot \sqrt{N} = \frac{1}{4} \cdot \alpha \cdot \sqrt{N} \tag{3}$$

in which $\Delta\mu = \mu_2 - \mu_1$ is the difference in the electrophoretic mobilities of the separated DNA species, and μ is the average mobility of the two species. In **Eq. 2**, the selectivity term is independent from the capillary length. However, because of the dynamic nature of DNA molecules (changes in DNA conformation), the selectivity is a function of the applied electric field strength *(15)*. In general, selectivity can decrease significantly by increasing the applied electric field strength to more than few hundred V/cm.

1.2.1. Extra-Column Dispersion

The extra-column effects due to the original size of the injected sample, σ^2_{inj}, and the finite size of the detection spot, σ^2_{det}, are independent of the electric field strength and the separation time. In miniaturized systems, where the contribution of time dependent dispersions is minimized by the short analysis time, the extra-column dispersion effects can be dominant *(16)*. The contribution of LIF to the extra-column dispersion can also be minimized by use of a detection spot size of 10–50 μm. On the other hand, sample injection varies greatly and can have a significant impact on the resolution. It is, therefore, important to inject only a very narrow injection plug, if high-speed DNA separations are to be achieved with sufficient resolution. Since the DNA sample is typically introduced by electromigration, a narrow injection plug can be generated by use of a short injection time. Injecting from a desalted solution can further help to sharpen the injected band by sample stacking *(17)*. In microfabricated channels, where the starting zone dimensions are given by the shape of the injection loop, a narrow injection plug can be achieved by a proper microdevice design *(18)*.

The effect of injection length and size of the detection window on the minimum migration path X needed for the total separation of two zones is schematically shown in **Fig. 1**. Here, the separation of a faster moving component 2 from a slower moving species 1 is depicted in the distance-time scheme. If all other sources of dispersion are neglected, the total width of a separating zone is given by the length of a sample plug injected into the separation capillary. It would seem that both components injected as a zone of a length L_{S1}, are completely separated at a separation distance A. However, the distance-time record of a detector response (shaded areas) at this region shows that the zones are not resolved completely in time dimension. It is evident that the length of a detection window, L_D, contributes to the total width of the zone detected at a given position in time. Therefore, a longer separation path X_B is needed for the total time-separation of both zones. This situation is depicted in region B. Here, the rear boundary of the faster migrating zone 2 is leaving, and the front boundary of slower migrating zone 1 is entering the detection window at the same time. If the lengths of a detection window and an injection plug are reduced to lower values of L_{D2} and L_{S2} (region C), the minimum separation path X_C and the migration time t_C are reduced as well. The

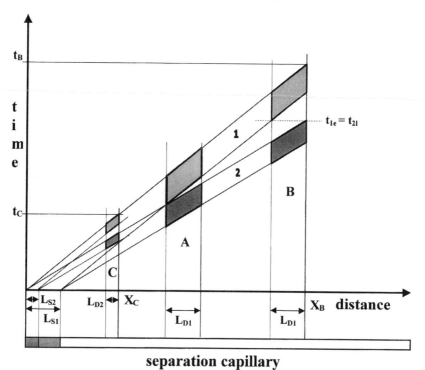

separation capillary

Fig. 1. Effect of lengths of injection zone and detection window on the minimum migration path, space-time scheme of the separation. *li, ld* injection zone and detection window length, respectively.

minimum separation distance can be evaluated using the above mentioned equality between the times t_{2l} and t_{1e}, at which zones 2 and 1 are leaving and entering the detection window, respectively.

$$t_{2l} = t_{1e} \tag{4}$$

Both times can be expressed as ratios of the migration paths and the respective velocities. The distances between the injection point of the capillary and the rear boundaries of the zones are taken as the migration paths. Then, with respect to **Eq. 1**, we can write:

$$\frac{X}{E\mu_2} = \frac{X - (L_D + L_S)}{E\mu_1} \tag{5}$$

This relationship can be rearranged to:

$$X = (L_D + L_S)/\alpha \tag{6}$$

Hence, the minimum migration path needed for complete separation of two compounds with similar mobility values is proportional to the sum of the lengths of injec-

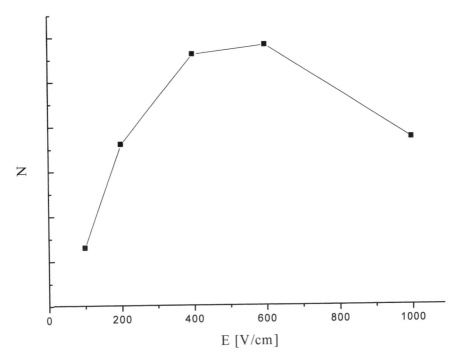

Fig. 2. The dependence of the separation efficiency N on the applied electric field strength E for the 603-bp fragment of the ΦX174/*Hae*III digest. Experimental conditions: 50 μm id × 3-cm capillary (DB-1, J&W Scientific, CA), 1% methyl cellulose in a background electrolyte of 40 m*M* Tris-HCl, 40 m*M* TAPS, pH 8.4. Capillary temperature is 22°C.

tion plug and the detection window, and is inversely proportional to the separation selectivity.

1.3. Diffusion Dispersion

Since the contribution of diffusion is directly proportional to the time of analysis, faster separation will result in a lower longitudinal diffusion and, consequently, in higher separation efficiencies. The faster the separation times, the less important the effects of diffusion are on the separation efficiencies.

1.4. Thermal Dispersion

In analogy to the diffusion dispersion, the thermal dispersion is also a function of the analysis time. But, in contrast to the diffusion dispersion, thermal dispersion scales with the 6th power of the applied electrical field strength, *(19)*. Thus, when increasing the separation speed by higher electric field strength, two opposing forces are at work. As the applied field strength increases, the diffusion band broadening decreases; however, the thermal dispersion increases. When the separation efficiency is plotted against the applied electric field strength an optimum region of the separation field strength can be found. This is shown in **Fig. 2**. In practice, depending on the size of the DNA

Table 1
System Optimization Parameters

Parameter	Optimization
σ^2_{det}	Focus light to have small detection spot
σ^2_{inj}	Inject narrow sample plug (inject for short time/use sample stacking)
σ^2_{therm}	Use of short effective capillary length (1–10 cm, depending on resolution needed)
σ^2_{diff}	Application of optimum electric field strength (300–800 V/cm)
σ^2_{other}	High frequency data acquisition system, use of coated capillaries

fragments and the conductivity of the separation matrix this optimum will typically be between 300 and 800 V/cm. It should be noted, that in separations with high selectivity (large differences in sizes of the separated DNA fragments), higher field strength can be used to further increase the separation speed.

1.5. Other Sources of Dispersion

Other sources of dispersion relate to adsorption of the analyte onto the capillary wall, to electroosmotic mixing, to nonhomogeneity of the electric field strength due to the sample matrix ions, or to insufficient speed of the acquisition of data. Fortunately, the negatively charged DNA molecules mostly do not adsorb significantly on the column walls, and most of the hydrophilic surface coatings developed for CE *(20)* will eliminate the adsorption completely. The surface coating and the use of viscous separation media also eliminate electroosmosis. Electromigration dispersion is typically minimized by sample cleanup and by short injection times. Data acquisition with more than 100 data points per second is adequate for recording peak widths down to 0.1 s, with a total analysis time of few seconds. Most of the standard A/D boards will be sufficient for the data collection.

1.6. Dynamic Structure of DNA

For argument, the DNA fragment is assumed to be a static molecule. However, its molecular orientation under the influence of high electric field in entangled polymer solution can change its diffusion coefficient and its electrophoretic mobility. This effect can be significant especially in case of larger DNA fragments *(15)*. In practice, optimum conditions for fast separations can easily be determined experimentally. **Table 1** lists the parameters, which can be optimized to achieve high-speed DNA separations in entangled polymer solutions.

2. Materials

2.1. Capillary Electrophoresis

1. Fast capillary, fused silica (effective length): 3 cm × 50 µm (id), (DB-1, J&W Scientific, CA).
2. Capillary 6.3: 12.1 cm × 50 µm id.
3. Constant denaturant capillary electrophoresis (CDCE): Capillary, 7 cm × 50 µm (id), coated with poly(vinyl alcohol) (PVA).

4. 1% methylcellulose in 40 mM Tris-base, 40 mM TAPS, pH 8.4 background electrolyte.
5. Electrophoresis buffer: 40 mM Tris-base, 40 mM TAPS, pH 8.4, 1 µg/mL ethidium bromide (EtBr).
6. Standard DNA: ΦX174/*Hae*III digest.
7. DNA size standard (Boehringer VIII).
8. (CA)$_{-18}$ microsatelite repeat polymorphism in the *FcERIβ* gene.
9. Mitochondrial DNA.
10. Agarose sieving medium: 4% solution of Agarose BRE (FMC Rockland, ME) in 0.1 M Tris-base, 0.1 M TAPS, pH 8.4 and 7 M urea.
11. LPA Separation matrix: 4% linear polyacrylamide (LPA) in 50 mM Tris-base, 50 mM TAPS, pH 8.4.
12. SYBR Green II dye.
13. Fluorescein at a concentration of 10^{-8} M.
14. 1X TBE buffer: 89 mM Tris-base, 89 mM boric acid, 1 mM EDTA, pH 8.8.

2.2. Chip Electrophoresis

1. Electrophoresis chip with integrated electrochemical detection.
2. Sieving medium: 0.75% (w/v) hydroxyethyl cellulose in 1X TBE buffer, pH 8.8 with 1 µM EtBr.
3. DNA: *Rsa*I-digested HFE amplicons.
4. Standard DNA: ΦX174/*Hae*III digest.
5. *Salmonella* PCR product.

3. Methods
3.1. Examples of Fast DNA Separations

1. This section illustrates the potential of CE for ultra-fast separations. CE decreases the analysis time typically by ten times compared to standard slab-gel electrophoresis. Depending on the application, the miniaturized CE and microfabricated electrophoresis systems can further decrease the analysis time to tens of seconds or less *(16)*.

3.2. Double-Stranded DNA

1. **Figure 3** shows the separation of the DNA size standard ΦX 174/*Hae*III, which contains 11 DNA fragments ranging from 72 to 1352 bp. In this example, a coated capillary was used to prevent band broadening by DNA-wall interactions. A high frequency data acquisition system allowed a sampling rate of 100 Hz, ensuring a sufficient number of points to define the bands, which are 0.1–0.2 s wide. In order to achieve resolution of the closest migrating fragments (271 bp/281 bp) on a 3-cm long column, a very narrow sample plug is injected with the aid of a fast switching high-voltage power supply *(16)*. A sample plug width of less than 100 µm can be injected into the capillary during a 100-ms electromigration injection. Under these conditions, baseline resolution of the 271-bp/281-bp bands is achieved in less than 30 s. The insert of **Fig. 3** shows the deleterious effect of a longer injection time on the peak resolution.
2. The speed of CE can be applied to the development of DNA diagnostic methods for molecular identification of hereditary diseases or cancer. The ultra-fast CE systems utilizing short electrophoresis columns (length of several centimeters or less) will allow for a variety of high-throughput DNA diagnostics methodologies *(21–23)*.
3. An example of the application of short capillaries in clinical diagnostics is shown in **Fig. 4**. *FcERIβ* is a high affinity glycoprotein receptor for IgE located on chromosome 11

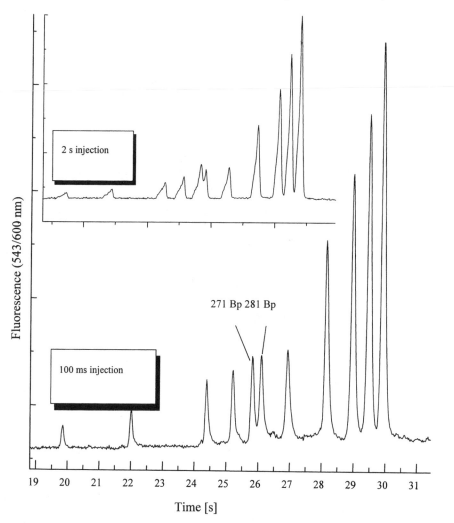

Fig. 3. Separation of ΦX174/*Hae*III digest with 3-s and a 100-ms electrokinetic injection. Experimental conditions: 50-μm id capillary (DB-1, J&W Scientific, CA). Separation distance: 3 cm. Separation matrix: 1% methyl cellulose in 40 m*M* Tris-base, 40 m*M* TAPS, pH 8.4, 1 μg/mL EtBr. Field strength 800 V/cm. LIF detection (excitation 543 nm/emission 600 nm).

(11q13), and variability of *FcERIβ* gene causes differences in excess of IgE responses, which is a typical feature of atopies such as allergic rhinitis, bronchial asthma, dermatitis, and food allergies. One of the genetic variants identified in the *FcERIβ* locus is a $(CA)_{-18}$ microsatelite repeat polymorphism in intron 5 of the gene. The analysis of the short tandem repeat polymorphism in the *FcERIβ* gene of an heterozygous individual is presented here, covering the $(CA)_{-18}$ repeat which ranges from 112 to 132 bp. The analysis is performed at an electric field strength of 256 V/cm, in a capillary with an effective length of

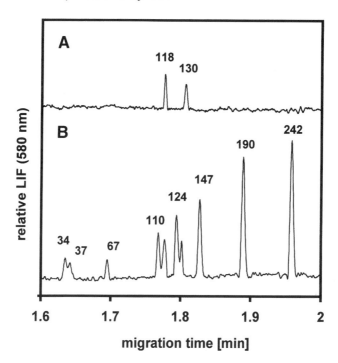

Fig. 4. Detection of $(CA)_{-18}$ microsatelite repeat polymorphism in *FcERIβ* gene of a heterozygous individual. LIF detection at 580 nm with fluorescein as an intercalator at a concentration of 10^{-8} *M*. The separation conditions are 2% Agarose BRE (FMC) in 0.1 *M* Tris-base, 0.1 *M* TAPS, pH 8.4. The electrophoresis is at room temperature; injection time is 3 s at 289 V/cm; electrophoresis is at 256 V/cm; capillary 6.3 (12.1) cm, 50 μm id. The sizes of PCR specific products of the sample (**A**) were evaluated using an addition of DNA size standard (Boehringer VIII) (**B**).

6.3 cm in less than 2 min. The respective polymerase chain reaction (PCR) products of sizes 118–130 bp (record A) are identified using an addition of DNA size standard (Boehringer marker VIII) (record **B**).

3.3. Partially Melted DNA

1. Although there is a difference in the migration behavior of dsDNA, partially melted DNA and ssDNA, most of the system optimization procedures apply to all three types of separations. One application, which takes advantage of differential melting of DNA and is used for the detection of point mutations, is CDCE *(24)*. DNA fragments which contain a mutation usually melt more readily under the influence of a denaturant (heat or urea). Partially melted fragments exhibit a different structure, and therefore, a different migration behavior through an entangled polymer solution.
2. **Figure 5** illustrates the method of screening for an appropriate temperature for the CDCE analysis of a point mutation in a mitochondrial DNA fragment. The sample contains wildtype DNA, mutant DNA, and the two heteroduplex combinations created by boiling-

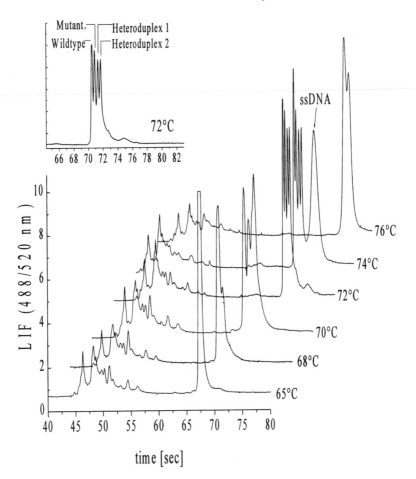

Fig. 5. CDCE of mitochondrial DNA. Capillary 50 μm id × 7 cm coated with PVA. Separation matrix: 4% LPA in 50 m*M* Tris-TAPS, pH 8.4. Field strength 600 V/cm (I = 12 μA). LIF detection (excitation 488 nm/emission 520 nm).

reannealing steps between the two molecules. The CDCE analysis can be performed in less than 80 s using a 7-cm long capillary and high electric field strength (*E* = 600 V/cm). In this example, the temperature screening tests yielded a temperature at which all four different species were separated. The short analysis times helps to rapidly find the optimum temperature for CDCE analysis.

3.4. Single-Stranded DNA

1. DNA sequencing is one area of DNA electrophoresis where the shortening of the analysis time is of key importance. The capillary length cannot be shortened as much as in the previous examples, as extremely high-fragment resolution is required. **Figure 6** shows that up to 300 bases can be separated at 600 V/cm in less than 190 s in a 7-cm long capillary, under optimized separation conditions.

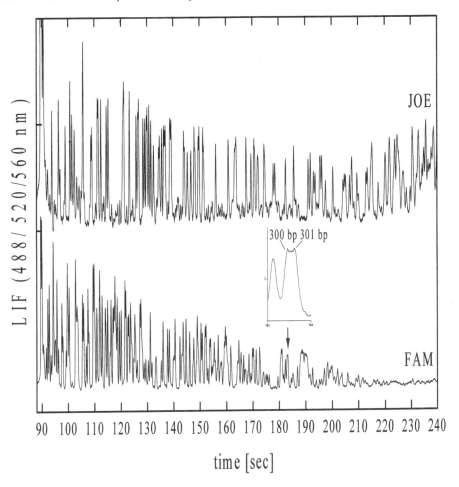

Fig. 6. Fast separation of a sequencing reaction mixture (M13mp18 template). Capillary 50 μm id (PVA coating) × 7 cm. Separation matrix: 3% LPA, 50 mM Tris-TAPS, pH 8.4. Field strength 600 V/cm, (I = 7 μA). LIF detection at (excitation 488 nm/emission 520 nm and 560 nm).

2. Polymorphism in the short tandem repeats, consisting from three adjacent repeat regions of CT, CA, and GC and situated between 979 and 1039 position of *Endothelin 1* gene, seems to play a role in transcription regulation. Even though polymorphism does not exactly change the amino acid sequence of the encoded protein, it may have effects on the dynamics of gene expression. Endothelin is a potent vasoconstrictor and, therefore, its gene variability probably impacts the origin of hypertension. The effect of an increased selectivity on the minimum migration path is demonstrated in **Figs. 7** and **8**. These show fast separations under denaturing conditions of DNA fragments amplified from *Endothelin 1* gene of heterozygous individuals.
3. The higher separation selectivity of ssDNA fragments enables the use of a capillary with an effective length of 2.5 cm (total length 5 cm), without a loss of resolution. As a result,

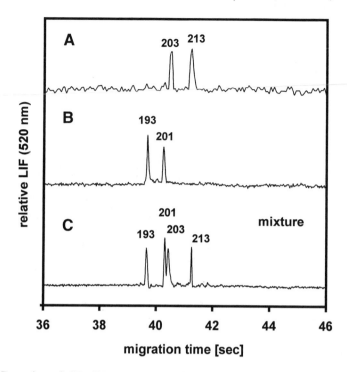

Fig. 7. Detection of CT, CA, GC dinucleotide short tandem repeats polymorphism in *Endothelin 1* gene. Panels (**A**) and (**B**): separation of fragments amplified from genomes of heterozygous individuals. Panel (**C**): mixture of samples A and B. Electrophoresis at 600 V/cm and 60°C in a capillary of 2.5 (5) cm long, 50 μm id, PVA coated. Sieving medium: 4% sol of Agarose BRE (FMC Rockland, ME) in 0.1 *M* Tris-base, 0.1 *M* TAPS, pH 8.4, 7 *M* urea. Sample is denatured in 0.01 *M* NaOH at room temperature and stained with SYBR Green II. Injection is for 5 s at 600 V/cm. LIF detection: excitation with argon-ion laser at 488 nm, collected emission is 520 nm.

the resolution needed for the analysis of the dinucleotide repeat polymorphism, can be completed in 42 s. The fragments of lengths of 203 and 213 (panel A) and 193 and 201 nt (panel B) are mixed (panel C) to confirm the resolution between fragments differing by two nucleotides occurs. Heterozygous alleles with the repeats that differ by a single dinucleotide unit were not available (**Fig. 7**).

4. To increase the speed and resolution yet further, the capillary is held at a temperature of 60°C. This lowers the viscosity of the separation medium and increases its denaturing ability, whereas the thermal energy protects DNA molecules to be stretched under the effect of electric field. Thus, electric field strengths of up to 600 V/cm can be applied. The elevated temperature is conveniently controlled by a 1-cm long heater made of electrically conductive rubber, which also serves as the capillary holder.
5. It is very difficult to achieve denaturation of the DNA fragment carrying the short tandem repeat region of the *Endothelin 1* gene, as the gene has 55% GC pairs. The complementary fragments easily recombine to form heteroduplexes.

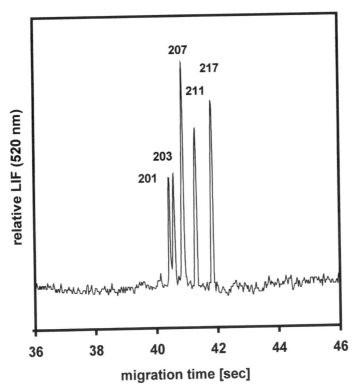

Fig. 8. Fast sizing of a homozygous fragment (203 nt) by using two heterozygous fragments of sizes 207, 217 and 201, 211 nt, respectively. All other conditions are as described in **Fig. 7**.

6. Dimethylsulfoxide or dimethylformamide, which are generally used for DNA denaturation, are not sufficient for the disruption of short tandem repeats in GC rich regions, and a stronger denaturing agent such as NaOH is used. A solution of 0.01 *M* NaOH proved to denature the fragments satisfactorily at room temperature. Another advantage of the presence of NaOH in the sample solution is the possibility to use the electrophoretic stacking technique for the injection of a very sharp zone onto the capillary. Based on the isotachophoretic principle, the slower migrating DNA fragments are stacked behind the zone of highly mobile and conductive OH⁻ anions, and form a much narrower zone than it would for corresponding to the injection times without the ions *(25)*. Thus, as shown in **Figs. 7** and **8**, even with an injection time of 5 s at the same electric field strength as during the electrophoretic run, the separation window of all fragments is only 2 s.

7. The DNA size-standards are not suitable for calibration, since the migration of fragments amplified from *Endothelin 1* gene is strongly affected by their sequence. Even using strong denaturing conditions, GC rich ssDNA fragments migrate anomalously. Therefore, it is necessary to use known sequence fragments amplified from the *Endothelin 1* genes as size markers for the calibration of unknown samples. In **Fig. 8**, the size of a homozygous fragment is determined to be 203 nt, by using two heterozygous samples with fragments of sizes 207, 217, and 201, 211 nt, respectively.

A

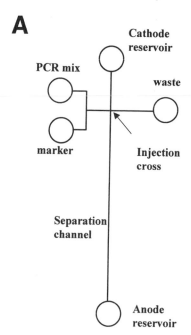

Fig. 9. (**A**) Schematic of a microchip used for sizing of PCR products with a DNA marker. (**B**) (a) Electropherogram of the sizing ladder; (b) Electropherogram of PCR products; (c) Electropherogram of the PCR product mixed equally with the sizing ladder. The microchip was filled with 3% (v/v) linear PDMA, the separation length was 2.5 cm. Field strength: 127 V/cm. The numbers denote fragment sizes in base pairs. Adapted with permission from **ref. 28**. Copyright (1998) American Chemical Society *(continued on opposite page).*

8. The use of denaturing electrophoresis in short capillaries with LIF detection resulted in 20-fold reduction in analysis time, when compared to commercially available CE systems.

3.5. DNA Separations on Microfabricated Devices

1. All previous examples utilize a short piece of a fused silica capillary as the separation column. Microfabrication technologies common in electronics are becoming increasingly important for a new generation of analytical instrumentation. Although the first microfabricated instrument (gas chromatograph) was developed some two decades ago *(26)*, it is the current advancement in protein and DNA research that command the development of adequate analytical instrumentation capable of high-throughput analysis of minute amounts of complex samples. Attempts to integrate several of the sample preparation/analysis steps led to the establishment of a new an analytical field, that of the micro Total Analytical Systems – "µTAS" *(27)*. Mass production of such integrated systems may lead to disposable devices made of inexpensive plastic materials, simplifying routine operation and eliminating sample carry-over or sample cross contamination. In addition, multiple channels on a chip open up the possibility for high-throughput analysis on a microscale. Although the term microdevice (chip, microchip) implies miniaturized size of the separation channel, it should be stressed that since the separation principle remains

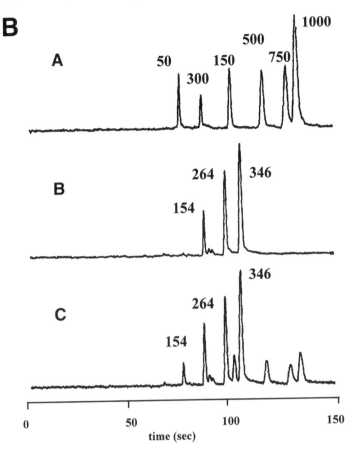

Fig. 9. *(continued)* (**B**).

the same, the speed of analysis will be the same, regardless of the type of the separation channel used (capillary or chip).

2. The main advantage of the chip technology is the ease of fabrication of a large number of channels. In comparison to standard instrumentation where all the connections between the different fluid paths are typically made with fluid fittings, zero dead volume channel junctions can easily be microfabricated. For example, precisely defined amounts of the sample can easily be introduced using a simple double T structure microfabricated as an integral part of the separation channel. The resulting significant size reduction will lower analysis cost by increased analysis speed and reduce the consumption of both the sample and separation media. Moreover in principle, sample preparation procedures can also be integrated on the same chip. This is illustrated in **Fig. 9**. **Figure 9A** shows a monolithic microchip device where the steps of cell lysis, multiplex PCR amplification, and electrophoretic analysis are executed sequentially *(28)*. **Figure 9B** shows the separation of a 500-bp region of bacteriophage lambda DNA and 154-, 264-, 346-, 410-, and 550-bp regions of *E. coli* genomic and plasmid DNAs, amplified using a standard PCR protocol.

The electrophoretic analysis of the products is executed in <3 min following completion of the amplification steps, and the product sizing is demonstrated by proportioning the amplified product with a DNA sizing ladder.

3. Since the microfabrication in glass may not be the most cost effective approach for mass production, there is much interest in alternative materials and fabrication procedures. One promising direction is injection-molding using inexpensive plastic materials *(29)*. The strategy for producing the devices involves solution-phase etching of a master template onto a silicon wafer. Then for the consecutive step, a nickel injection mold insert is made from the silicon master by electroforming in nickel. High-resolution separations of dsDNA fragments with total run times of less than 3 min are demonstrated with good run-to-run and chip-to-chip reproducibility. It is expected that injection molded devices could lead to the production of low-cost, single-use electrophoretic chips, suitable for a variety of separation applications including DNA sizing, DNA sequencing, random primary library screening, and/or rapid immunoassay testing. **Figure 10A** shows the complete separation chip with electrodes, electrical connectors, and buffer reservoirs. **Figure 10B** shows the separation of the fragments of a *Hae*III digest of ΦX174 DNA in less than 2.5 min in such a sealed plastic chip.

4. Although the development of microfabricated CE systems is mainly focused on the separation "chip" itself, it is clear that substantial changes to the external parts of the instrumentation will be also necessary, especially changes to the detector. Current advances in solid-state lasers will bring substantial miniaturization of LIF detection. The recently described electrochemical detector, which can be microfabricated as an integral part of the chip itself, is also a promising development *(30)*.

5. Photolithographic placement of the working electrode just outside the exit of the electrophoresis channel provides high-sensitivity electrochemical detection with minimal interference from the separation electric field. Indirect electrochemical detection is used for high-sensitivity DNA restriction fragment and PCR product sizing. These microdevices match the size of the detector to that of the microfabricated separation and reaction devices, bringing to reality the "lab-on-a-chip" concept. **Figure 11** shows the electrophoresis chip with integrated electrochemical detection and corresponding separation of *Salmonella* PCR product (shaded) with 500 pg/mL ΦX174/*Hae*III restriction digest obtained on this chip.

6. As previously mentioned, one of the advantages of the microfabrication is the possibility to make a large number of channels of a variety of shapes with zero dead volume connections. Typical application of this potential may be a CAE microplate that can analyze 96 samples in less than 8 min *(31)*. This CAE chip has been produced by the bonding of 10-cm diameter micromachined glass wafers to form a glass sandwich structure. The microplate with 96 sample wells and 48 separation channels permitted the serial analysis of two different samples on each capillary. Individual samples are addressed with an electrode array positioned above the microplate.

7. The detection of all lanes with high temporal resolution is achieved by using a laser-excited confocal fluorescence scanner. **Figure 12** shows the microdevice used for electrophoretic separation with fluorescence detection for the diagnosis of hereditary hemochromatosis. Electropherograms represent two sequential separations of *Rsa*I-digested *HFE* amplicons. The three genotypes are characterized by a single peak at 140 bp corresponding to the 845G type, a single peak at 111 bp corresponding to the 845A type, and the heterozygote type that exhibits both the 140- and 111-bp peaks.

8. Electrophoretic separation in a short narrow capillary allows extremely fast separations with resolution comparable to, or better than standard slab gel or CE. On the top of the

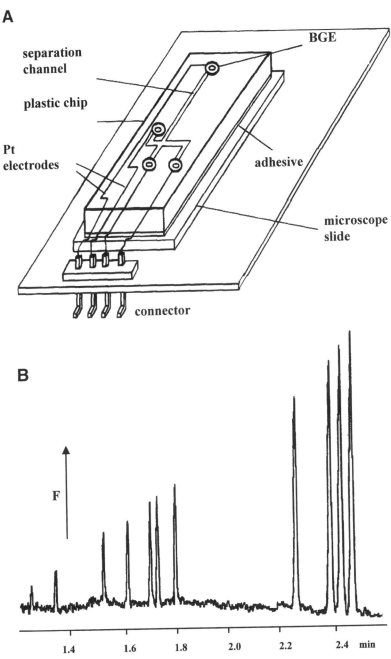

Fig. 10. (**A**) Completed separation chip with electrodes, electric connectors and buffer reservoirs. (**B**) Separation of a *Hae*III digest of ΦX174 RF DNA fragments in a sealed plastic chip. Adapted with permission from **ref. 29**. Copyright (1997) American Chemical Society.

A

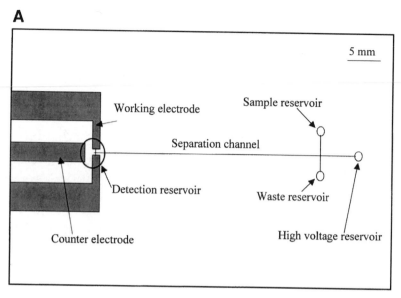

B

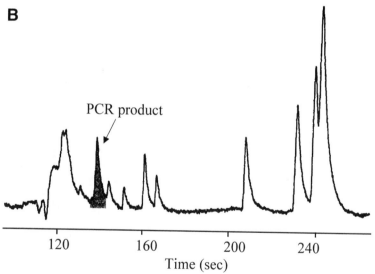

PCR product

Time (sec)

Fig. 11. **(A)** Diagram of an electrophoresis chip with integrated electrochemical detection. **(B)** Separation obtained on this chip of a *Salmonella* PCR product (shaded) along with 500 pg/mL of ΦX174/*Hae*III restriction digest. Adapted with permission from **ref. *30***. Copyright (1998) American Chemical Society.

analysis speed, the current technology also provides sensitive LIF detection for all critical applications. The maturity of the CE has currently been confirmed by its application for the accelerated completion of the Human Genome Project. It can be also anticipated

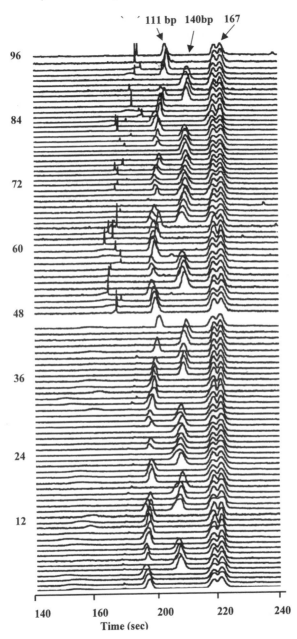

Fig. 12. ACE separations of two sequential "48-samples" of *Rsa*I-digested *HFE* amplicons. The three genotypes correspond to a single peak at 140 bp (845G type), 111 bp (845A type) and two peaks at 140 bp and 111 bp (heterozygote). Samples were separated in 10-cm long channels using 0.75% (w/v) hydroxyethyl cellulose in 1X TBE buffer with 1 μ*M* EtBr. Field strength 300 V/cm. Adapted with permission from **ref. *31***. Copyright (1998) American Chemical Society.

that future development of these microfabricated devices will enable extremely rapid analytical and diagnostic tools, integrating the fast electrophoretic separation methods with microscale sample processing.

Acknowledgments

The preparation of this chapter has been supported by Grants No. 203/00/0772 and 301/97/1192 of the Grant Agency of the Czech Republic, and with support from the Barnett Institute for Contribution #768.

References

1. Shapiro, A. L., Vinuela, E., and Maizel, J. V. (1967) Molecular weight estimation of polypeptide chains by electrophoresis in SDS-polyacrylamide gels. *Biochem. Biophys. Res. Commun.* **28,** 815.
2. O'Farrell, P. H. (1975) High resolution two-dimensional electrophoresis of proteins. *J. Biol. Chem.* **250,** 4007–4021.
3. Sanger, F., Nicklen, S., and Coulson, A. R. (1977) DNA sequencing with chain-terminating inhibitors. *Proc. Natl. Acad. Sci. USA* **74,** 5463–5467
4. Maxam, A. M. and Gilbert, W. (1977) A new method for sequencing DNA. *Proc. Natl. Acad. Sci. USA* **74,** 560–654.
5. Andrews, A. T. (1986) *Electrophoresis: Theory, Technique, and Biochemical and Clinical Applications,* 2nd ed., Clarendon Press, Oxford.
6. Rickwood, D. and Hames, B. D. (eds.), (1990) *Gel Electrophoresis of Nucleic Acids,* 2nd ed., Oxford University Press, Oxford, New York.
7. Deyl, Z. (ed.), (1979) Electrophoresis: Part A: techniques. *J. Chromatogr. Library* **18,** Elsevier, Amsterdam.
8. Hjertén, S. (1967) Free zone electrophoresis. *Chromatogr. Rev.* **9,** 122–219.
9. Virtanen, R. (1974) Zone electrophoresis in a narrow-bore tube employing potentiometric detection. *Acta Polytech. Scand.* **123,** 1–67.
10. Mikkers, F. E. P., Everaerts, F. M., and Verheggen, Th. P. E. M. (1979) High performance zone electrophoresis. *J. Chromatogr.* **169,** 11–20.
11. Jorgenson, J. W. and Lukacs, K. D. (1981) Zone electrophoresis in open tubular glass capillaries. *Anal. Chem.* **53,** 1298–1303.
12. Karger, B. L., Chu, Y., and Foret, F. (1995) Capillary electrophoresis of proteins and nucleic acids. *Annu. Rev. Biophys. Biomol. Struct.* **24,** 579–610.
13. Grossbach, U. (1965) Acrylamide gel electrophoresis in capillary columns. *Biochim. Biophys. Acta* **107,** 180–182.
14. Hyden, H., Bjurstam, K., and McEwen, B. (1966) Protein separation at the cellular level by micro disc electrophoresis. *Anal. Biochem.* **17,** 1–15.
15. Slater, G. W., Mayer, P., and Grossman, P. D. (1995) Diffusion, Joule heating, and band broadening in capillary gel electrophoresis of DNA. *Electrophoresis* **16,** 75–83.
16. Mueller, O., Minarik, M., and Foret, F. (1998) Ultrafast DNA analysis by capillary electrophoresis/laser-induced fluorescence detection. *Electrophoresis* **19,** 1436–1444.
17. Chan, K. C., Muschik, G. M., and Issaq, H. J. (1997) High-speed electrophoretic separation of DNA fragments using a short capillary. *J. Chromatogr.* **695,** 113–115.
18. Woolley, A. T., Sensabaugh, G. F., and Mathies, R. A. (1997) High-speed DNA genotyping using microfabricated capillary array electrophoresis chips. *Anal. Chem.* **69,** 2181–2186.

19. Foret, F., Deml, M., and Bocek, P. (1988) Capillary zone electrophoresis: quantitative study of the effects of some dispersive processes on the separation efficiency. *J. Chromatogr.* **452,** 601–613.

20. Hjertén, S. (1985) High-performance electrophoresis: elimination of electroendosmosis and solute adsorption. *J. Chromatogr.* **347,** 191–198.

21. Liu, S., Shi, Y., Ja, W. W., and Mathies, R .A., (1999) Optimization of high-speed DNA sequencing on microfabricated capillary electrophoresis channels. *Anal. Chem.* **71,** 566–573.

22. Schmalzing, D., Adourian, A., Koutny, L., Ziaugra, L., Matsudaira, P., and Ehrlich, D. (1998) DNA sequencing on microfabricated electrophoretic devices. *Anal. Chem.* **70,** 2303–2310.

23. Kleparnik, K., Mala, Z., Havac, Z., Blazkova, M., Holla, L., and Bocek, P. (1998) Fast detection of a $(CA)_{18}$ microsatelite repeat in IgE receptor gene by capillary electrophoresis with laser induced fluorescence detection. *Electrophoresis* **19,** 249–255.

24. Khrapko, K., Hanekamp, J., Thilly, W., Belenky, A., Foret, F., and Karger, B. L. (1994) Constant denaturant capillary electrophoresis (CDCE) - a high resolution approach to mutational analysis. *Nucleic Acids Res.* **22,** 364–369.

25. Bocek, P., Deml, M., Gebauer, P., and Dolnik, V. (1988) *Analytical Isotachophoresis.* VCH Verlagsgesellschaft, Weinheim.

26. Terry, S. C., Jerman, J. H., and Angell, J. B. (1979) A gas chromatographic analyzer fabricated on a silicon wafer. *IEEE Trans. Electron. Devices* **26,** 1880–1886.

27. Manz, A., Harrison, D. J., Verpoorte, E. M. J., Fettinger, J. C., Paulus, A., Ludi, H., et al. (1992) Planar chips technology for miniaturization and integration of separation techniques into monitoring systems. Capillary electrophoresis on a chip. *J. Chromatogr.* **593,** 253–258.

28. Waters, L. C., Jacobson, S. C., Kroutchinina, N., Khandurina, J., Foote, R. S., and Ramsey, J. M. (1998) Microchip device for cell lysis, multiplex PCR amplification, and electrophoretic sizing. *Anal. Chem.* **70,** 158–162.

29. McCormick, R. M., Nelson, R. J., Alonso-Amigo, M. G., Benvegnu, J., and Hooper, H. H. (1997) Microchannel electrophoretic separations of DNA in injection-molded plastic substrates. *Anal. Chem.* **69,** 2626–2630.

30. Woolley, A. T., Lao, K. Q., Glazer, A. N., and Mathies, R. A. (1998) Capillary electrophoresis chips with integrated electrochemical detection. *Anal. Chem.* **70,** 684–688.

31. Simpson, P. C., Roach, D., Woolley, A. T., Thorsen, T., Johnston, R., Sensabaugh, G. F., et al. (1998) High-throughput genetic analysis using microfabricated 96-sample capillary array electrophoresis microplates. *Proc. Natl. Acad. Sci. USA* **95,** 2256–2261.

3

Fast DNA Fragment Sizing in a Short Capillary Column

Haleem J. Issaq and King C. Chan

1. Introduction

Capillary electrophoresis (CE) is an efficient and fast microseparation technique that is well suited for the separation of nucleic acids. CE is routinely used for the separation of double-stranded DNA fragments; single-stranded DNA fragments and polymerase chain reaction (PCR) products in our laboratory. CE is a versatile tool for separating nucleic acids in molecular biology *(1)*. CE has the advantages of automation, small sample requirement, fast and efficient separation, real-time detection, and negligible buffer waste in comparison to slab-gel electrophoresis. However, the major drawback of CE is low amount of sample throughput because samples are analyzed sequentially. This problem is resolved by using a capillary array electrophoresis (CAE) *(2,3)*. Since conventional CE systems are equipped with a single capillary, a practical way to increase throughput is to minimize the analysis time per sample. This has been accomplished by using a range of an effective length capillary as short as 7 cm and field strength as high as 2000 V/cm for the separations of small ions, drugs, and proteins *(4–7)*.

Recently, we were able to achieve fast separation of the wild-type and mutant PCR products of the TGF-β_1 gene in 60 s using a 7-cm effective length capillary and 560 V/cm *(8)*. However, high field strength degraded the resolution of large DNA fragments *(9)* and may cause migration anomalies *(10)*. Alternatively, one can use a shorter capillary length with low field strength. We describe here the fast sizing of DNA fragments using 1-, 2-, and 7-cm effective length capillaries *(11)*. Fast DNA separation can be achieved with capillaries with an effective length of only 1–2 cm. The low electric field associated with such short capillaries extends the range of DNA fragment sizes that may be separated by this means, which is useful if the analysis of large DNA fragments is required.

2. Materials
2.1. Apparatus

1. All experiments are performed with a home-built CE laser-induced fluorescence (LIF) system, similar in design to that used previously *(12)*. See **Fig. 1** for a detailed diagram.

From: *Methods in Molecular Biology, Vol. 163:*
Capillary Electrophoresis of Nucleic Acids, Vol. 2: Practical Applications of Capillary Electrophoresis
Edited by: K. R. Mitchelson and J. Cheng © Humana Press Inc., Totowa, NJ

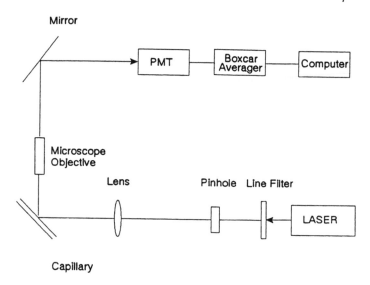

Fig. 1. A sketch of the LIF-CE system. Reprinted from **ref. *12***, pp. 1877–1890, by courtesy of Marcel Dekker.

The CE system comprises: High-voltage power supply (Glassman, Whitehouse Station, NJ, USA); Photomultiplier tube (PMT) (Oriel, Stratford, CT, USA); Picoammeter (Keithley, Cleveland, OH, USA); Data collection by the Beckman 406 data acquisition module; argon-ion laser (ILT, Salt Lake City, UT, USA).

2. DB-17 coated fused silica capillaries, 50 μm id, are purchased from J&W (Folsom, CA, USA) and are cut to size.
3. DNA separation buffer is from Sigma (St. Louis, MO, USA).
4. YO-PRO-1 dye is from Molecular Probe (Eugene, OR, USA).
5. 20- and 100-bp DNA ladders are from Gensura (Del-Mar, CA, USA).
6. Buffers are from Fisher Scientific (Pittsburgh, PA, USA). Sigma DNA separation buffer is used undiluted, or diluted with 40% water to 60% separation buffer.

3. Methods
3.1. Capillary Preparation

1. A 15–20-cm DB-17 capillary is cut and a narrow window is made at 1, 2, or 7 cm from one end to form the detection window of the 1-, 2-, and 7-cm effective length capillaries.
2. A detection window is created in the capillary by burning the polyimide coating with a gentle flame, then washing the area with methanol.
3. Each capillary is then rinsed with water for 5 min, with buffer for 2 min, and then filled with the replaceable gel solution.

3.2. Fast CE

1. The DNA separation buffer is diluted with deionized-distilled water to 60%, after which Yo-Pro-1 is added to 1 μ*M*.
2. The concentrations of the 20- and 100-bp ladders are 0.5 and 0.25 ng/μL, respectively.

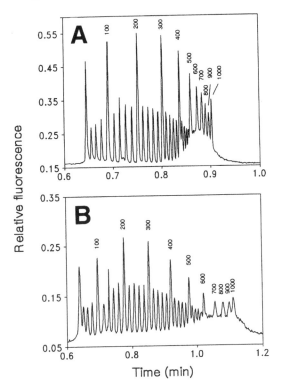

Fig. 2. Separations of a 20–1000-bp ladder using: (**A**), a 7-cm effective length capillary at 556 V/cm; and (**B**), a 2-cm effective length capillary at 185 V/cm. Reprinted from Chan, K. C., Muschik, G. M., and Issaq, H. J. High-speed electrophoretic separation of DNA fragments using a short capillary. Copyright (1997), pp. 113–115, with permission of Elsevier Science.

3. The samples are injected electrokinetically, typically for 5 s, at 0.01 kV/cm for the 1 or 2 cm separation capillary and at 0.04 kV/cm for the 7-cm separation capillary.
4. The capillary is flushed after a few injections with the gel buffer.
5. The capillary is mounted on a *x-y* translational stage for precise movement.
6. The argon-ion laser beam is focused onto the capillary with a bi-convex lens. Fluorescence is collected at 90° from the excitation beam with a 10X microscopy objective.
7. After passing through a 488-nm interference filter, the fluorescence is detected by a PMT. The PMT current is monitored by the picoammeter. The output from the picoammeter is fed to a Beckman 406 data acquisition module. The rate of data sampling is 20 Hz with a rise time of 0.1 s.
8. The electric field for CE separation is up to 30 kV.

3.3. Results

3.3.1. DNA Separation

1. A mixture of DNA fragments in a 20–1000-bp ladder can be resolved on the three different effective length capillaries, 7, 2, and 1 cm (total capillary length 20 cm each), using the 60% diluted DNA separation buffer.

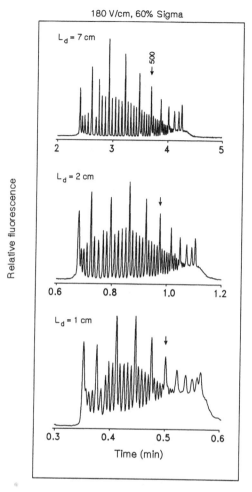

Fig. 3. Electropherogram of the separation of a 20-bp and 100-bp DNA standard solution in 10 mM Tris-borate buffer containing 0.1 mM EDTA, pH 8.0, and 1 μL/mL YO-PRO-1 intercalting fluorescent dye (1 mM in DMSO). Detection was achieved via an argon-ion laser λ_{ex} = 488-nm and a 520-nm interference filter. A 50 μm id DB-17 coated fused silica capillary of 7-, 2-, and 1-cm length filled with Sigma DNA replaceable gel buffer was used with an applied voltage of 180 V/cm. Reprinted from **ref. 11**.

2. The 7-cm capillary is run at 556 V/cm and allows the separation of the 20–1000-bp ladder in about 1 min (**Fig. 2A**).
3. However, the 2-cm effective length capillary is run at 185 V/cm and provides both a similar speed of migration and a better resolution of the larger DNA fragments (**Fig. 2B**). This is because larger fragments are stretched in electric fields and migrate according to a "biased reptation" mode, in which their mobilities are less dependent on molecular sizes at high field strength (*9*). In addition, column heating degrades the DNA separations at higher fields, as our home-built system does not have capillary cooling.

4. The 1-cm effective length capillary is run at 180 V/cm and is also able to separate the DNA ladder standards (**Fig. 3**). However, the peak resolution is not as good as when 7- and 2-cm capillaries are used, but it is expected to improve if the duration and field used for sample injection phase and the separation phase of the run are optimized.

Acknowledgments

This project has been funded in whole or in part with Federal funds from the National Cancer Institute, National Institutes of Health, under Contract No. NO1-CO-56000. By acceptance of this article, the publisher or recipient acknowledges the right of the United States. Government to retain a nonexclusive, royalty-free license and to any copyright covering the article. The content of this publication does not necessarily reflect the views or policies of the Department of Health and Human Services, nor does mention of trade names, commercial products, or organizations imply endorsement by the United States Government.

References

1. Righetti, P. G. (ed.). (1996) *Capillary Electrophoresis in Analytical Biotechnology.* CRC Press, Boca Raton, Florida.
2. Ueno, K. and Yeung, E. S. (1994) Simultaneous monitoring of DNA fragments separated by electrophoresis in a multiplexed array of 100 capillaries. *Anal. Chem.* **66,** 1424–1431.
3. Mathies, R. A. and Huang, X. C. (1992) Capillary array electrophoresis: an approach to high speed, high-throughput DNA sequencing. *Nature* **359,** 167–169.
4. Lausch, R., Scheper, T., Reif, O.-W., Schlosser, J., Fleischer, J., and Freitag, R. (1993) Rapid capillary gel electrophoresis of proteins. *J. Chromatogr. A* **654,** 190–195.
5. Hjertén, S., Valtcheva, L., Clenbring, K., and Liao, J.-L. (1995). Fast, high-resolution (capillary) electrophoresis in buffers designed for high field strengths. *Electrophoresis* **16,** 584–594.
6. Altria, K. D., Kelly, M. A., and Clark, B. J. (1996) The use of a short-end injection procedure to achieve improved performance in capillary electrophoresis. *Chromatographia* **43,** 153–157.
7. Bert, L., Robert, F., Denoroy, L., and Renand, B. (1996) High speed separation of subnanomolar concentrations of noradrenaline and dopamine using capillary zone electrophoresis with laser-induced fluorescence. *Electrophoresis* **17,** 523–525.
8. Chan, K. C., Muschik, G. M., Issaq, H. J., Garvey, K. J., and Generlette, P. L. (1996) High speed screening of polymerase chain reaction products by capillary electrophoresis. *Anal. Biochem.* **243,** 133–139.
9. Slater, G. W., Desruisseaux, C., and Hubert, S. J. (2001) DNA separation mechanisms during capillary electrophoresis, in *Capillary Electrophoresis of Nucleic Acids,* Vol. 1 (Mitchelson, K. R., and Cheng, J., eds.), Humana Press, Totowa, NJ, pp. 27–41.
10. Berka, J., Pariat, Y. F., Mueller, O., Hebenbrock, K., Heiger, D. N., Foret, F., and Karger, B. L. (1995) Sequence dependent behavior of double-stranded DNA in capillary electrophoresis. *Electrophoresis* **16,** 377–388.
11. Issaq, H. J., Chan, K. C., and Muschik, G. M. (1997) The effect of column length, applied voltage, gel type, and concentration on the CE separation of DNA fragments and polymerase chain reaction products. *Electrophoresis* **18,** 1153–1158.
12. Chan, K. C., Janini, G. M., Muschik, G. M., and Issaq, H. J. (1993) Pulsed UV laser-induced fluorescence detection of native peptides and proteins. *J. Liq. Chromatogr.* **16,** 1877–1890.

II

PRACTICAL APPLICATIONS FOR MUTATION DETECTION

4

Detection of DNA Polymorphisms Using PCR-RFLP and Capillary Electrophoresis

John M. Butler and Dennis J. Reeder

1. Introduction

The most fundamental difference between two individuals is DNA variation at the nucleotide level. This DNA variation, often referred to as a single nucleotide polymorphism or SNP, can occur approximately once every several hundred base pairs. The ability to accurately determine the DNA sequence at specific sites throughout an organism's genome is important in a variety of applications including disease detection, differentiation of microorganisms, agricultural genetics and molecular breeding, and human identification (i.e., forensic and paternity testing). A number of laboratory techniques have been developed in recent years to analyze DNA rapidly and reliably for the purpose of mutation detection. One of these methods involves the use of DNA restriction enzymes to generate DNA molecules of varying length depending on the presence or absence of restriction sites. The length variations in DNA molecules caused when the restriction enzymes do their cutting are called restriction fragment-length polymorphisms (RFLP). Another molecular procedure that allows one to work with small amounts of DNA is the polymerase chain reaction (PCR), a method for amplifying and enriching the sample within a targeted region of a DNA sequence. When the two procedures are combined, the method is referred to as PCR-RFLP. A separation step is required to resolve and measure the length of the restriction fragments from one another after the enzymatic reactions (**Fig. 1**), and this essential step is where capillary electrophoresis (CE) offers a number of advantages.

Traditionally, PCR-RFLP fragments are separated using polyacrylamide gel electrophoresis (PAGE). In recent years, CE has matured into a reliable technique that offers many advantages over PAGE including speed, sensitivity, high resolution, and automation. Using pumpable polymer solutions and intercalating dyes combined with laser-induced fluorescence (LIF), CE has proven effective for separating and detecting PCR-RFLP fragments *(1)*. CE methods have been reported for separating PCR-RFLP fragments from mRNA *(2)*, DNA from bacteriophages *(3)*, and oncogenes *(4)*, as well

From: *Methods in Molecular Biology, Vol. 163:*
Capillary Electrophoresis of Nucleic Acids, Vol. 2: Practical Applications of Capillary Electrophoresis
Edited by: K. R. Mitchelson and J. Cheng © Humana Press Inc., Totowa, NJ

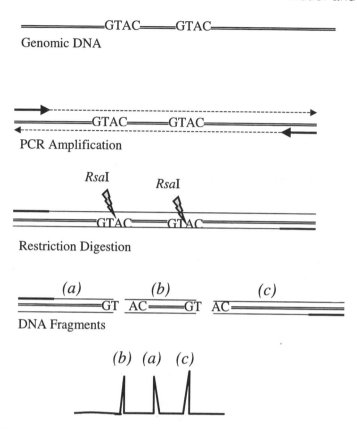

Fig. 1. Diagram showing the steps in processing of a PCR-RFLP sample, using the restriction enzyme *Rsa*I as an example.

as to differentiate plasmids which have been digested with various restriction enzymes *(5)*. In addition, *apoE* genotyping *(6)*, bacterial identification *(7,8)*, disease diagnosis *(9,10)*, and mitochondrial DNA typing *(1)* have each been performed using CE separation of the PCR-RFLP fragments.

A PCR-RFLP sample has a limited number of peaks compared to a sequencing reaction and typically does not require single-base resolution, and often resolution of approx 10–20 bp is more than sufficient. The goal with PCR-RFLP samples is to make comparisons of the peak pattern and to differentiate between samples that have generated different sized restriction fragments because of their polymorphisms at one or more nucleotide positions at the enzyme cutting site. PCR-RFLP serves as a screening/confirmation method, rather than a primary method of polymorphism discovery, such as DNA sequencing.

Prior to performing any laboratory work, *"in silico"* studies may be accomplished using programs such as GeneRunner or Mac Vector to predict the location of restriction sites within the sequence of interest. These computer programs will typically test a DNA sequence against 200–300-restriction endonuclease recognition sequences.

Most restriction enzyme recognition sites are 4, 6, or 8 nt in length. If the polymorphic SNP site of interest creates or eliminates a restriction site, then an appropriate restriction enzyme may be used to generate a series of DNA fragments with a pattern that can be distinguished between wild-type and mutant DNA sequences. Several hundred different restriction enzymes are commercially available from a variety of sources, often at costs of only a few cents per reaction (*see* **Note 1**).

One example of a nucleotide variation of diagnostic value that can be detected with PCR-RFLP is a test for hemochromatosis. Hemochromatosis is a common hereditary iron regulation disorder that causes iron overload in tissues and leads to liver disease, diabetes, and other clinical complications. This genetic disease is easy to treat by routine blood donation. Many individuals who have a family history of this disorder desire to know if they are homozygous for the gene, and thus susceptible to the deleterious effects that usually appear in the fifth or sixth decade of life. However, because of uncertainties about the prevalence and extent of penetrance of the mutations, genetic testing on a massive scale is not recommended at this time *(11)*. The *hemochromatosis* gene is 348 amino acids in length and a simple cysteine-to-tyrosine mutation at amino acid 282 (C282Y) is thought to cause the onset of the disease *(12)*. The recognition site of the restriction enzyme *Rsa*I (GT/AC) is altered by a single G-to-A base change at nucleotide position 845 in the *hemochromatosis* gene on chromosome 6. With a 307-bp PCR product, *Rsa*I digestion of a normal allele results in two fragments of 167 bp and 140 bp. A mutant allele, on the other hand, yields three fragments (167, 111, and 29 bp) whereas heterozygous individuals have four fragments (167, 140, 111, and 29 bp). Thus, a PCR-RFLP test involving *Rsa*I can provide an effective diagnostic test for hemochromatosis *(10,13)*.

The PCR-RFLP method described in this chapter was recently demonstrated as an effective tool for differentiating various mitochondrial DNA (mtDNA) types *(1)*. Human mitochondrial DNA contains a highly polymorphic control region, also called the D-loop, which is used in identity testing *(14)*. Samples are typically sequenced to compare the variation across approx 600 bases of the control region. Dozens of SNP sites exist in the mtDNA control region (*see* **Note 2**). Screening methods, such as PCR-RFLP, have the advantage of being less expensive and rapid (*see* **Note 3**), and reduce the need to fully sequence differing samples in order to allow exclusion (*see* **Notes 4** and **5**).

2. Materials
2.1. Equipment
1. CE is performed on a Beckman P/ACE® 5500 CE instrument with a Laser Module 488 nm argon-ion laser (Beckman Instruments, Fullerton, CA). P/ACE Station software version 1.0 is used for instrument control and data collection.
2. Heat block (set at 37°C).
3. Thermal cycler PE9700 (PE Applied Biosystems) for PCR amplification steps. The cycler may also be used for isothermal restriction enzyme digestion steps.

2.2. Reagents and Accessories
1. Capillary: 27 cm × 50 µm id CElect-N coated capillary (Supelco, Bellefonte, PA), or a 27 cm × 50 µm id DB-17 coated capillary (J&W Scientific, Folsom, CA). The distance from

injection to detection is 20 cm because the detection window is located 7 cm from the outlet end of the capillary in a Beckman capillary cartridge.

2. Sieving buffer: 1% w/v hydroxyethyl cellulose (2% HEC = 80–125 mPa.s) (Aldrich, Milwaukee, WI) in 100 mM Tris-base, 100 mM borate, 1 mM EDTA pH 8.3 (Tris-borate-EDTA) solution. The HEC (5 g) is typically dissolved by stirring overnight at room temperature in 500 mL of the Tris-borate-EDTA solution.

3. Intercalating dye: 500 ng/mL YO-PRO-1 (Molecular Probes, Eugene, OR) (*see* **Note 6**). A 12-µL aliquot of 1 mM YO-PRO-1 is added to 15 mL of the HEC polymer solution in an aluminum foil covered tube. The dye is light sensitive and a possible carcinogen and should be treated appropriately.

4. Methanol, high performance liquid chromatography (HPLC) grade.

5. CE vials: 4-mL wide-mouth vials with threads for CE instrument.

6. Beckman sample caps: silicon rubber caps for the 4-mL vials.

7. Sample vials: 0.2 mL MicroAmp® Reaction tubes (Perkin-Elmer, Norwalk, CT).

8. Float dialysis membranes, 0.025 µm VSWP (Millipore, Medford, MA).

9. Deionized water.

2.3. Molecular Biology Materials

1. Restriction enzymes: A variety of enzymes are available from New England Biolabs as well as other molecular biology supply houses (*see* **Note 1**). For example, *Rsa*I, which cuts between GT/AC, is available from 18 different suppliers.

2. PCR reagents: buffers, polymerase, dNTPs, MgCl$_2$, and tubes (available from a variety of sources, including PE Applied Biosystems, Promega Corporation, etc.).

3. PCR primers: synthetic oligonucleotides specific to the DNA target region of interest.

4. DNA size standards: 20-bp ladder (GenSura Laboratories, Inc., Del Mar, CA) or a φX174 *Hae*III DNA restriction digest (Sigma, St. Louis, MO)

5. Internal DNA standard to be mixed with sample: a 200-bp DNA fragment (QS-200) at a concentration of 100 ng/µL (GenSura, Del Mar, CA). Alternatively, a 150-bp DNA fragment from BioVentures (Murfreesboro, TN) may be used.

2.4. Software and Sequence Information

1. GeneRunner® (version 3.03; Hastings Software, Inc., Hastings, NY) is used to view the DNA sequence information and perform *"in silico"* studies with the 287 commercially available restriction enzymes listed in the default restriction search table.

2. DNA sequence information may be downloaded into GeneRunner from GenBank (http://www2.ncbi.nlm.nih.gov/cgi-bin/genbank). The Anderson reference sequence for the 16,569-bp mtDNA genome may be accessed via the accession number M63933.

3. Previously observed sequence polymorphisms are inserted into the appropriate location in the reference sequence and stored as separate mtDNA types. In the case of other genetic loci, wild-type and mutant alleles may be saved as separate sequence files for testing against the GeneRunner restriction enzyme search table.

3. Methods

1. Identify the potential restriction site(s) within the sequence of interest and choose the restriction enzyme(s). Use GeneRunner or some other DNA sequence evaluation software to locate the restriction enzyme cut sites and to predict the expected fragment sizes.

2. Amplify DNA using PCR: In the example used here, the hypervariable region I (HV1) from the control region of mtDNA is amplified via the PCR using the L15997 and H16395

primers and PCR conditions as previously described *(14)*. These primers result in a 436-bp PCR product spanning positions 15978–16413 of the human mtDNA genome.

3. Digest PCR products with restriction enzyme(s): Take a portion of the PCR product and dilute into the restriction enzyme buffer along with the appropriate enzyme. A 1–5-µL aliquot of the PCR product works well with 5–10 U of restriction enzyme depending on the quantity of target material produced by the PCR reaction. Reaction volumes of 10 µL are effective. Heat the sample at 37°C for 30 min to perform the digestion.

4. Prepare sample for CE: A 1:50 dilution of the sample in deionized water is an effective means of sample preparation with LIF detection *(15,16)*.

5. Float dialysis is also useful for removing PCR buffer salts from the sample if a weak signal is observed in the CE *(17)*. Add a 1-µL aliquot of the digestion mixture to 49 µL deionized water. An internal standard, such as a 150-bp DNA product, may be added for use in adjusting migration times between samples provided that it does not interfere with the analysis and can be resolved from any restriction fragments of interest (**Fig. 2**). Mix the sample well by drawing it into and out of the pipet tip several times to ensure a uniform DNA concentration. Place the sample vial (0.2-mL tube) on a spring inside a 4-mL wide-mouth vial. Screw a silicon rubber cap on the 4-mL vial (to prevent evaporation) and load the samples into the auto-sampler.

6. Prepare the CE capillary and buffers: Prepare the capillary by cutting it to the desired length and removing approx 5 mm of the polyamide coating for the detection window. Place the capillary in an LIF capillary cartridge (Beckman). The cartridge will allow liquid coolant to flow around the capillary and maintain a constant temperature environment.

7. Fill three 4-mL vials with the HEC buffer containing the intercalating dye. Two buffer vials are used as the inlet and outlet vials during electrophoresis, whereas the third vial is used to fill the capillary with fresh separation media between each run. Rinse the capillary with methanol for 1 min between each run helps increase the longevity of the capillary.

8. Separate digestion products on CE: Enter the sample names in the collection software spreadsheet corresponding to the samples and their positions in the auto-sampler tray. Program the method for CE analysis. Inject each sample at 2 kV for 5 s.

9. Apply a separation voltage of 5 kV (185 V/cm) if a high-resolution separation is desired or 15 kV (556 V/cm) if a more rapid separation with lower resolution is more desirable (*see* **Note 7**).

10. A 20-bp ladder should be analyzed prior to running samples in order to determine the resolution over the size range of interest as well as to correlate the migration times to DNA sizes.

11. Perform data analysis: Compare the CE electropherograms for each sample to assess peak patterns (*see* **Note 8**). The Beckman P/ACE Station software has a "Supercompare" program that permits multiple electropherograms to be overlaid and viewed together.

4. Notes

1. A valuable source of information on restriction enzymes is REBASE: The Restriction Enzyme Database, which is maintained by New England Biolabs. REBASE may be found on the World Wide Web at http://rebase.neb.com/rebase/. This database contains a comprehensive list of over 3000 known restriction enzymes along with their recognition sequences. However, only ~500 restriction enzymes are available from the 20 suppliers listed in REBASE.

2. The example used to illustrate the technique is for mtDNA, but the technique is much more general and may be applied to any SNP where a restriction site is either eliminated or created provided that the restriction fragments may be resolved in the CE.

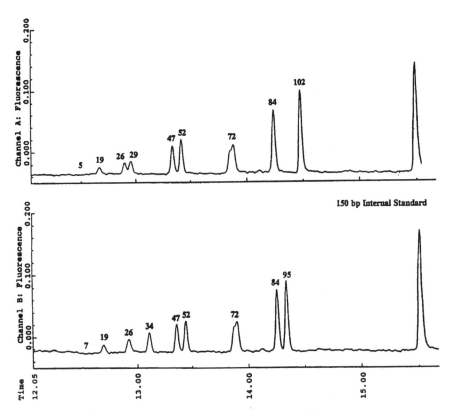

Fig. 2. CE electropherograms comparing two mtDNA PCR products from the HV1 region. The numbers refer to expected DNA fragment sizes in bp. Restriction enzyme digestion is with *Rsa*I for 30 min at 37°C. An internal standard of 150 bp is added to the samples to align the two electropherograms. The 19-, 26-, 47-, 52-, 72-, and 84-bp peaks remain constant between the two samples. Variation between these two samples is observed with the 29-bp fragment in the upper electropherogram, and the 34-bp fragment in the lower one, as well as a 102-bp fragment in the upper electropherogram and a 95-bp fragment in the lower one. The smallest peaks of 5 bp and 7 bp are not detected (*see* **Note 6**).

3. The recent advent of microchannel CE in the form of microchips and capillary array microfabricated devices suggests that PCR-RFLP analysis will be able to be performed at a much faster rate using shorter separation distances than conventional CE instruments *(10)*.

4. Although sequence analysis is the most comprehensive method of detecting DNA polymorphisms, it is relatively time consuming and expensive. Sequencing is also inefficient if one is only trying to evaluate a single nucleotide or even a limited number of nucleotide positions out of a 500-nt sequencing read. More rapid tests, such as the PCR-RFLP technique described here, the dot blot assay *(18)*, or minisequencing *(19)* assays, provide alternative methods for screening samples (at a fixed number of polymorphisms defined by assay probes) and can reduce the cost and time of analysis compared to DNA sequencing.

5. Multiple restriction enzymes may be used sequentially or in combination to distinguish a greater number of DNA types than a single restriction enzyme.
6. Intercalating dyes work well for detecting DNA restriction fragments over a wide range of sizes. A lower sensitivity limit of ~20 bp exists due to a limited number of intercalating sites in short DNA molecules. However, the inability to see the smaller fragments does not adversely impact the PCR-RFLP assay as variation in larger fragments also can be seen. Other fluorescent labeling techniques, such as incorporation of fluorescent dye-labeled primers, are not as effective for PCR-RFLP since only the variation in a single fragment (i.e., the one closest to the end) may be seen.
7. Rapid CE results are possible for separating restriction fragments with widely differing sizes (*1*). The resolution can be tailored to the minimum size difference requirement between DNA fragments. For example, if the expected fragment sizes after digestion are 100 and 150 bp, then the time of separation can be reduced by increasing the CE voltage, since a resolution of only 50 bp is needed to distinguish between the two fragments.
8. DNA sizing information is not always a necessity with PCR-RFLP. Pattern recognition (i.e., 2 peaks vs 3 peaks) can often be more important than the measured DNA size in base pairs. An internal DNA sizing standard, such as a 150-bp fragment, may be added to samples to adjust for migration time differences between CE runs over time if accurate sizing information is desired.

Acknowledgments

The method described here was developed while John M. Butler was a National Research Council postdoctoral fellow in the Biotechnology Division of the National Institute of Standards and Technology. The commercial manufacturers are identified only for the purpose of describing the method, and inclusion does not imply endorsement by the National Institute of Standards and Technology.

References

1. Butler, J. M., Wilson, M. R., and Reeder, D. J. (1998) Rapid mitochondrial DNA typing using restriction enzyme digestion of polymerase chain reaction amplicons followed by capillary electrophoresis separation with laser-induced fluorescence detection. *Electrophoresis* **19**, 119–124.
2. Nathakarnkitkool, S., Oefner, P. J., Bartsch, G., Chin, M. A., and Bonn, G. K. (1992) High-resolution capillary electrophoretic analysis of DNA in free solution. *Electrophoresis* **13**, 18–31.
3. Klepárník, K., Malá, Z., Doskar, J., Rosypal, S., and Boček, P. (1995) An improvement of restriction analysis of bacteriophage DNA using capillary electrophoresis in agarose solution. *Electrophoresis* **16**, 366–376.
4. Ulfelder, K. J., Schwartz, H. E., Hall, J. M., and Sunzeri, F. J. (1992) Restriction fragment length polymorphism analysis of ERBB2 oncogene by capillary electrophoresis. *Anal. Biochem.* **200**, 260–267.
5. Maschke, H. E., Frenz, J., Belenkii, A., Karger, B. L., and Hancock, W. S. (1993) Ultrasensitive plasmid mapping by high performance capillary electrophoresis. *Electrophoresis* **14**, 509–514.
6. Schlenck, A., Visvikis, S., O'Kane, M., and Siest, G. (1996) High-resolution separation of PCR product and gene diagnosis by capillary gel electrophoresis. *Biomed. Chromatogr.* **10**, 48–50.

7. Bull, T. J., Shanson, D. C., and Archard, L. C. (1995) Rapid identification of mycobacteria from AIDS patients by capillary electrophoresis profiling of amplified SOD gene. *J. Clin. Pathol. Mol. Pathol.* **48,** M124–M132.

8. Avaniss-Aghajani, E., Jones, K., Chapman, D., and Brunk, C. (1994) A molecular technique for identification of bacteria using small subunit ribosomal RNA sequences. *BioTechniques* **17,** 144–149.

9. Del Principe, D., Iampieri, M. P., Germani, D., Menichelli, A., Novelli, G., and Dallapiccola, B. (1993) Detection by capillary electrophoresis of restriction length polymorphism: analysis of a polymerase chain reaction-amplified product of the DXS 164 locus in the dystrophin gene. *J. Chromatogr.* **638,** 277–281.

10. Simpson, P. C., Roach, D., Woolley, A. T., Thorsen, T., Johnston, R., Sensabaugh, G. F., and Mathies, R. A. (1998) High-throughput genetic analysis using microfabricated 96-sample capillary array electrophoresis microplates. *Proc. Natl. Acad. Sci. USA* **95,** 2256–2261.

11. Burke, W., Thomson, E., Khoury, M. J, McDonnell, S. M., Press, N., Adams, P. C., et al. (1998) Hereditary hemochromatosis - Gene discovery and its implications for population-based screening. *J. Am. Med. Assoc.* **280,** 172–178.

12. Feder, J. N., Gnirke, A., Thomas, W., Tsuchihashi, Z., Ruddy, D. A., Basava, A., et al. (1996) A novel MHC class I-like gene is mutated in patients with hereditary haemochromatosis. *Nat. Genet.* **13,** 399–408.

13. Lynas, C. (1997) A cheaper and more rapid polymerase chain reaction restriction fragment length polymorphism method for the detection of the HLA-H gene mutations occurring in hereditary hemochromatosis. *Blood* **90,** 4235–4236.

14. Wilson, M. R., Polanskey, D., Butler, J. M., DiZinno, J. A., Replogle, J., and Budowle, B. (1995) Extraction, PCR amplification, and sequencing of mitochondrial DNA from human hair shafts. *BioTechniques* **18,** 662–669.

15. Butler, J. M., McCord, B. R., Jung, J. M., Lee, J. A., Budowle, B., and Allen, R. O. (1995) Application of dual internal standards for precise sizing of polymerase chain reaction products using capillary electrophoresis. *Electrophoresis* **16,** 974–980.

16. Devaney, J. M. and Marino, M. A. (2001) Purification methods for preparing polymerase chain reaction products for capillary electrophoresis analysis, in *Capillary Electrophoresis of Nucleic Acids,* Vol. 1 (Mitchelson, K. R., and Cheng, J., eds.), Humana Press, Totowa, NJ, pp. 43–49.

17. McCord, B. R., Jung, J. M., and Holleran, E. A. (1993) High resolution capillary electrophoresis of forensic DNA using a non-gel sieving buffer. *J. Liq. Chromatogr.* **16,** 1963–1981.

18. Stoneking, M., Hedgecock, D., Higuchi, R. G., Vigilant, L., and Erlich, H. A. (1991) Population variation of human mtDNA control region sequences detected by enzymatic amplification and sequence-specific oligonucleotide probes. *Am. J. Hum. Genet.* **48,** 370–382.

19. Tully, G., Sullivan, K. M., Nixon, P., Stones, R. E., and Gill, P. (1996) Rapid detection of mitochondrial sequence polymorphisms using multiplex solid-phase fluorescent minisequencing. *Genomics* **34,** 107–113.

5

High Resolution Analysis of Point Mutations by Constant Denaturant Capillary Electrophoresis (CDCE)

Konstantin Khrapko, Hilary A. Coller, Xiao-Cheng Li-Sucholeiki, Paulo C. André, and William G. Thilly

1. Introduction

1.1. Purpose and Main Features of CDCE

In recent years, the need for techniques capable of detecting and identifying point mutants has increased dramatically in the fields of genomics, cancer research, and molecular diagnostics. The large arsenal of methods for mutation detection ranges from direct sequencing and array hybridization to allele-specific polymerase chain reaction (PCR). CDCE *(1)* holds a unique position within this list of techniques due to its high flexibility. Having set up a CDCE system, one can use it to solve a wide range of tasks. At one extreme, CDCE is capable of detecting mutants at very low frequencies (as low as 10^{-6}) *(2,3)*, in which it rivals the high sensitivity mutation detection methods like allele specific PCR. At the other extreme, CDCE is capable of identifying mutants with precision comparable to that of direct sequencing *(4)*. In addition, the instrument for CDCE does not need to be dedicated to one purpose. In fact, most capillary DNA electrophoresis systems, such as DNA sequencing capillary instruments, can be adapted for CDCE separations.

1.2. Theory of CDCE: Principle of Separation

Separation of point mutants by CDCE is based on the different melting behavior of DNA fragments with different mutations. A DNA fragment suitable for separation of mutants by CDCE should consist of two isomelting domains, one with a low and the other with a high melting temperature. The melting of such a fragment proceeds through an equilibrium state between the completely unmelted form and the partially melted intermediate, in which the low melting domain is completely denatured, whereas the high melting domain is still completely helical. This partially melted intermediate form has a significantly reduced electrophoretic mobility compared to

From: *Methods in Molecular Biology, Vol. 163:*
Capillary Electrophoresis of Nucleic Acids, Vol. 2: Practical Applications of Capillary Electrophoresis
Edited by: K. R. Mitchelson and J. Cheng © Humana Press Inc., Totowa, NJ

fully duplex molecules. Consequently, as the melting proceeds (e.g., with increasing temperature), the average mobility of the fragment decreases since the equilibrium shifts toward the partially melted intermediate and the DNA fragment spends more time in the low-mobility state (*see* **Note 1**). It appears that the melting behavior of such a fragment is highly sensitive to single nucleotide changes in the low melting domain. Hence, at a given temperature, the melting equilibrium of a fragment carrying a mutation will be shifted compared to the wild-type sequence. This, in turn, will result in a changed electrophoretic mobility of the mutant fragment, and the two species will be potentially separable by electrophoresis. For a comprehensive discussion of the theory of CDCE, *see* **ref. 5**. Historically, CDCE evolved from denaturant gradient gel electrophoresis *(6)* and constant denaturant gel electrophoresis *(7)*.

1.3. Overview of the Procedure

In accordance with the principle outlined above, the first step of the procedure is to choose an appropriate target sequence with the required biphasic melting profile. The target sequence should be either PCR amplified, or cut out of genomic DNA by restriction enzymes. To separate mutants and wild-type molecules, the mixture of DNA fragments is run through a polyacrylamide-coated capillary filled with replaceable linear (noncrosslinked) polyacrylamide matrix. The actual separation of mutants takes place in a portion of the capillary maintained by a water jacket at the temperature appropriate for the establishment of the partial melting equilibrium. The DNA fragments pass through a LIF detector and are recorded as individual peaks in a separation profile, after being separated in the heated zone. Further, the mutant molecules may be individually collected by fractional elution at the anode end of the capillary.

2. Materials
2.1. Selection of a Target DNA Sequence

1. MacMelt®/WinMelt® software (for Macintosh and Windows, respectively), Bio-Rad (Hercules, CA).

2.2. PCR Amplification

1. A light-driven thermocycler (Idaho Technology, Idaho Falls, ID).
2. *Pfu* thermostable DNA polymerase (Stratagene, La Jolla, CA).

2.3. Capillaries and Separation Matrix

1. Fused-silica capillaries: of 75 µm id and 375 µm od (Polymicro Technologies, Phoenix, AZ). Capillaries can be prepared in large batches and used as needed to save time and minimize the influence of batch-to-batch variability. Approximately 12 m of fused silica capillary coiled on a bobbin for convenience is used.
2. Dry methanol, high performance liquid chromatography (HPLC) grade.
3. [3-(methacryloyloxy)propyl]trimethoxysilane (Sigma, St Louis, MO).
4. 30% acrylamide solution: filter through a 0.2-µm filter and store at 4°C.
5. Freshly prepared, 10% ammonium persulfate, electrophoresis grade.
6. *N,N,N',N'*-tet TEMED, electrophoresis grade.
7. Stainless steel thin needles (#7186) and pipeting needles (#7957) (Popper & Sons, New Hyde Park, NY).

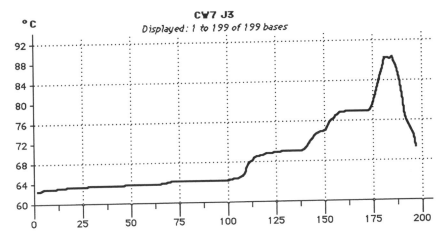

Fig. 1. A typical melting profile of a DNA fragment suitable for CDCE separation. A fragment of human mitochondrial genome (base pair 1 corresponds to the genomic position 10,011).

8. 10-mL glass syringes: clean, dry, and chilled at 4°C.
9. 100-μL high pressure-lok syringes, clean and dry: (#160025WB, VICI Precision Sampling, Baton Rouge, LA).
10. Teflon tubing of 1/16" od and 0.010" id (Bodman, Aston, PA).

2.4. Instrument

1. High-voltage power supply: model CZE-1000R, (Spellman, Plainview, NY).
2. Neslab Instruments model EX-111 Constant Temperature Circulator, microprocessor-controlled with external temperature sensor (Neslab Instruments, Portsmouth, NH).
3. Stainless steel tubing, (Small Parts, Miami Lakes, FL have a wide collection of steel tubing and will cut it to specifications). Steel tubing with the inner diameter slightly larger than the outer diameter of the capillary is necessary. For example, 22-gauge tubing for 375-μm capillary.
4. MP100 data acquisition system, (Biopack Systems, Goleta, CA).

2.5. Electrophoresis

1. Electrophoresis buffer (1X TBE): 89 mM Tris-base, 89 mM borate, pH 8.4, 1 mM EDTA, filter through a 0.2-μm filter.

3. Methods
3.1. Selection of a Target DNA Sequence

1. The goal of this step is to find or create a suitable CDCE target sequence, i.e., a DNA fragment with an appropriate biphasic melting profile, which can be successfully used to detect mutations.
2. The melting profile of any given sequence can be calculated using MacMelt/WinMelt software (Bio-Rad), which is based on the original program developed at MIT (*5*). The melting profile of a typical "good" CDCE target plotted by MacMelt is shown in **Fig. 1**. The fragment consists of about 110 bp with relatively low thermal stability (the low melting domain, bp 0–110) adjacent to a stretch with a higher thermal stability (the high melting domain, bp 111–199). It is desirable that the melting profile of the low melting domain

remains flat or descends gently and monotonically outward, away from the high melting domain. "Humps" and "pits" on the low melting domain profile should be avoided (*see* **Note 2**).

3. If a high melting domain does not naturally occur adjacent to the target sequence of interest, it can be artificially attached during PCR by the choice of appropriate primers. For example, one primer contains an additional non-monotonous 5' GC-rich sequence region in addition to the actual genomic annealing priming sequence at the 3' end of the primer. The additional GC region is then incorporated into the extended PCR product to create an artificial high melting domain (clamp). Alternatively, the clamp domain can be ligated to the target sequence using specially designed adapters *(1)*. The commonly used 40-bp clamp sequence is d(CGC CCG CCG CGC CCC GCG CCC GTC CCG CCG CCC CCG CCC G). In general, the addition of the clamp to a sequence makes its melting profile much flatter than that of the sequence without a clamp. In fact, many sequences, which alone do not look at all as an isomelting domain, will be turned into an isomelting domain upon the addition of a clamp.

3.2. PCR Amplification

1. In most situations, the DNA samples for CDCE are prepared by PCR amplification, although other approaches are required in certain important cases *(1)*.

2. The reaction conditions for PCR amplification of the DNA fragments deserve special attention. There are two types of PCR amplification artifact, each of which create a problem for CDCE separation and the interpretation of peaks.

3. First, PCR is known to introduce DNA polymerization (misincorporation) errors into the amplified DNA. This becomes a problem when one tries to detect mutants occurring at low frequencies, because the PCR-generated mutants create a background that makes it impossible to detect the original low-frequency mutations encoded in the genomic template *(8)*.

4. Secondly, PCR generates significant amounts of "modified" products, which are distinct from the classic primer-to-primer double-stranded PCR product. Examples of these "modified" species include fragments with additional nontemplate 3'-adenosine nucleotides *(9)* and fragments with incomplete strands. Some of these "modifications" will alter the temperature at which the low melting domain disassociates, and thus will be separated from the main wild-type peak during CDCE and be detected, although they do not carry any mutations. In practice, these "modified" wild-type fragments create a background signal in those regions of the separation profile, where mutant peaks are expected to appear. This decreases both the sensitivity of mutant detection and the efficiency of fraction collection.

5. It is worth noting that these "modified" species usually go unnoticed because their mobility in conventional electrophoresis is very similar to that of "normal" PCR products. Therefore, even a PCR product that appears as a perfect single band in a conventional gel might be unsuitable for CDCE.

6. Both of the above-mentioned PCR problems can be minimized by use of thermostable *Pfu* DNA polymerase, which is known to have the lowest error rate of all the thermostable DNA polymerases. In addition, we found that under PCR conditions described below (2–4-fold higher concentration of the enzyme and a particular PCR temperature profile), *Pfu* is capable of producing very low levels of the "modified" PCR products.

3.2.1. PCR Amplification Conditions

1. PCR is performed in 10-µL sealed glass capillaries.
2. A light-driven thermocycler (Idaho Technology, Idaho Falls, ID), which is capable of extremely high ramping rates, is used in conjunction with native *Pfu* thermostable DNA polymerase (Stratagene, La Jolla, CA).
3. The PCR buffer supplied by the manufacturer includes: 20 mM Tris-HCl, pH 8.0, 10 mM KCl, 6 mM (NH$_4$)$_2$SO$_4$, 2 mM MgCl$_2$, and 0.1% Triton X-100. The reaction mix also has added 100 µg/mL BSA, 0.2 µM of each primer, 0.1 mM dNTPs, and 0.1 U/µL of *Pfu* DNA polymerase.
4. The cycle consists of a denaturation step at 95°C, an annealing step at a temperature determined for each primer pair according to conventional protocols, and an extension step at 72°C, 10 s each. After the desired number of cycles, the samples are incubated at 72°C for 2 min and then at 45°C for 30 min. Under such conditions it should be possible, by applying extra PCR cycles, to "exhaust" the primers, i.e., incorporate most of the primer molecules into the PCR product. In our hands, these conditions result in DNA products of about 10^{11}copies/µL.
5. The high melting domain amplification primer is labeled with a fluorescent tag, usually fluorescein or tetramethylrhodamine (TMR). It is important that the label is incorporated into the product via the high melting domain primer. In our hands, labeling of the low melting domain primer with fluorescein results in a doubling of all peaks in the separation profile. Apparently, the fluorescein residue exists in two isomeric forms which differentially affect the melting temperature of the low melting domain and, hence, the CDCE mobility of the fragment.

3.3. Capillaries and Separation Matrix

3.3.1. Coating of the Capillaries

1. The inner surface of the capillaries must be coated with linear polyacrylamide chains to prevent electroosmotic flow which otherwise severely limits the efficiency of separation.
2. The coating procedure is adapted from the method of Hjertén *(10)*. During the procedure, the inner surface of the capillary is first activated by NaOH/HCl treatment to produce free silanol groups. The silanol groups then serve as anchors for the attachment of a silane bearing methacryl groups, [3-(methacryloyloxy)propyl]trimethoxysilane.
3. Further, a subsequent chain reaction of acrylamide polymerization is performed inside the capillary. The surface-attached methacryl groups are incorporated into the growing polyacrylamide chains along with the acrylamide monomers, so that the polyacrylamide chains become covalently bound to the inner surface of the capillary.
4. Capillaries can be prepared in large batches to minimize the influence of batch-to-batch variability. Approximately 12 m of fused silica capillary coiled on a bobbin is filled with 1 M NaOH and incubated for 1 h. The capillary is then washed sequentially with 1 M HCl, then with dry methanol, then rinsed once with [3-(methacryloyloxy)propyl]trimethoxysilane and then left to react overnight with fresh [3-(methacryloyloxy)propyl]trimethoxysilane, and finally the capillary is washed clean with methanol.
5. Immediately following the methanol rinse, the capillary is filled with a fresh, ice-cold 6% acrylamide solution in TBE containing 0.1% TEMED and 0.025% ammonium persulphate, and then left to polymerize for 1 h at room temperature. The coated capillary can be

stored in a refrigerator for at least a year; and pieces of the appropriate length (about 30 cm) can be cut and used as needed. Further information concerning coating of capillaries is described in **ref. *11***.

3.3.2. Fluidics: Syringes and Fittings

1. The replacement of the reagents during capillary coating procedures and the replacement of the separation matrix between electrophoresis runs (*see* **Subheading 3.3.3.**) requires relatively high pressures.
2. We use 100-μL syringes with teflon-lined plungers and blunt needles equipped with custom fittings to attach the syringes to the capillaries. The fittings are made of Teflon tubing (Bodman, Aston, PA) with an inner bore diameter that is smaller than the outer diameter of the capillary to be attached. For example, a 1/16 in. od, 0.01 in. id tubing is appropriate for the commonly used 375-μm od capillary.
3. A piece of tubing (5 cm long) is heated on a weak flame until tender and the syringe needle is quickly inserted about 2 cm deep into the bore of the tubing. Then the other end of the tubing is heated and a spare piece of capillary is inserted into it about 1 cm deep. After the tubing cools, the capillary piece is pulled out. The expanded bore enables the capillaries to be inserted and attached to the syringe.

3.3.3. Separation Matrix

1. The capillaries are filled with a fluid polyacrylamide matrix, which must be replaced before each run. It has been determined that a polyacrylamide matrix with longer chains provided a higher resolution. The goal of the following procedure is to synthesize very long linear polyacrylamide chains.
2. To achieve this goal, the concentration of the chain initiator, ammonium persulfate, is kept very low, so that the number of initiated chains is small. In addition, the reaction mixture is completely deoxygenated by bubbling argon gas to prevent oxygen molecules from terminating the growing polyacrylamide chains. These two factors, in conjunction with low polymerization temperature, result in the synthesis of extremely long polyacrylamide chains. The resulting solution, although it contains only 5% polyacrylamide, is an extremely viscous gel-like substance.
3. The matrix is prepared as follows: A 5% acrylamide solution in TBE is placed into a flask sealed with multiple layers of parafilm and placed on an ice-bath. The solution is deoxygenated for 10 min by intense argon bubbling via a long needle inserted through the parafilm. The flask is kept at positive argon pressure thereafter to avoid any contact of the solution with air.
4. The TEMED and ammonium persulfate are added to 0.03% and 0.003%, respectively, by inserting pipets through parafilm.
5. The solution is thoroughly mixed and is immediately dispensed into 10-mL glass syringes (flushed with argon) by sucking with long needles inserted through the parafilm.
6. The syringes are left overnight at 4°C to allow acrylamide to polymerase. This basic matrix polymer can be stored in the 10-mL syringes at 4°C for at least a year.
7. The matrix is dispensed as needed from the 10-mL syringes into 100-μL U6K syringes, which are used to replace the matrix in capillaries.
8. The recipe above describes a basic matrix containing only 1X TBE. Components such as sodium borate can be added to improve resolution (*see* **Note 3**), or urea/formamide can be included to decrease the optimal separation temperature (*see* **Subheading 3.5.4.**). The additives should be added to the cocktail prior to the polymerization step.
9. If denaturants are present in the matrix polymer, it should be stored at a temperature just above its freezing point (e.g., –15°C for 20% urea, 20% formamide). The low tempera-

ture decreases the rate of degradation of the denaturants. The accumulation of degraded denaturants would cause a gradual increase in the ion concentration and in the conductivity of the matrix, and therefore would raise the optimal separation temperature as ions stabilize DNA structure (*see* **Note 4**).

3.4. Capillary Electrophoresis Instrumentation

A CDCE instrument consists of a CE apparatus, a constant temperature circuit, and a detector, all assembled on an optical breadboard. The all-inclusive cost of parts required for the assembly is below $25,000. A full description of the instrument is beyond the limits of this communication. Anyone interested in using our setup is welcome to contact the corresponding author for detailed building instructions.

3.4.1. Capillary Electrophoresis

1. The capillary is positioned horizontally 3–4 in. above the breadboard and held by the water jacket and the detector. The grip is loose and the capillary can be easily removed and replaced when it is worn out (*see* **Subheading 3.5.3.**).
2. The ends of the capillary are bathed in two buffer reservoirs filled with the electrophoresis buffer. Each buffer reservoir (Weaton) is a short glass tube with threads for septa caps on each end. One septum is permanently pierced with a short piece of a pipet tip, which facilitates removal of the capillary from the reservoir to inject a sample or elute a fraction.
3. The buffer reservoir is quite an unusual vessel: something like a bottle with two caps on the opposite ends. The ends of the capillary are bathed in two buffer reservoirs filled with the electrophoresis buffer (about 5 mL). The reservoir should have two entrances, one for the capillary and one for the electrode, preferably on the opposite ends. A short glass tube with a thread for a septum cap on each end is used. One septum is pierced with a short piece of a pipet tip. The tip is left in the septum and is used to insert/remove the capillary into/from the reservoir to inject a sample or collect a fraction. The electrode (a piece of platinum wire) is inserted through the other septum. Alternatively, a primitive buffer reservoir can be made by piercing opposite ends of a plastic tube.
4. The reservoirs are held by horizontal three-prong clamps attached to vertical swiveling rods. The clamp with the reservoir can be easily rotated in the horizontal plane to expose the tip of the capillary for sample injection, or for fraction collection.
5. The other septum of each reservoir is pierced with a short piece of platinum wire, which serves as an electrode. The platinum wire is connected to the power supply via a crocodile connector.
6. Persons who are seriously going to build an instrument should contact Professor W. G. Thilly for details.

3.4.2. Water Jacket

1. As described in **Subheading 1.**, the actual separation of mutants takes place in a portion of the capillary heated by a water jacket connected to a Neslab EX-111 constant temperature circulator.
2. To make the "jacket," cut a rectangular polystyrene foam block to the desired length (e.g., $15 \times 5 \times 1$ cm^3 for a 15-cm jacket). Wrap silicon tubing around the longer side and both of the two shorter sides of the block, and fix it in place with tape. Although the commonly used white polystyrene foam is appropriate, we prefer construction-grade blue or pink foam to be used for thermal insulation of walls for its superior mechanical strength.

3. Take a piece of steel tubing with the inner diameter slightly larger than the outer diameter of the capillary being used. The piece of steel tubing should be about 1 cm longer than the foam block.

4. Pierce the silicon at one bending point, push the steel tubing inside the silicon tubing along the long side of the block until it reaches the other bending point, and force it to pierce the silicon again. Clamp the jacket to a firm support, insert the capillary into the steel tubing and connect the silicon tubing to a water bath.

5. To correct for possible oscillations of the ambient temperature, the external temperature sensor of the circulator should be inserted into the water stream inside the tubing via a T-shaped connector as close as possible to the jacket.

3.4.3. Detector

1. The capillary is illuminated by a vertical 488-nm argon-ion laser beam. Emitted light is collected at right angles to both the beam and the capillary by a microscope objective, split into two equal beams by a beamsplitter, and directed into two photomultiplier tubes through appropriate sets of filters. The beam path is closed so the instrument can be operated at ambient light.

2. A combination of a 520 ± 10-nm bandpass and a 515-nm long pass filter is used for fluorescein, and for TMR a single 580 ± 10 nm bandpass filter is employed.

3. The signals from the photomultiplier tubes are recorded by a computerized data acquisition system (e.g., MP100 from Biopack Systems, Goleta, CA).

4. It is important that the portion of the capillary illuminated by the laser is stripped of the outer polyimide protective coating, which otherwise absorbs laser light and creates a huge fluorescent background. The coating can be easily removed by gently cutting it off with a razor blade to make a short (~0.5 cm long) stretch of bare fused silica. **NOTE:** A capillary is rather brittle without the polyimide coating, and it should be handled with extreme care to prevent breakage.

3.5. Electrophoresis

1. A typical separation of mutants is performed in 30 cm long by 75-μm id capillaries at about 200 V/cm.

2. A typical current is on the order of 10 μA, but it depends on several variables such as the ionic composition of the buffer, the temperature and length of the jacket, capillary diameter, and so on.

3. During electrophoresis at a constant voltage, the current generally decreases with time, presumably because the concentrations of ions in the capillary changes. The current is restored to the original value after replacement of the matrix.

3.5.1. Loading the Sample

1. DNA is loaded into the capillary by electroinjection. The DNA solution (as little as a few microliters) is put into a 0.6-mL centrifuge tube.

2. Then the cathode-side tip of the capillary is removed from its buffer reservoir and is dipped into the DNA solution together with a short platinum wire, which is used as an electrode. The other tip of the capillary is still in its reservoir connected to the power supply.

3. The platinum wire is clamped to the tube by the cathode crocodile connector of the power supply and a current pulse is applied to inject the DNA.

4. **NOTE:** *To avoid an electrical shock (not dangerous, though painful), make sure that the "ground" output of the power supply is used for injecting and the high-voltage output is positive and connected to the other end of the capillary.*

5. The amount of DNA injected is proportional, to a first approximation, to the amount of electricity in the pulse (current × time), and to the relative concentration of DNA vs other anions present in the solution.

6. Anions compete with DNA molecules for the injection current, thus higher anion concentrations result in the electrokinetic injection of proportionally smaller amounts of DNA, even if the DNA concentration is kept constant.

7. PCR reactions can be used directly for sample injection without purification: the salt concentration in the PCR buffer allows for the injection of a convenient amount of about 10^8 copies for a 30 s × 1 μA pulse. A PCR reaction contains about 10^{11} copies/μL, which is much more than necessary for a single injection.

8. Alternatively, the PCR reaction solution can be diluted ~10 times with water, which will not decrease the injection efficiency since the DNA/anion ratio does not change. Since only a small fraction of DNA in the sample is injected into the capillary (about 0.1%) with each injection, multiple injections can be performed from such a sample. There is only a slight decrease in loading efficiency after several samplings, which can be compensated for by increasing the duration of the loading pulse.

3.5.2. Loading of Large Amounts of DNA

1. In certain cases (e.g., loading of the TMR-labeled standard mutants, *see* **Subheading 3.6.2.**), it is necessary to load much larger amounts of DNA.

2. Desalting of the DNA sample (e.g., by ethanol precipitation and redissolving into water) results in injections of about 10^{10}–10^{11} copies of a fragment, because many fewer anions are now competing with DNA for the injection current.

3. To precipitate DNA from a PCR reaction: Add 3 vol of ethanol to the reaction and mix well, hold the mixture at room temperature for 10 min. Centrifuge the mixture at 12,000*g* for 3 min to pellet the DNA, remove the supernatant, centrifuge the tube again briefly and carefully remove any residual supernatant, then dissolve the pellet in 1 vol of water.

4. The desalted sample can be injected in the same way as described in **Subheading 3.5.1.**

3.5.2.1. LOADING OF LARGE AMOUNTS OF DNA FROM TUBING

1. However, if it is important to quantitatively load the entire DNA sample into the capillary, a "loading from tubing" procedure should then be used.

2. About 1 μL of the desalted sample should be pipeted into a piece of Teflon tubing (~1 cm long) with an id exactly matching the od of the capillary. Such a "loading tubing" can be made by forcing a spare piece of capillary into a piece of Teflon tubing softened by heating on a weak flame. The piece of capillary is then pulled out, leaving the tubing with a bore of exactly the required id.

3. The separation capillary is then inserted into the tubing with the sample, avoiding a bubble between the capillary and the sample. Then the capillary is pushed inside the tubing until the sample moves all the way to the other side of the tubing, thus displacing all the air from the bore. The capillary with the tubing is inserted into the buffer reservoir. After an injection pulse of about 10 μA for 30 s, the tubing is removed and the electrophoretic separation is performed as usual.

3.5.3. Stability of the Capillary

1. A capillary normally withstands about a hundred injections. However, multiple injections of very large amounts of DNA (over 10^{11} copies for a 75-μm id capillary), or the presence of some impurities in DNA samples results in the premature "wearing" of the capillary.

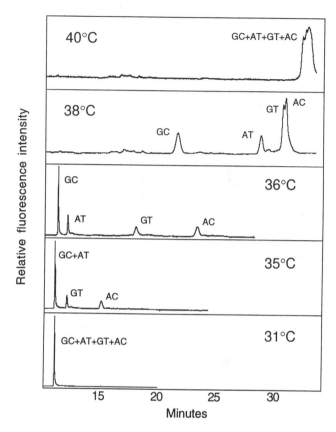

Fig. 2. Effect of temperature on a typical CDCE separation. Conditions: 1X TBE, 20% urea, 20% formamide, 10-cm jacket, 250 V/cm. DNA fragment as in **Fig. 1.**

2. "Failure of the capillary" manifests itself as a sudden increase in electrical resistance and (in most cases) the appearance of bubbles inside the injection end of the capillary.
3. Although sometimes the capillary recovers from a failure after matrix replacement, the safest option is to change the capillary once it fails. If injection of a sample repeatedly causes a decrease in the current, the injection pulse should be decreased or wide bore capillaries should be used *(3)*.

3.5.4. Optimization of Separation Conditions

1. This subheading describes the procedure to determine the optimal temperature for separation of mutants occurring in the low melting domain of a given target DNA fragment with a biphasic melting profile.
2. The effect of temperature on a typical CDCE separation is illustrated in **Fig. 2.** The sample consists of four DNA species differing in one base pair. Two of the four DNA variants are perfect duplexes which contain either GC or AT base pair at a certain position within the

low melting domain. The other two species are heteroduplexes, which contain the mismatches GT or AC in the same position.

3. The melting profile of the GC fragment is also shown in **Fig. 1**. The GC to AT change results in destabilization of the low melting domain. However, mismatches in the duplex cause even more significant destabilization.

4. As shown in **Fig. 2**, all four species migrate as a single peak at 31°C because none of the fragments are partially melted for any significant proportion of the time during their migration through the heated zone. Therefore, all of the fragments migrate at a speed characteristic of double-stranded DNA.

5. However at 35°C, the two heteroduplexes containing AC and GT mismatches are partially melted, such that the mobilities of the two heteroduplexes are decreased and they separate from the two homoduplexes which remain unmelted.

6. At 36°C, all four DNA species are at a "partial melting equilibrium" (although to different extents) and are separated nicely from each other.

7. At 38°C, the partial melting equilibrium of the two heteroduplexes is completely shifted toward the "partially melted" form. The mobilities of the two species are therefore very low and are barely able to be distinguished from each other.

8. Finally at 40°C, all four species are partially melted most of the time, and all have very similar electrophoretic mobilities, corresponding to the mobility of the partially melted form of the DNA fragment.

9. **Figure 2** also illustrates a general feature of CDCE separations. As the temperature in the heated zone of the capillary increases, the mobility of a DNA fragment undergoes a sharp decrease that reflects a transition from the unmelted to the partially melted form. Thus, the optimal temperatures for the separation of any set of mutants lies within this range of "transitional temperatures," at which the mutants are in transition from the duplex to the partially melted form.

10. The melting temperature of the low melting domain, as provided by the MacMelt software, gives a fair estimate of the transitional temperature, assuming that the separation matrix only contains TBE.

11. If separation matrix also contains denaturants, the transition temperature should be corrected by decreasing approx 0.63°C/1% of added urea or formamide. For example, the theoretical transition temperature of the GC fragment shown in **Fig. 2** is about 63°C. The matrix used for separation contained 20% urea and 20% formamide, hence the optimal separation temperature is expected to be $63 - (0.63 \times 20) - (0.63 \times 20) = 38°C$.

12. The estimated transition temperature is a good starting point for the optimization of the procedure, and to refine the separation conditions. In order to determine both the temperature range of the transition, and the corresponding electrophoretic mobilities of the partially melted fragments, a series of test CDCE separations are performed at several temperatures above and below the estimated transition temperature.

13. If mutant DNAs are available, the optimal separation temperature can be determined directly by selecting the best separation pattern among those obtained at different temperatures. For example, the optimal separation temperature for the set of mutants shown in **Fig. 2** appears to be between 36°C and 38°C, i.e., about 1°C below the predicted temperature.

14. If only the wild-type DNA fragments are available for initial testing, a good approximation of the "optimal separation temperature" can be made. This is the temperature at which the mobility of the wild-type decreases by 1/4 of the difference between the mobilities of its unmelted and partially melted forms.

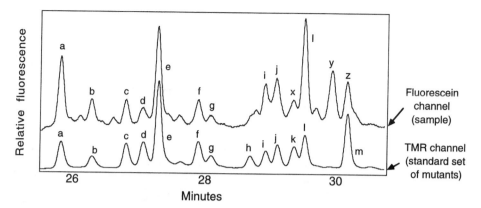

Fig. 3. Identification of mutants in complex separation profiles. Conditions: 15-cm jacket, 1X TBE +30 mM sodium borate pH 8, at 71°C, 150 V/cm. DNA fragment as in **Fig. 1.**

3.5.5. Quantification and Sensitivity

1. The relative amounts of DNA molecules in each peak of a separation profile can be determined by comparing the areas under the corresponding peaks.
2. If absolute numbers of molecules per peak are required, the instrument must be calibrated by injecting a known amount of labeled DNA using the "loading tubing" approach (*see* **Subheading 3.5.2.1.**).
3. An aliquot of a fluorescently labeled DNA sample is subjected to gel electrophoresis, stained (e.g., by ethidium bromide) and the copy number is quantified by comparison to conventional DNA standards (e.g., commercially available DNA size markers).
4. Another aliquot of the same sample with a known number of DNA copies is then quantitatively injected into the capillary and electrophoresed under nondenaturing conditions, so that the DNA appears as a single peak.
5. The area under the peak is measured and the conversion coefficient from the area (in V × s) to copy number is calculated. The coefficient can be used to calculate the copy number for any peaks given that the DNA fragments are labeled with the same dye, and that detector parameters are unchanged.

3.6. Identification of Mutants

1. CDCE is capable of resolving complex mixtures of mutants as shown in **Fig. 3**. It is therefore essential to be able to unambiguously identify each of the multiple peaks of a separation profile.
2. Initially, the mutant molecules are individually isolated by fraction collection, PCR amplified, and DNA sequenced.
3. As more samples are analyzed and more mutants are isolated and sequenced, it may become clear that the same specific mutant molecules appear in different samples. Consequently, more efficient modes of identification of the mutants than isolation and sequencing are acceptable. For example, identification by comigration and by hybridization with known standards is possible.

3.6.1. Isolation of Mutants by Fraction Collection

1. The procedure for the collection of CE fractions is a mirror image of loading a sample (*see* **Subheading 3.5.1.**). The anode buffer reservoir is moved away from the capillary tip and the tip is dipped into a 0.5-mL centrifuge tube containing 5 µL of 0.1X TBE, 0.1 mg/mL BSA.
2. A platinum wire is also dipped in the tube and used as a positive electrode.
3. The wire is held in place by a crocodile connector, which grips both the tube and the wire and represents the positive output of the power supply.
4. It is preferable to invert the polarity of the power supply so that the positive output is grounded and the negative output is under high negative potential. The separation is resumed for a specific time interval, then the tube is changed to a new one and another fraction is collected, and so on.
5. To determine the timing of fraction collection, the "cold migration speed" of the DNA fragment in the unheated portions of the capillary. This can be measured by running a sample with the water jacket turned off. The "cold speed" is the same for all fragments irrespective of the mutations they carry. The migration speed for a 200-bp fragment for electrophoresis at 200 V/cm should be about 1 cm/min.
6. Using the value of the "cold speed," the timing for a series of collections can be calculated based on the time it will take the DNA to migrate from the detector to the end of the capillary. Usually, several sequential fractions are collected, PCR amplified, and then analyzed by CDCE. The amplified fraction with the highest enrichment of the mutant is then selected for further CDCE fractionation, or for direct DNA sequencing.

3.6.2. Identification by Comigration

1. Mutant DNA fragments, which have been purified and sequenced, are labeled with TMR via the PCR primer tag and mixed together to form a "standard set" of mutants.
2. The standard set is then coinjected with a fluorescein-labeled sample under study and both samples are independently observed on a single CDCE separation by means of a two-wavelength detector. The mutants in the fluorescein-labeled sample are then identified, based on their comigration with the previously isolated, TMR-labeled mutants (**Fig. 3**).
3. If a peak in the sample does not have a counterpart in the standard set (e.g., peak "y" in the sample profile in **Fig. 3**), then it may be a novel mutation. It should be isolated, sequenced and then added to the "standard set" of mutants.
4. To create a TMR-labeled mutant, the fluorescein-labeled DNA fragment is diluted at least 1000-fold and reamplified with a TMR-labeled primer. The detector setup (*see* **Subheading 3.4.3.**) is optimized for the detection of fluorescein, so the sensitivity for TMR is relatively low, and hence more DNA has to be injected to obtain a clear signal.
5. To increase the amount of DNA injected, the standard set is desalted by ethanol precipitation and injected using the "injection tubing" procedure (*see* **Subheading 3.5.2.1.**).
6. The fluorescein-labeled sample is loaded immediately after the standard set, directly from the PCR reaction (*see* **Subheading 3.5.1.**) and both samples are subjected to a CDCE separation. The mobility shifts induced by the two fluorescent residues attached to the DNA are very similar, and any two differentially labeled DNA fragments carrying the same mutation both migrate at the same speed. This allows the identification of the mutants by direct alignment of the peaks of the two profiles.

3.6.3. Identification by Hybridization

1. Comigration of a known peak with a known standard is not an absolute proof of identity. Two different mutants could comigrate leading to false identification. In order to provide

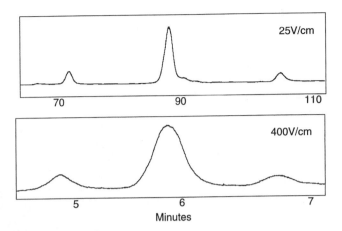

Fig. 4. Performing CDCE separations involves a trade-off between efficiency and speed. Conditions: 5-cm jacket, 0.25X TBE. DNA fragment as in **Fig. 1.**

a more rigorous test of peak identity, a procedure based on on-column hybridization of sample peaks with the mutants of the standard set must be applied.

2. For example, hybridization of the sample shown in **Fig. 3** revealed that peaks x and z in the sample profile were not identical to the comigrating standard mutants k and m, respectively.

3. The CDCE instrument used for "identification by hybridization" includes two consecutive heating jackets.

4. The sample and the standards are coseparated in the first jacket as for identification by comigration, and the resulting set of peaks is stopped inside the second jacket. The temperature in the second jacket is increased initially to denature the DNA, and then the temperature is decreased to permit reannealing of single strands. Finally, the test sample is coelectrophoresed through the rest of the second jacket to the detector, along with an excess of the standard mutant set.

5. Two heteroduplexes between the "sample mutant" and the "standard mutant" will form during the hybridization stage of the electrophoresis if a "mutant peak" in the sample is not identical to the "standard mutant" with which it comigrates. Such heteroduplexes are always less stable than the homoduplexes from which they were derived, and they move to a different position during the posthybridization portion of CDCE separation. This results in the "disappearance" of the sample peak from the spectrum. This approach is as precise as sequencing, with an additional advantage of not requiring isolation of individual mutants prior to identification. More details on the identification of mutants by hybridization can be found in *(4)*.

4. Notes

1. The factors determining separation efficiency of CDCE have been studied in detail in *(5)*. **Figure 4** shows the dramatic decrease of separation efficiency of a triplet of mutants as the speed of separation is increased 16 times by increasing the electric field strength. Unexpectedly, this decrease of efficiency is not a result of Joule heating of the capillary at higher field strengths. Instead, the physical basis for such a trade-off between the speed

and the quality of CDCE separation is in the limiting slow kinetics of the partial melting equilibrium. In other words, the more individual melting-reannealing events the DNA fragments undergo although in the heating zone, the higher is the separation efficiency. Consequently, in addition to decreasing the speed of separation, the efficiency can also be improved by increasing the rate of melting-reannealing reaction. This can be achieved in several ways, as discussed in **Notes 2–4**.

2. Nonmonotonic features of the melting profile of the low melting domain ("humps" and "pits") indicate that partial melting/reannealing of the fragment may be unusually slow. We believe that melting of the DNA duplex domain is normally initiated at its free end, and then proceeds inward as the duplex gradually "unzippers." According to this model, a "hump" in the melting would make a potential barrier for unzippering, and a "pit" would make a barrier for rezippering, which would inhibit one or the other process. Fragments bearing either of these two features are expected to form very broad peaks at the transition temperature. Some peaks are so broad that in a series of test separations performed at increasing temperatures the peak virtually "disappears" as temperature approaches the transition temperature of the low melting domain, and then the peak "reappears" at much higher temperatures. If such a peak is detected, the target sequence should be redesigned.

3. The DNA renaturation rate is proportional to the cube of ionic strength *(12)*. Hence, the separation efficiency is observed experimentally to increase with the ionic strength of the electrophoresis buffer. We recommend adding about 30 mM sodium borate, pH 8.0 to the separation matrix and to the electrophoresis buffer if a high resolution of mutants is essential. Addition of sodium ions also increases the thermostability of DNA and consequently requires an increase of separation temperature by about 6°C.

4. A high concentration of urea/formamide increases the microviscosity of the medium, and this would be expected to slow the rate of DNA melting-reannealing. The efficiency of separation in the presence of denaturants significantly decreases compared with TBE only. The addition of 20% urea and 20% formamide results in about a 2.5-fold decrease in separation efficiency.

Acknowledgments

This work was supported in part by NIH grant CA77044 to WGT.

References

1. Khrapko, K., Hanekamp, J. S., Thilly, W. G., Belenki, A., Foret, F. and Karger, B. L. (1994). Constant denaturant capillary electrophoresis (CDCE): a high resolution approach to mutational analysis. *Nucleic Acids Res.* **22**, 364–369.
2. Khrapko, K., Coller, H., André, P. C., Li, X.-C., Foret, F., Belenky, A., Karger, B. L., and Thilly, W. G. (1997) Mutational spectrometry without phenotypic selection: human mitochondrial DNA. *Nucleic Acids Res* **25**, 685–693.
3. Kim, A., Li-Sucholeiki, X.-C., and Thilly, W. G. (2001) Applications of constant denaturant capillary electrophoresis and complementary procedures: measurement of point mutational spectra, in *Capillary Electrophoresis of Nucleic Acids*, Vol. 2 (Mitchelson, K. R., and Cheng, J., eds.), Humana Press, Totowa, NJ, pp. 175–189.
4. Khrapko, K., Coller, H. A., Hanekamp, J. S., and Thilly, W. G. (1998) Identification of point mutations in mixtures by capillary electrophoresis hybridization. *Nucleic Acids Res* **26**, 5738–5740.
5. Khrapko, K., Coller, H., and Thilly, W. (1996) Efficiency of separation of DNA mutations by constant denaturant capillary electrophoresis is controlled by the kinetics of DNA melting equilibrium. *Electrophoresis* **17**, 1867–1874.

6. Lerman, L. S. and Silverstein, K. (1987) Computational simulation of DNA melting equilibrium and its application to Denaturing Gradient Gel Electrophoresis. *Meth. Enzymol.* **155,** 482–499.

7. Hovig, E., Smith-Sorenson, B., Brogger, A., and Borresen, A.-L. (1991). Constant denaturant gel electrophoresis, a modification of denaturing gradient gel electrophoresis. *Mutation Res.* **262,** 63–71.

8. André, P., Kim, A., Khrapko, K., and Thilly, W. G. (1997) Fidelity and mutational spectrum of *Pfu* DNA polymerase on a human mitochondrial DNA sequence. *Genome Res.* **7,** 843–852.

9. Hu, G. (1993) DNA polymerase-catalyzed addition of non-template extra nucleotides to the 3' end of a DNA fragment. *DNA Cell Biol.* **12,** 763–770.

10. Hjertén, S. (1985) High performance electrophoresis. Elimination of electroendosmosis and solute adsorption. *J. Chromatogr.* **347,** 191–198.

11. Chiari, M. and Cretich, M. (2001) Capillary coatings: choices for capillary electrophoresis of DNA, in *Capillary Electrophoresis of Nucleic Acids*, Vol. 1 (Mitchelson, K. R., and Cheng, J., eds.), Humana Press, Totowa, NJ, pp. 125–138.

12. Studier, F. W. (1969) Effects of the conformation of single-stranded DNA on renaturation and aggregation. *J. Mol. Biol.* **41,** 199–209.

6

Point Mutation Detection
by Temperature-Programmed Capillary Electrophoresis

Cecilia Gelfi, Laura Cremoresi, Maurizio Ferrari, and Pier Giorgio Righetti

1. Introduction

Small alterations in DNA sequence of genomic DNA lead to many human diseases, such as cancer, diabetes, heart disease, atherosclerosis, cystic fibrosis, Alzheimer disease, Duchenne muscular dystrophy, and various thalassemias. These alterations in DNA sequence include many types of mutations and polymorphisms, such as substitutions of one or several nucleotides, deletions or insertions of some larger sequences, differences in variable number of tandem repeats (VNTR), and the genomic instability of microsatellite repeat *(1)*. The diagnosis of human diseases by DNA polymorphism analysis has very important applications in the fields of genetic and medical research, clinical chemistry and forensic science. Because a large portion of sequence variations in the human genome is caused by single base changes, any method used to detect mutations or polymorphisms must be capable of detecting single-base substitutions.

Until recent times, slab-gel electrophoresis has been the standard method for analysis of polymerase chain reaction (PCR) products for the screening of genetic diseases, but the technique is time consuming, labor-intensive, and too unwieldy for precise quantitation. It is particularly vexing that many manipulations are still required to produce the final diagnostic data following the electrophoretic step, such as gel-slab staining (and destaining in silvering techniques), photographing, and signal densitometry. It is for these reasons that capillary zone electrophoresis (CZE) is rapidly emerging as a unique tool for analysis of gene defects *(2)*. CZE represents a fully automated version of such an electrophoresis method. CZE has additional valuable features including the on-line detection and quantitation of analytes, minute sample requirements, automatic sample injection, large analyte handling capability in the case of capillary arrays *(3)*, and drastically reduced analysis times in the case of microfabricated chips *(4)*,

From: *Methods in Molecular Biology, Vol. 163:*
Capillary Electrophoresis of Nucleic Acids, Vol. 2: Practical Applications of Capillary Electrophoresis
Edited by: K. R. Mitchelson and J. Cheng © Humana Press Inc., Totowa, NJ

and/or when using isoelectric buffers *(5)*. This chapter discusses a method our group has developed for precise temperature-programmed CZE runs under nonisocratic conditions that identify point mutations in genomic DNA. First, we will explain two important aspects of the technique: (a) the properties of sieving liquid polymers, as routinely adopted in CZE for all DNA separations, and (b) the basic concepts underlying denaturing gradient gel electrophoresis (DGGE), which is at the heart of our nonisocratic methodology.

1.1. Sieving Liquid Polymers

Modern CZE technologies are based solely on the use of liquid polymers that are prepared in the absence of any crosslinker. A variety of different polymer networks have been described, mainly for separation of dsDNA. Such materials include methyl cellulose (MC), hydroxyethyl cellulose (HEC), hydroxypropyl cellulose, hydroxypropyl methyl cellulose (HPMC), poly(ethylene glycol) (PEG), poly(ethylene oxide) (PEO), poly(vinyl alcohol), liquid or liquid/solid agarose and glucomannan, as reviewed in *(6)*. These polymers are typically used at concentrations in the 0.1–1.0% range, the lower levels (0.1–0.2%) being preferred for large DNA fragments (e.g., 5000–50,000 bp) and the higher amounts (0.7–1.0%) being adopted for resolving shorter DNAs (e.g., 100–500 bp). However, polyacrylamide appears to be the best matrix by far for achieving the best separation of short DNA fragments and for DNA sequencing. In 1990, Heiger et al. *(7)* first proposed the use of polyacrylamide at low and 0% crosslinking levels for these purposes. However, liquid linearpolyacrylamide suffers from many drawbacks. First, the 6–10% polymer solution necessary to achieve DNA sieving is extremely viscous and is pumped into capillaries only with difficulty. Second, polyacrylamide rapidly hydrolyses to produce polyacrylate at the pH values (pH 8.2) and at the high temperatures used for detection of DNA point mutations. "Hot spots" develop along the capillary length, producing air bubbles that cut off the electric current. Moreover, the reproducibility of analyte migration cannot be guaranteed.

Our group has proposed several remedies: viscosity can be drastically decreased by producing "fluidized" or "short-chain" (mol wt of ca. 250,000, Mn of ca. 55,000) polyacrylamides, by polymerizing the monomer in presence of a chain-transfer agent (e.g., 3% 2-propanol) at high temperatures (70°C) *(8)*. Hydrolysis-resistant *N*-substituted acrylamide monomers can eliminate degradation typical of normal acrylamide. We synthesized a series of Ω-hydroxy, *N*-substituted acrylamides, notably *N*-acryloylaminopropanol (AAP) and *N*-acryloylaminobutanol (AAB) *(9–11)*. Additionally, we proved that the performance in DNA separations was even ameliorated in mixtures of these monomers, such as 5% AAP spiked with 1% AAB *(12)*. Moreover, in a recent report *(13)*, we have compared the performance of our novel polyacrylamides with that of celluloses, at both room temperature and at 60°C. Contrary to previously held beliefs, which attributed best separations in polyacrylamides to small DNA fragments (typically in the 50–1000-bp size range) and in celluloses to large DNA fragments, it was shown that celluloses can also achieve fine sieving of short DNA sizes provided they are used at much higher concentrations than previously reported, e.g., in the case of HEC, typically used at 0.2–0.8% concentrations, levels of 3% produce excellent patterns, at 25°C, in the 50–600-bp size range. If separations are conducted at 60°C, sieving is lost in most liquid polymers. However, sieving is fully restored if the concentration of HEC is raised to 6%, and that of HPMC to above 1%. Among all the polymers investigated, an 8% sol of poly(AAP) was found to offer the best performance and the highest theoretical plate numbers in the 25°C–60°C interval. Such separations at high temperatures are necessary when dealing with detection of point mutations in temperature-programmed CZE.

1.2. The DGGE Technique

Among the different methods available for revealing DNA point mutants, DGGE, as originally described by Myers et al. *(14)*, has become one of the most popular techniques. DGGE involves the electrophoresis of dsDNAs in a polyacrylamide gel containing increasing concentrations of denaturing agents such as formamide, urea or temperature *(15)*, and exploits the melting properties of the DNA molecule as a separation mechanism. DNA melts in domains at a temperature (T_m) which is dependent on the nucleotide sequence of the fragment under study. At the point along the gradient where the DNA fragment undergoes partial denaturation, its electrophoretic mobility will be strongly retarded as a consequence of the entanglement of branched DNA molecules with the polyacrylamide matrix. This will permit the separation of two dsDNA molecules differing by a single nucleotide in the domain with the lowest T_m, owing to the fact that rather than melting in a continuous zipper-like manner, most DNA fragments melt in a stepwise process within a very narrow range of denaturing conditions. However, DNA segments differing by base changes within the domain with the highest T_m cannot be resolved in this way, since reaching the highest T_m would bring about complete strand dissociation. Complete strand dissociation can be overcome by introducing a GC-rich domain (GC-clamp) at the 5' end of one of the two amplification primers during PCR. The DNA from an individual heterozygous for a mutation or polymorphism will show four bands on DGGE. The two faster migrating bands are the homoduplexes, corresponding to the wild-type (W_t/W_t) and the mutant (M/M) alleles. The two additional bands that migrate at a slower rate in the gel consist of the two heteroduplexes (W_t/M) formed by reassorting of strands during PCR. As they are electrophoresed down the length of a gel containing a denaturing gradient, they melt differentially, so as to be resolved into the typical four-zone pattern (*see* **Notes 1** and **2**). It should be noted that the two heteroduplexes, due to the mismatch reverberating along a more extended domain in the double helix, are the first ones to melt *(14)*. The melting behavior can be predicted by using the computer algorithm developed by Lerman and Silverstein *(15)*. **Figure 1** gives an example of the melting profile of a point mutation in exon IIIB of the *globin* gene that causes thalassemia. The lower melting region has a very flat profile (which is very advantageous from an analytical point of view, since it means a very sharp melting point and thus the generation of a narrow electrophoretic zone) with a T_m value of 73.3°C. The high melting region (i.e., the first 44-base region) is the G-C clamp extremity, which impedes total dissociation into single-stranded filaments (which would result in a complete loss of resolution).

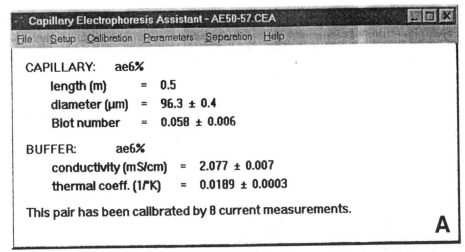

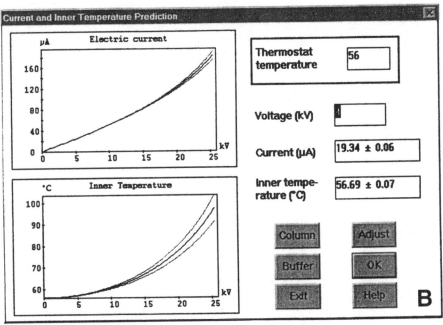

Fig. 1. Method for generating a temperature gradient in a capillary. (A) Input parameters for a specific run, divided into capillary parameters (length, diameter and Biot number) and buffer parameters (conductivity and thermal coefficient). (B) Graphs for predicting current and temperature inside the capillary lumen corresponding to given voltage ramps. The upper graph links a particular voltage profile to the corresponding current generated in the capillary, having the parameters shown in the upper panel. The lower graph links a voltage ramp to a precise temperature generated in the capillary lumen. The thin lines above and below the central, thicker line represent the standard deviation. The windows to the left exemplify a single point prediction along the curve to the right: given an outside thermostat temperature of 56°C, at a voltage of 4 kV and at a current of 19.34 µA, the temperature inside the capillary lumen will be 56.69°C.

2. Materials

2.1. Polymerase Chain Reaction

1. DNA: 1 µg of human genomic DNA.
2. 2 U of DyNAzyme II DNA Polymerase (Finnzymes, OY).
3. 30 pmol/µL of each GC-clamped primer (Boehringer-Mannheim).
4. 200 µM of each of 4 dNTPs (Pharmacia).
5. PCR is performed in a final volume of 100 µL. The PCR buffer mixture contains 10 mM Tris-HCl (pH 8.8 at 25°C), 1.5 mM MgCl$_2$, 150 mM KCl, 0.1% Triton X-100, 200 µM each dNTP, 30 pmol of each primer, 1 µg of genomic DNA template and 2 U of DyNAzyme II DNA polymerase.
6. Thermal cycler (Omnigene, Hybaid).

2.2. Agarose Gel Electrophoresis

1. Agarose: 2 g/100 mL.
2. 50X TAE buffer: 242 g Tris-base, 57.1-mL glacial acetic acid, 37.2 g Na$_2$EDTA 2H$_2$O, H$_2$O to 1 L). Working solution is: 1X.
3. 10 mg/mL ethidium bromide. Working concentration: 0.5 µg/mL.
4. Loading buffer: 70% 1X TAE, 30% glycerol, 0.1% bromophenol blue, 0.1% xylene cyanol.
5. DNA molecular size standard (Boehringer-Mannheim).

2.3. Purification of PCR Products

1. Centricon 30 spin-filters (Amersham).

2.4. Capillary Zone Electrophoresis

1. Capillary: 100 µm id, 375 µm od, 50–60 cm long (Polymicro Technologies) (*see* **Note 3**).
2. TGCE buffer: 8.9 mM Tris-base, 8.9 mM borate, pH 8.3, 1 mM EDTA, 10 mM NaCl, 6 M urea (*see* **Notes 4** and **5**). After preparation, the buffer is passed through a 0.22-µm filter membrane (Millipore).
3. Acryloylaminopropanol (Bio-Rad): 8 g/100 mL is dissolved in filtered thermal gradient capillary electrophoresis (TGCE) buffer.
4. HEC: 27,000 Mr (6 g/100 mL) (Polysciences) is dissolved in filtered TGCE buffer.

2.5. Computer Software

1. Dedicated computer programs are necessary to set up the thermal programmed capillary electrophoresis (TPCE) method. The Melt-87 program of Lerman and Silverstein *(15)* (Bio-Rad Laboratories) is used for calculating the melting profiles of DNAs on the basis of their base composition (and sequence).
2. The program of Bello et al. *(16)* (Bio-Rad Laboratories) is used for calculating buffer electric conductivity (λ), thermal coefficient of conductivity (α), and temperature increments in the capillary lumen is also necessary.
3. Whereas the first program can be run on any personal computer, the other program is adapted to the CZE instrumentation available from Bio-Rad (Bio Focus 2000 or 3000 instruments), thus it might not perform well if used on other CZE equipment.

3. Methods

3.1. DNA Extraction

1. Genomic DNA is extracted from nucleated blood cells by proteinase K digestion and phenol-chloroform extraction according to standard procedures.

3.1.1. DNA Samples

1. DNA specimens are obtained from cystic fibrosis (CF) patients. All patients are heterozygous for one of the following mutations in the *CFTR* gene:
2. The S1251N (G → A transition at position 3884 in exon 20) *(17)*,
3. The 1717-1G → A *(18)*,
4. The G542X (G → T at 1756) *(19)*,
5. The 1784delG in exon 11 *(20)*,
6. The polymorphism in exon 14a, T854T (C2694 T/G) *(21)*.
7. These samples are analyzed in parallel with normal controls. The DGGE conditions, including the GC-clamped primer sequences, the denaturant gradient, the time and the voltage of the electrophoretic separation are as described *(22)*.
8. For increased resolution, heteroduplexes are generated at the end of the PCR reaction.

3.1.2. Melting Profile Determination

1. This is performed using the Melt 87 program. It is essential that the specific primer pair chosen is designed to produce a homogeneous melting domain for the entire region of the fragment that is to be analyzed. The position of a GC clamp at the 5'-end or at the 3'-end of the molecule is related to the melting profile, and depending in which zone the high (temperature) melting sequence its presence improves the homogeneity of the melting domain.
2. In order to provide information on the reliability of the system, we suggest that practitioners simulate the melting profiles generated by known or even unknown mutations, through the use of the computer programs.
3. The simulated mutations should be located at different positions within the target fragment, and should be compared to the melting profiles obtained in presence and in absence of the base change. This also allows evaluation of the difference in melting behavior introduced by the sequence alteration, and allows easy prediction of the temperature gradient required for differentiation of the mutations.
4. Note that in general, conservative transversions (G ↔ C and A ↔ T) do not significantly affect the melting profile of the mutated fragment with respect to the wild-type sequence, since they do not alter the overall base composition of the DNA fragment. In fact, these are the most difficult sequence alterations to identify, and very often their detection is based solely on the presence of the different heteroduplex species.

3.1.3. Polymerase Chain Reaction

1. 1 μg of genomic DNA is amplified in a final volume of 100 μL containing 10 mM Tris-HCl (pH 8.8 at 25°C), 1.5 mM MgCl$_2$, 150 mM KCl, 0.1% Triton X-100, 200 μM each dNTP, 2 U of DyNAzyme II DNA polymerase and 30 pmol of each primer.
2. The standard amplification protocol consists of an initial denaturation step of 94°C for 5 min, followed by 30 rounds of thermal cycling (94°C for 1 min, 55°C for 1 min and 72°C for 1 min), and a final incubation at 72°C for 10 min. For increased resolution, heteroduplexes are generated at the end of each PCR session by heating PCR products for 5 min at 94°C and then allowing the DNA to slowly reanneal by maintaining the temperature at 56°C for 1 h.

3.1.4. Agarose Gel Electrophoresis of PCR Products

1. In order to check amplification, 10 μL of the PCR product are loaded onto a 2% agarose gel containing ethidium bromide. The gels are run at 50 V for 45 min in TAE buffer and are photographed by ultraviolet (UV) transillumination.

3.1.5. PCR Sample Purification

1. The PCR products are desalted and concentrated by ultrafiltration by centrifugation for 15 min at 3000*g* through an anisotropic Centricon 30 membrane. A volume of 5–10 µL is recovered.

3.2. Capillary Zone Electrophoresis

3.2.1 Capillary Coating

1. The capillaries are precoated by a Silane coating procedure. Reagents are prepared in 1-mL disposable vials provided with Swagelok tee which is held in place with the aid of copious amounts of Teflon tape. The side of the tee is connected with Teflon tubing to a nitrogen cylinder and the top is connected to a 1/16 in. adapter, which is linked to the capillary (*see* **Note 6**).
2. All the reagents are flushed through the capillary under nitrogen pressure at 20 psi. A 4–5 m length capillary is first flushed with 0.1 *M* NaOH for 30 min and then with water for another 30 min.
3. A 4% solution of γ-methacryloxypropyltrimethoxysilane in a 1:1 mixture of glacial acetic acid and water is prepared, and the capillary is flushed with this solution for 20 min. The reaction is allowed to proceed for 1 h. The capillary is then flushed with water for 10 min.
4. One mL of 3%T solution of AAP monomer is carefully degassed and 4 µL of freshly prepared 10% ammonium persulphate and 1 µL of TEMED are added. The capillary is filled with this solution by flushing for 2–3 min and allowed to react for 3 h. Water is flushed through the capillary to replace the polymer and the coated capillary is washed for 10 min with distilled water and then for 10 min with running buffer. Alternatively, an AAP coated capillary is commercially available from Bio-Rad Laboratories.

3.2.2. Electroosmotic Flow (EOF) Check

1. In order to evaluate the coating of the capillary, EOF is measured by monitoring the elution time of 20 m*M* acrylamide by absorbance at 214 nm. The sample is run with 20 m*M* Tris-acetate, pH 8.0 running buffer, and an applied electric field of 400 V/cm. The formula used for EOF evaluation is $(L \times Lw)/(V \times T)$, in which: L = total capillary length, Lw= length to the window, V = total voltage, T = elution time of acrylamide peak. A successful coating has an EOF of at least 1×10^{-6} cm^2 V^{-1}s^{-1}.

3.2.3. Preparation of Liquid Sieving Polymer

1. The AAP sieving liquid polymer is prepared by diluting to 8% final concentration a 50% stock solution of AAP monomer in TGCE buffer, then the solution is carefully degassed for 15 min.
2. TEMED (2 µL/mL) and 10% freshly prepared ammonium persulphate (4 µL/mL) are added.
3. The capillary is filled with this solution by flushing for 2–3 min, and is allowed to polymerize for 1 h at 25°C.
4. The conductivity of the polymer is then tested by applying a field of 100 V/cm. This check is made for the absence of air bubbles that can be generated during the polymerization step.
5. For those who prefer not to work with acrylamide, it is possible to replace polyacrylamides with celluloses, especially HEC. We prefer short-chain HEC, such as the 27,000 Dalton. Considering the high temperatures used in TGCE, 6% HEC should be used to achieve proper sieving. This solution is very viscous however, and to facilitate capillary filling the solution should be heated at 60°C so as to lower its viscosity prior to injection.

6. The cellulose is dissolved in TGCE filtered buffer and is then carefully sonicated to eliminate all the air bubbles formed during the dissolving step prior to use in electrophoresis.

3.3. Temperature-Programmed Capillary Electrophoresis

1. How is a temperature gradient generated inside a capillary and how is it optimized for detection of DNA point mutations? We took advantage of a computer program (*16*) developed in CZE for predicting the inner capillary temperature as a function of a number of experimentally measured parameters.
2. The success of such a technique is dependent on the following critical parameters: the stability of the viscous sieving polymer, the capillary length, the choice of buffers with the correct conductivity, and the choice of thermal gradients with the right slope. We will briefly review here the thermal theory underlying the present methodology. The temperature increments (ΔT) produced inside the capillary by given voltage gradients (E, in V/cm) can be calculated according to:

$$\Delta T = \lambda_o \, E^2 \, d^2/4\chi \tag{1}$$

where λ_o is the buffer specific electric conductivity at a reference temperature (25°C), d is the capillary diameter and χ the thermal conductivity of the buffer solution. All thermal theories assume the buffer conductivity to be linearly dependent on temperature, where:

$$\lambda = \lambda_o[1 + \alpha(T - To)] \tag{2}$$

2. Here α is the temperature coefficient of conductivity, and T is the temperature inside the capillary (*16*). It is seen from **Eq. 1**, that within a given experiment the most direct way for generating a temperature gradient is via voltage ramps, as all other parameters, once chosen, remain constant. Thus, the experimental parameters needed for predicting the temperature increments linked to voltage ramps are the following: the capillary diameter, its total length, the electric current values (μA) linked to a given applied voltage, the buffer electric conductivity (λ_o, in mS/cm), and its thermal coefficient of conductivity (α, in 1/°K).

3.3.1. Determination of Buffer Electric Conductivity (λ_o)

1. The capillary is filled with the buffer containing 8% AAP, 8,9 mM TBE, 10 mM NaCl, 6 M urea and allowed to polymerize for 1 h as described in **Subheading 3.2.3.**
2. After polymerization, the external temperature is fixed at 45°C and the voltage is increased from 4 to 20 kV to allow the electric conductivity of the buffer to be determined. At least 10 conductivity values are required for use in the thermal program for λ_o calculation. The same concept is applied to a buffer containing HEC or other sieving liquid polymers.

3.3.2. Determination of the Thermal Coefficient of Conductivity (α)

1. Five mL of AAP dissolved in TGCE buffer is polymerized in a glass tube and equilibrated at 20°C. A conductimeter electrode is then coupled with a thermocouple and is introduced into the glass tube and the temperature is increased in increments of 5°C until 60°C.
2. The conductivity data are collected and inserted into the CZE.exe (Bio-Rad) for the α calculation. One needs to calculate those parameters every time the buffer composition is changed.
3. Total Capillary Length: Typically 60 cm, with an id of 100 μm is recommended.

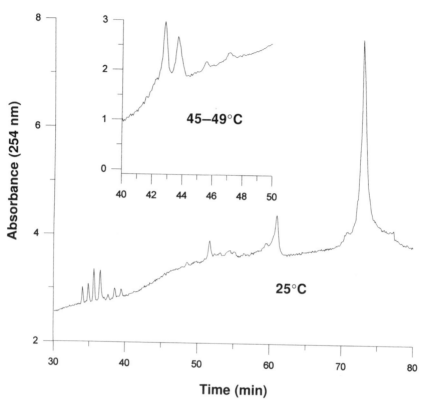

Fig. 2. Electropherograms of a PCR-amplified sample from a CF patient exhibiting two polymorphisms in exon 14a (V868V, T854T/other) on the same chromosome. The CZE was run in a Water's Quanta 4000E unit equipped with a 60 cm long, 100 μM id capillary, coated with a covalent layer of 6% poly(AAP) and filled with an 8.9 mM TBE buffer, pH 8.3, 1 mM EDTA, 10 mM NaCl in 6 M urea and 8% poly(AAP) as a liquid sieving polymer matrix. Lower tracing: a constant temperature run at 45°C, and 6 kV. Upper electropherogram: temperature gradient run, from 45 to 49°C, with increments of 0.2°C/min, obtained with a voltage ramp (6-kV start, 22-kV end settings). Note that in the upper tracing, the expected four-peak pattern is obtained, whereas in the lower tracing the single peak is an envelope of the four bands. The early eluting peaks represent unpurified primers and GC-clamp fragments. Detection is by UV absorbance at 254 nm. Reprinted from **ref. 23**.

3.3.3. Generation of Thermal Gradients

1. Given the input parameters calculated above, the dedicated software allows precise determination of the capillary inner temperature (*see* **Fig. 1**, upper panel) *(16)*. Graphs can then be easily constructed linking first voltage ramps with electric current (in μA) and then voltage ramps with temperature gradients (*see* **Fig. 1**, lower panel).
2. First of all, it is essential to determine the precise melting temperature of the fragment to be analyzed using the Lerman program *(15)* as shown in **Fig. 3**. The melting temperature represents the middle point of the gradient, which will span half degree over and half

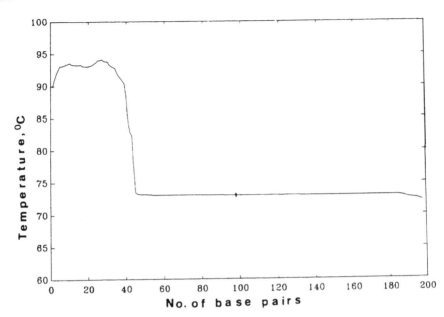

Fig. 3. Melting profile of fragment IIIB (200 bp) carrying a C → T mutation (as indicated by the black dot) in codon 39 of the β-*globin* gene. The melting profile is calculated with the Lerman and Silverstein program *(15)*. Note the sharp T_m value of 73.3 and the very flat profile of the low-melting region. The high melting region to the left of the diagram represents the GC clamp domain.

    degree under the melting temperature. Then the temperature gradient is obtained by increasing the voltage step by step to give an increment of 0.1°C in the temperature with each step; this is usually achieved by a total voltage increment of 100 V (*see* **Note 4**).

3. The typical experimental parameters adopted for most runs are: 100 μm id, 375 μm od, 50-cm long capillaries, coated with a hydrolytically stable poly(AAP) *(10)* and filled with a 8.9 m*M* Tris-base, 8.9 m*M* borate, 1 m*M* EDTA, 10 m*M* NaCl, 6 *M* urea buffer, pH 8.3, containing 6% poly(AAP) as sieving liquid polymer, polymerized *in situ* at 25°C. The experimentally measured physico-chemical parameters of this buffer are: $\alpha = 0.019 \pm 0.00035°K^{-1}$ and $\lambda_o = 0.985 \pm 0.0035$ S/cm.

4. A number of rules should be followed for TGCE: (a) The DNA sample should be injected in the capillary using conditions that are maintained (by combined chemical and thermal means) just below the expected T_m value, (b) in general, the temperature ramp should be rather narrow (~1–1.5°C), and (c) the sweep slope should be very gentle, typically 0.05°C/min.

5. A combination of the constant denaturant concentration (6 *M* urea), the low ionic strength of the buffer (8.9 m*M* TBE), and the correct outside temperature platform at which the capillary is equilibrated, combine to produce a denaturant (chemical and thermal) plateau. This combination of conditions serves to bring the DNA fragments very close (and just below) their respective T_m values at the start of the run (*see* **Note 5**).

6. As soon as the voltage ramp is generated by the programmed temperature increments, it produces a sudden decrement in the mobility of the duplexes which start unwinding, thus

allowing optimal resolution of the DNA analyte into the characteristic four-band pattern of homoduplexes and heteroduplexes along the migration path. Contrary to what is routinely performed in gel-slab operations (where ΔT's along the migration path as high as 15°C are typical), optimum separation is achieved in all cases within a very narrow temperature range (1–1.5°C). The most frequently employed gradients of denaturants used in DGGE have also been reproduced in the TGCE mode for the different classes of DNA sequence: 10–60% denaturant for low-melters; 20–70% denaturant for intermediate low-melters; 30–80% denaturant for intermediate high-melters, and 40–90% denaturant for high-melters. The technique was successfully applied to point mutation detection in CF *(23,24,26)* and in thalassemia *(25)*.

3.6. Separation of Low-Melters to High-Melters by Nonisocratic CZE

1. **Figure 2** shows the electropherograms of a PCR-amplified sample of the *CFTR* gene of a CF patient exhibiting two polymorphisms (V868V, T854T/other) in exon 14a (V868V; T854T) *(21)*. The constant-temperature run (lower profile) gives a single peak, representing an envelope of four duplexes. The four duplexes are fully resolved in the temperature-gradient run shown in the upper tracing. Note that in this case, the M/M has the highest melting point, thus it is the first eluting peak because it is the last one to unwind along the electrophoretic track. This is a typical example of resolution of a low melter, since the upper run occurs in a temperature interval from 45° to 49°C. It should be noted, however, that the real T_m is considerably higher because the presence of 6 *M* urea weakens the hydrogen bonds of the double helix and thus the apparent T_m is lowered by as much as 12°C (2°C/U of urea molarity). This is one of the earliest experimental runs we made, and it was performed in a naive way. As we later learned, it is not wise to perform a separation in such a wide temperature interval, since quite often point mutants have rather minute ΔT_m, thus they can only be separated in quite shallow T gradients.

2. **Figure 4** shows how important it is to optimize a TGCE run for developing the correct four-peak pattern. Here an intermediate- to high-melter in the *CFTR* gene (an S1251N mutation in exon 20) is being examined *(17)*. Panel A shows a control run in the absence of a temperature gradient, but at a T value (57°C) very close to the T_m: only one broad peak is obtained, surely indicative of sample heterogeneity, but of no diagnostic value. If a too broad T interval is adopted (e.g., panel B, 57° to 60°C) with a sweep rate which is too steep (0.25°C/min), although embracing the T_m value the two homoduplexes are poorly resolved. In contrast, the two heteroduplexes are clearly visible in the 68–70 min time window. Panel C shows that when the same temperature interval (57°–60°C) is used, but with a gentle slope (0.125°C/min), the two homoduplexes are well resolved, and now the two heteroduplexes fuse into a single peak eluting at 67 min. If the temperature interval is in the wrong range (panel D, 57–58°C interval, centered on the low side of the T_m) partial resolution of the two homoduplexes and good resolution of the two heteroduplexes is achieved. When the T gradient has the inflection point just right on the T_m value (panel E), optimum resolution of the four peaks is engendered.

3. **Figure 5** shows the analysis of a set of intermediate low-melting fragments, amplified from CF patients heterozygous for different mutations in exon 11 of the *CFTR* gene: 1717-1G → A (panel A); G542X (G → T at 1756; panel C) and 1784delG (panel D) with their respective normal control (panel E). All samples of the mutated gene heteroduplexes exhibit the characteristic four-peak profile, compared to a single band in the control. As shown in the temperature profile of panel B, these mutated genes are intermediate-low-melters, with T_m's in the 56.5–57.8°C range *(20)*.

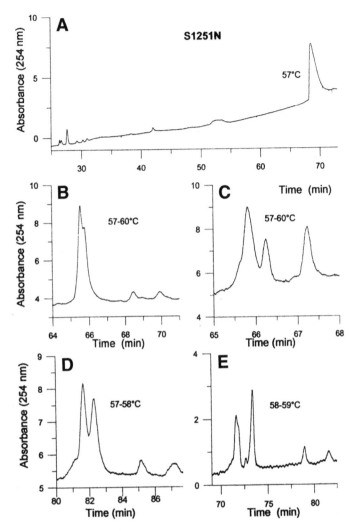

Fig. 4. Results of CZE runs under suboptimal conditions, as applied to the analysis of the S1251N mutation in exon 20 of the *CFTR* gene. Sample injection: electrokinetic, 3 s at 4 kV. The standard run (58–59°C temperature gradient, 3–4 kV voltage ramp, 0.05°C/min sweep rate) is shown in panel E. (**A**) Injection at a constant plateau of 57°C and run in the absence of a temperature gradient. (**B,C**) 57–60°C gradient, with a sweep rate of 0.25°C/min (**B**) or 0.125°C/min (**C**), over a 3–7 kV voltage ramp. (**D**) Same as (**E**), but in the 57–58°C interval. Reprinted from **ref. *24***.

4. Notes

1. Among the different mutation screening methods, DGGE has not yet gained wide application despite satisfying most of criteria for an optimal technique. DGGE is accurate, inexpensive, easy to perform, and does not require the use of radioactively labeled mate-

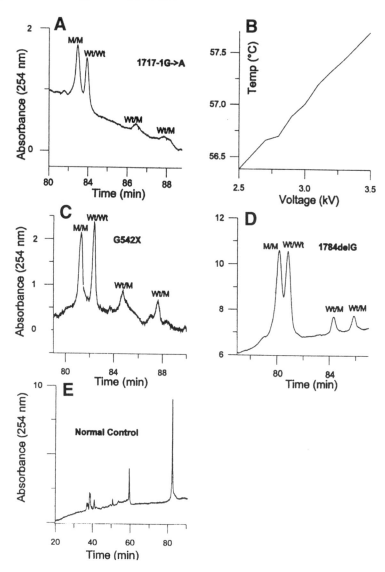

Fig. 5. CZE analysis of 3 mutants in exon 11 of the *CFTR* gene: 1717-1G → A (**A**); G542X (**C**) and 1784delG (**D**); the normal control (NC) is in panel **E**. All other conditions as in **Fig. 4**, except that the starting temperature plateau was 56.5°C. (**B**) Plot of the temperature profile over the applied voltage ramp. Reprinted from **ref. 24**.

rial. The unpopularity is mostly because of the need that the denaturant slope and running times must first be optimized for each of the different DNA fragments to be examined. This requires more time than for other scanning techniques. Actually, although the primer location, the electrophoresis run times, and the denaturant slopes can be predicted for every DNA fragment by computer modeling using algorithms developed by Lerman, optimal conditions still need to be empirically determined.

2. Because the migration time needed to resolve the homoduplexes often greatly exceeds the time required for separating the heteroduplexes, the latter often produce less sharp or even blurred bands that are scarcely distinguishable from the background fluorescence after ethidium bromide staining. The difficulty in detecting the heteroduplex peaks is particularly crucial in permitting the detection of mutations involving conservative nucleotide transversions. In conservative transversion mutations, the homoduplex species often comigrate leading to false negative results. In order to solve this problem, we developed a modified version of conventional DGGE by introducing a second gradient of porosity over the denaturing one (double gradient DGGE or DG-DGGE) *(26)*. In this system, the additional porosity gradient is able to suppress band broadening even during prolonged time runs, while maintaining the zone-sharpening effect. We also demonstrated the enhanced power of DG-DGGE as compared to conventional DGGE in even resolving homoduplexes, which appear as a unique band in DGGE, and in the detection of heteroduplexes that are almost indistinguishable over the ethidium bromide background. Both conventional DGGE and DG-DGGE are time-consuming (running times range from a minimum of 4 h at high voltage to 15–17 h at low voltage) and scarcely automatable. Moreover, a minimal gel-to-gel variability can also affect resolution, and is due to the fact that gels are manually cast. The purity of reagents and quality of primers are crucial for reproducibility of the DGGE technique and can seriously affect the routine application of the method.

3. Temperature programmed CE allows fast automated molecular scanning of large numbers of samples. We recommend the use of large bore capillaries, which facilitate the creation of temperature gradients in the capillary lumen, while also permitting a higher detection sensitivity at 254 nm using the intrinsic DNA UV absorbance.

4. The addition of NaCl salt is also recommended to increase the buffer conductivity and allow the creation of a temperature increment.

5. Urea is added so as to lower the melting point of the DNA fragments (6 *M* urea results in a decrement of ca. 12°C).

6. The inner capillary coating is necessary for reproducible runs since it eliminates the electroendoosmotic flux. Proper coating procedures require good skills in organic chemistry and are quite complex and laborious. We recommend the purchase of commercially available coated capillaries.

Acknowledgment

This work is supported by grants from AIRC (Associazione Italiana Ricerca sul Cancro) and by Telethon Fondazione Onlus, grant No. E. 0893 (Rome, Italy).

References

1. Kaplan, J. C., and Delpech, M. (1993) *Biologie Moleculaire et Medicine*, Flammarion, Paris.
2. Righetti, P. G., ed. (1996) *Capillary Electrophoresis in Analytical Biotechnology*, CRC Press, Boca Raton, pp. 431–437.
3. Wang, Y., Hung, S. C., Linn, J.F., Steiner, G., Glazer, A. N., Sidransky, D., and Mathies, R. A. (1997) Microsatellite-based cancer detection using capillary electrophoresis and energy-transfer fluorescent primers. *Electrophoresis* **18,** 1742–1749.
4. Effenhauser, C. S., Bruin, G. J. M., and Paulus, A. (1997) Integrated chip-based capillary electrophoresis. *Electrophoresis* **18,** 2203–2213.

5. Righetti, P. G., Gelfi, C., Perego, M., Stoyanov, A. V., and Bossi, A. (1997) Capillary zone electrophoresis of oligonucleotides and peptides in isoelectric buffers: theory and methodology. *Electrophoresis* **18**, 2145–2153.

6. Righetti, P. G. and Gelfi, C. (1996) Capillary electrophoresis of DNA, in *Capillary Electrophoresis in Analytical Biotechnology* (Righetti, P. G., ed.), CRC Press, Boca Raton, pp. 431–476.

7. Heiger, D. N., Cohen, A. S., and Karger, B. L. (1990) Separation of DNA restriction fragments by high performance capillary electrophoresis with low and zero cross-linked polyacrylamide using continuous and pulsed electric fields. *J. Chromatogr.* **516**, 33–44.

8. Gelfi, C., Orsi, A., Leoncini, F., and Righetti, P. G. (1995) Fluidified polyacrylamides as molecular sieves in capillary zone electrophoresis of DNA fragments. *J. Chromatogr. A* **689**, 97–107.

9. Simò-Alfonso, E., Gelfi, C., Sebastiano, R., Citterio, A., and Righetti, P. G. (1996) Novel acrylamido monomers with higher hydrophilicity and improved hydrolytic stability. I: synthetic route and product characterization. *Electrophoresis* **17**, 723–731.

10. Simò-Alfonso, E., Gelfi, C., Sebastiano, R., Citterio, A., and Righetti, P. G. (1996) Novel acrylamido monomers with higher hydrophilicity and improved hydrolytic stability. II: properties of *N*-acryloyl amino propanol. *Electrophoresis* **17**, 732–737.

11. Gelfi, C., Simò-Alfonso, E., Sebastiano, R., Citterio, A., and Righetti, P. G. (1996) Novel acrylamido monomers with higher hydrophilicity and improved hydrolytic stability. III: DNA separations by capillary electrophoresis in poly(*N*-acryloyl amino propanol). *Electrophoresis* **17**, 738–743.

12. Simò-Alfonso, E., Gelfi, C., Lucisano, M., and Righetti, P. G. (1996) Performance of a series of novel N-substituted acrylamides in capillary electrophoresis of DNA fragments. *J. Chromatogr. A* **756**, 255–262.

13. Gelfi, C., Perego, M., Libbra, F., and Righetti, P. G. (1996) Comparison of the behaviour of N-substituted acrylamides and celluloses on double-stranded DNA separations by capillary electrophoresis at 25° and 60°C. *Electrophoresis* **17**, 1342–1347.

14. Myers, R. M., Maniatis, T., and Lerman, L. (1987) Detection and localization of single base changes by denaturing gradient gel electrophoresis. *Methods Enzymol.* **155**, 501–527.

15. Lerman, L. S. and Silverstein, K. (1987) Computational simulation of DNA melting and its application to denaturing gradient gel electrophoresis. *Methods Enzymol.* **155**, 482–501.

16. Bello, M. S., Levin, E. I., and Righetti, P. G. (1993) Computer-assisted determination of the inner temperature and peak correction for capillary electrophoresis. *J. Chromatogr.* **652**, 329–336.

17. Kälin, A., Dörk, T., Bozon, D., and Tümmler, B. (1992) A novel frame-shift mutation in exon 4 of the cystic fibrosis gene (435insA) demonstrates the ambiguity of restriction analysis for mutation screening. *Hum. Mol. Genet.* **1**, 545–546.

18. Guillermit, H., Fanen, P., and Ferenc, C. (1990) A 3' splice site consensus sequence mutation in the cystic fibrosis gene. *Hum. Genet.* **85**, 450–453.

19. Kerem, B., Zielenski, J., Markiewicz, D., Bozon, D., Gazit, E., Jahaf, J., et al. (1990) Identification of mutations in regions corresponding to the two putative nucleotide (ATP)-binding folds of the cystic fibrosis gene. *Proc. Natl. Acad. Sci. USA* **87**, 8447–8451.

20. Devoto, M., Ronchetto, P., Fanen, P., Telleria Orriolis, J. J., Romeo, G., Goossens, M., et al. (1991) Screening for non-deltaF508 mutations in five exons of the cystic fibrosis transmembrane conductance regulator gene in Italy. *Am. J. Hum. Genet.* **48**, 1127–1132.

21. Zielenski, J., Rozmahel, R., Bozon, D., Kerem, B., Grzelczak, Z., Riordan, J., et al. (1991) Genomic DNA sequence of the cystic fibrosis transmembrane conductance regulator gene. *Genomics* **10**, 214–228.

22. Brancolini, V., Cremonesi, L., Belloni, E., Pappalardo, E., Bordoni, R., Seia, M., et al. (1995) Search for mutations in pancreatic sufficient cystic fibrosis Italian patients: detection of 90% of molecular defects and identification of three novel mutations. *Hum. Genet.* **96,** 312–318.

23. Gelfi, C., Righetti, P. G., Cremonesi, L., and Ferrari, M. (1994) Detection of point mutations by capillary electrophoresis in liquid polymers in temporal thermal gradients. *Electrophoresis* **15,** 1506–1511.

24. Gelfi, C., Cremonesi, L., Ferrari, M., and Righetti, P. G. (1998) Temperature-programmed capillary electrophoresis for detection of DNA point mutations. *BioTechniques* **21,** 926–932.

25. Gelfi, C., Righetti, P. G., Travi, M., and Fattore, S. (1997) Temperature programmed capillary electrophoresis for the analysis of high-melting point mutants in thalassemias. *Electrophoresis* **18,** 724–731.

26. Cremonesi, L., Firpo, S., Ferrari, M., Righetti, P. G., and Gelfi, C. (1997) Double-gradient DGGE for optimized detection of DNA point mutations. *BioTechniques* **22,** 326–330.

7

Single-Nucleotide Primer Extension Assay by Capillary Electrophoresis Laser-Induced Fluorescence

Christine A. Piggee and Barry L. Karger

1. Introduction

1.1. Capillary Electrophoresis with Laser-Induced Fluorescence

One of the most sensitive methods of detection for capillary electrophoresis (CE) is laser-induced fluorescence (LIF). The reader is referred to a recent review (1) for a more detailed description of the technique, but in brief, laser irradiation is used to excite the fluorescent molecule, and the emission is detected at a distinct wavelength. Since most molecules do not exude native fluorescence, analytes can be derivatized with one of several commercially available fluorescent dyes either before, after or during the separation step. This leads to the major advantage of LIF, the low background and subsequent sensitivity, which affords a factor of 1000 or more improvement in detection limit compared to UV absorbance. One of the most popular applications of CE-LIF is the analysis of DNA either by specifically labeling one or more nucleotides with a fluorescent tag (e.g., on the primer or the dideoxy-terminator) or nonspecifically detecting DNA through intercalation or similar binding of dyes (2).

1.2. Single Nucleotide Primer Extension (SNuPE)

With recent progress in sequencing the human genome, the important link between genetic polymorphism and disease has been conclusively demonstrated. The most common variation found in the human genome is the single nucleotide polymorphism or point mutation. Currently, a considerable effort is underway to identify these mutations for use in further genetic studies (3).

A facile way to detect known point mutations is by means of SNuPE. **Figure 1** demonstrates the principle of this assay. A primer is chosen that will anneal immediately 5' to the position of putative mutation on the template, and then the complementary labeled base is incorporated by the polymerase to extend the primer (n nucleotides)

From: Methods in Molecular Biology, Vol. 163:
Capillary Electrophoresis of Nucleic Acids, Vol. 2: Practical Applications of Capillary Electrophoresis
Edited by: K. R. Mitchelson and J. Cheng © Humana Press Inc., Totowa, NJ

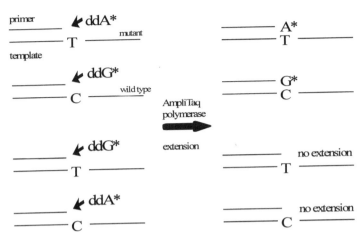

Fig. 1. The principle of SNuPE. The complementary primer anneals immediately upstream of the mutation site. Addition of the labeled dideoxy terminator complementary to the base at the mutation site will result in extension of the primer by one nucleotide to yield a fluorescent product that can be detected. Reprinted from *J. Chromatogr.* **781**, Piggee, C. A., Muth, J., Carrilho, E., and Karger, B. L. Capillary electrophoresis for the detection of known point mutations by single-nucleotide primer extension and laser-induced fluorescence detection. Copyright (1997), pp. 367–375, with permission from Elsevier Science.

by a single nucleotide ($n+1$ nucleotides). When the complementary base is not available, no extension occurs. Either regular deoxynucleotides (dNTPs) or dideoxynucleotide "terminators" (ddNTPs) can be employed for this primer extension. One advantage of using ddNTPs is that the primer can only be extended by a single nucleotide because the terminator is lacking the hydroxy group necessary to form the phosphodiester bond with the next molecule. If dNTPs were used for the extension when there are two or more of the same nucleotides in a row at the mutation site on the template (e.g., GG), then the possible extension products of a primer n nucleotides long would be both $n+1$ and $n+2$.

When the detection primer is labeled with biotin and immobilized on solid beads or in wells coated with avidin, this method is known as minisequencing, whereas when the extension reaction is performed in solution without immobilization, it is known as SNuPE. A comprehensive review of these techniques has recently been published *(4)*.

2. Materials
2.1. SNuPE Reaction
1. DNA template with and without known point mutation (*see* **Note 1**).
2. Primer for the SNuPE reaction which anneals next to mutation site (*see* **Note 2**).
3. Thermostable polymerase (*see* **Note 3**).
4. Dye-labeled terminators (dRhodamine Kit; Perkin-Elmer/ABI, Foster City, CA).
5. Thermocycler.

6. Microcentrifuge desalting spin columns (e.g., CentriSpin 10, Princeton Separations, Adelphia, NJ).
7. Microcentrifuge.

2.2. Capillary Electrophoresis

1. CE-LIF instrument with either 514- or 488-nm laser excitation (*see* **Note 4**).
2. Neutral, hydrophilic coated capillary (*see* **Note 5**) with separation length, l = 15–20 cm.
3. Separation buffer: 50 mM Tris-base, 50 mM TAPS, 2 mM EDTA and 3.5 M urea.
4. A 10% (w/v) solution of 360,000 mol wt polyvinylpyrrolidone (PVP) dissolved in separation buffer.

3. Methods

3.1. Single Dideoxyterminator Extension

1. For the SNuPE reaction, combine 10–100 ng of ds template, 10 pmol of primer, 1 μL of the appropriate labeled dideoxyterminator, and 2–8 U of Ampli*Taq* FS or other thermo-stable DNA polymerase in a 0.2-mL thin-wall PCR tube.
2. Insert the tube into the thermocycler and run a touchdown type temperature cycling program. The following program is a good place to start:
   a. 94°C for 1 min;
   b. 92°C for 30 s;
   c. 70°C for 40 s, with a decrease in temperature of –0.5°C/cycle for each subsequent cycle;
   d. cycle to step (2) 19×;
   e. 92°C for 30 s;
   f. 60°C for 40 s with an increase in temperature of +1 s/cycle for each subsequent cycle;
   g. cycle to step (5) 19×.
3. Because salts and other high electrophoretic mobility contaminants in the reaction (i.e., free dideoxyterminator dyes) interfere with electrokinetic injection in CE, the reaction must be desalted by using a microcentrifuge spin column (such as CentriSpin-10) as directed by the manufacturer. **Figure 2** illustrates the importance of reaction cleanup and the significant improvement in signal compared to the untreated reaction.

3.2. DNA Separation

1. The coated capillary column (100 or 75 μm id) should be installed in the instrument with a 15–20 cm length from injection to detection. Fill the capillary with the 10% PVP solution (*see* **Note 6**) and use plain separation buffer (TRIS/TAPS) in both electrode reservoirs.
2. Set the running voltage to 300 V/cm. Inject the desalted sample electrokinetically for 5 s at the running voltage.
3. Detection should be at 580 nm for a single detection wavelength instrument or 500–700 nm for the custom-built photodiode array instrument.

3.3. Multiplexing

1. In the SNuPE reaction, it is possible to detect more than one point mutation in a single reaction. Two or more primers of different lengths and similar annealing temperatures are chosen which anneal at different regions along the template.
2. An example of multiplexed SNuPE is shown in **Fig. 3**, where three primers (15, 20, and 25 nt long), are simultaneously extended along different regions of the template *(5)*. The additional primers are useful not only as internal standards but also as sizing standards.

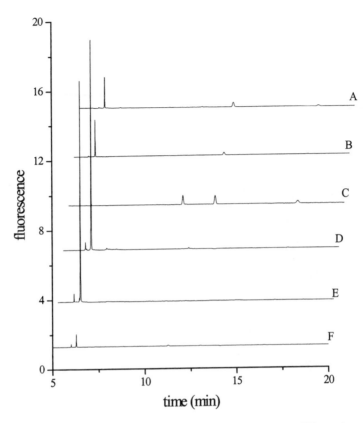

Fig. 2. Desalting of LHON SNuPE Reactions. Three separate SNuPE reactions using the same dye terminator (ddA) are used to detect the human mitochondrial mutations at the positions 3460, 11,778, and 14,459, which are associated with Leber's Hereditary Optic Neuropathy (LHON). 100-μm id neutral coated capillary, l = 15 cm, L = 25 cm, 10% PVP (mol wt 1×10^6) in 50 mM Tris/TAPS buffer, 300 V/cm, 514-nm excitation, and 560-nm emission. A 5-s electrokinetic injection of: (A–C), LHON SNuPE reactions of the 3460, 11,778, and 14,459 mutations before purification; (D–F), the same SNuPE fragments after purification by desalting. Reprinted from *J. Chromatogr.* **781**, Piggee, C. A., Muth, J., Carrilho, E., and Karger, B. L. Capillary electrophoresis for the detection of known point mutations by single-nucleotide primer extension and laser-induced fluorescence detection. Copyright (1997), pp. 367–375, with permission from Elsevier Science.

4. Notes

1. It is preferable to use templates with a few hundred bases as opposed to the entire DNA under investigation. Smaller templates can be amplified from the entire DNA by PCR, followed by treatment with shrimp alkaline phosphatase and exonuclease I to digest the primer and deactivate the unreacted nucleotides, and finally, clean-up with microcentrifuge spin columns to remove the enzymes (e.g., Ultrafree-MC Probind, Millipore) and salts (CentriSpin 10).

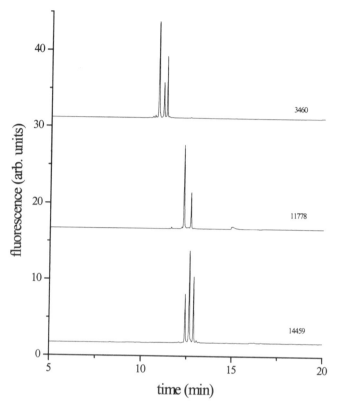

Fig. 3. Multiplexing with LHON SNuPE reactions. 100-μm id capillary, l = 15 cm, L = 25 cm, 10% PVP (mol wt 1 × 10^6) in 50 mM Tris/TAPS 2 mM EDTA 3.5 M urea buffer, at 300 V/cm (7.5 kV), 514-nm excitation, and 560-nm emission. A 5-s electrokinetic injection of the 3460, 11,778, and 14,459 SNuPE reactions containing additional longer and shorter primers. Eight out of the nine primers could be successfully detected. The primer with the lowest calculated annealing temperature would not extend even at very low (nonspecific) annealing temperatures. Equal amounts of all three primers were added to each reaction, so the affects of sequence-dependent extension can be seen in the differing amounts of extension product obtained. Reprinted from *J. Chromatogr.* **781**, Piggee, C. A., Muth, J., Carrilho, E., and Karger, B. L. Capillary electrophoresis for the detection of known point mutations by single-nucleotide primer extension and laser-induced fluorescence detection, Copyright (1997), pp. 367–375, with permission from Elsevier Science.

2. Primer design is crucial in these reactions to have a favorable annealing temperature. There are computer programs (such as "Oligo") available to calculate the reaction conditions for a designated primer, or the GC/AT method can be used to approximate annealing temperature. We generally use primers around 20 bases long. In addition, the use of a "touchdown" type thermocycling program improves the likelihood of eventually reaching the correct annealing temperature without having to run many different reactions, each having a single annealing temperature.

3. Some polymerases have been shown to be more efficient at incorporating dideoxy-terminators than others. Although extension efficiency can only be measured when there is a way to detect both the unextended and extended primer (e.g., mass spectrometry), it has been shown that ThermoSequenase more completely extended the primers *(6)*.

4. CE-LIF can be performed on laboratory-built instruments such as those described in references *(5)* and *(7)* or on the automated Beckman P/ACE® CE-LIF instrument.

5. The best CE performance is obtained during separations of DNA when the capillary has a neutral hydrophilic layer. Such coated capillaries can be purchased from companies such as Hewlett Packard and Beckman Coulter.

6. Replacement of the PVP before each run must be from the detector end to avoid bubbles that cause current disruption. Also, preconditioning of the matrix by applying the running voltage for a few minutes before injecting is sometimes helpful.

7. Electrokinetic injection of the sample is generally used when the capillary is filled with a replaceable polymer matrix.

8. The free dye terminator migrates slower than the average 20-mer primer. It is also helpful to realize that the terms "labeled-dideoxy-terminator" and "dye terminator" are used interchangeably.

9. Negative controls can be performed by using (one variable at a time) either, the incorrect template, the primer or dye terminator, or by omitting the template, the primer, the terminator or the polymerase although keeping the buffer concentration and final volume constant.

10. When the template concentration is too high, the double-stranded template competes with the primer for annealing. Try running reactions with several different template dilutions to find the optimum amount.

11. Small template impurities are sometimes detected, particularly if the template was generated by polymerase chain reaction (PCR).

12. The optimum number of cycles for the SNuPE reaction can be determined experimentally because past a certain point, nonspecific extension will occur at a higher rate than specific extension.

References

1. Tao, L. and Kennedy, R. T. (1998) Laser-induced fluorescence detection in microcolumn separations. *Trends Anal. Chem.* **17**, 484–491.

2. Righetti, P. G. and Gelfi, C. (1997) Recent advances in capillary electrophoresis of DNA fragments and PCR products. *Biochem. Soc. Trans.* **25**, 267–273.

3. Wang, D. G., Lipshutz, R., Lander, E. S., et al. (1998) Large-scale identification, mapping, and genotyping of single-nucleotide polymorphisms in the human genome. *Science* **280**, 1077–1082.

4. Syvänen, A.-C. (1999) From gels to chips: "minisequencing" primer extension for the analysis of point mutations and single nucleotide polymorphisms. *Hum. Mutat.* **13**, 1–10.

5. Piggee, C. A., Muth, J., Carrilho, E., and Karger, B. L. (1997) Capillary electrophoresis for the detection of known point mutations by single-nucleotide primer extension and laser-induced fluorescence detection. *J. Chromatogr.* **781**, 367–375.

6. Haff, L. and Smirnov, I. Genotyping multiple loci in DNA by MALDI-TOF mass spectrometry—the PinPoint assay. *Proceedings of the 45th ASMS Conference on Mass Spectrometry and Allied Topics*, Palm Springs, California, June 1–5, 1997.

7. Ruiz-Martinez, M. C., Berka, L., Belenkii, A., Foret, F., Miller, A. W., and Karger, B. L. (1993) DNA sequencing by capillary electrophoresis with replaceable linear polyacrylamide and laser-induced fluorescence. *Anal. Chem.* **65**, 2851–2858.

8

Amplification Refractory Mutation System Analysis of Point Mutations by Capillary Electrophoresis

Paola Carrera, Pier Giorgio Righetti, Cecilia Gelfi, and Maurizio Ferrari

1. Introduction

1.1. Point Mutation Detection: Significance and Technology

Mutation detection has become a very important chapter in medicine, following the development of molecular technology for the study of pathogenic mutations in human diseases. DNA technologies allow great advances in basic knowledge of the pathophysiology of disorders, adding new means of diagnosis to characterize the molecular defect and to correlate between genotype and phenotype. The field of molecular diagnosis is evolving rapidly, and currently, it broadly comprises genetic disease analysis, paternity testing, forensic studies, assessment of genetic risk, and so on. A number of molecular techniques have been developed during the last 10–15 yr that greatly improve diagnostic and prognostic capabilities, allowing carrier detection, and prenatal diagnosis.

Direct detection of mutations, and indirect detection by linkage analysis represent the two main approaches for molecular diagnosis. Direct detection is feasible when the affected gene is known and its structure is well defined. This approach is very powerful, particularly for those diseases in which the molecular defect shows little variability. In fact, the higher the genetic complexity and heterogeneity of mutations, the more complex the design of diagnostic approach. The direct approach is applied to routine clinical diagnosis after the definition of the different kind of mutations in the various populations, the establishment of correlations between genotype and phenotype, and the refinement of methods and testing protocols. Methods for mutation analysis differ with specific classes of mutations: extensive deletions, insertions and rearrangements, smaller deletions/insertions, or mutations such as triplet expansions or single base substitutions. Among these, PCR-based methods are suitable for the

From: *Methods in Molecular Biology, Vol. 163:*
Capillary Electrophoresis of Nucleic Acids, Vol. 2: Practical Applications of Capillary Electrophoresis
Edited by: K. R. Mitchelson and J. Cheng © Humana Press Inc., Totowa, NJ

detection of small mutations. Many different PCR-based procedures have been developed, devised either to detect defined mutations, or to search for and localize new mutations and polymorphisms. The latter mutation then being exactly identified by a further analysis step, such as direct sequencing.

1.2. Amplification-Refractory Mutation System (ARMS)

ARMS is one PCR-based method capable of detecting small deletion/insertions and point mutations in defined DNA fragments. This method was recently adapted to the simultaneous detection of the most common point mutations in the 21-hydroxylase *P450c21-B* gene, by combining the ARMS assay with capillary zone electrophoresis (CZE) for the analysis stages. Steroid 21-hydroxylase deficiency is a recessive inherited disease accounting for about 90% of Congenital Adrenal Hyperplasia (CAH) cases *(1)*. The *P450c21-B* gene, which is located on chromosome 6p21, encodes the 21-hydroxylase enzyme. The *P450c21-B* gene and a 98% homologous pseudogene form a tandem repeat adjacent to *C4B* and *C4A* genes, respectively. Diverse mutations of the active gene are known, such as whole gene deletion, gene conversion *(2,3)*, and more frequently a number of common point mutations, probably resulting from small-scale gene conversion events *(4)*. The molecular diagnosis protocol of this disease represents a good example to understand how a combination of methods for identifying different classes of mutations, is needed to reach a good feasibility. In particular, conventional Southern blot hybridization allows detection of gene deletion and larger gene conversion events, whereas single nucleotide point gene conversions can be scored by a variety of PCR-based methods, including the ARMS technique *(5–7)*. We describe the detection of known point mutations in the *P450c21-B* gene using a combination of a ARMS reaction with CZE in sieving liquid polymers. This protocol essentially depicts all of the critical issues of designing an ARMS assay.

1.3. Principles of ARMS

The ARMS method is based on the capability of oligonucleotides (oligos) to be extended to their 3' end by the enzyme *Taq* DNA-polymerase only when they are perfectly matched with the template *(8,9)*. Thus, in the presence of a mutant sequence, we can design two oligos that differ to their 3'-base, the first homologous to the wild-type sequence, the second oligo homologous to the expected mutant sequence. The two sequence-specific oligos are used in combination with another common primer in two separate PCR reactions, in which only the oligo with the correctly matched base will be extended in each PCR reaction. The specificity of the reaction is positively influenced by PCR reaction parameters, such as employing high stringent annealing temperatures. The insertion of a mismatch site at the penultimate 3'-base within the oligo further increases discriminatory amplification. Analysis of the PCR products allows inference of the genotype of an individual, with respect to the mutation analyzed as follows:

1. A normal subject only generates PCR products in the reaction containing the oligos specific for the wild-type sequence.

2. A homozygous mutant subject only generates PCR products in the reaction containing the oligos specific for the mutant sequence.
3. A heterozygous subject generates PCR products in both reactions, since it has a copy of both target sequences. Thus, the presence or absence of a particular PCR product indicates the presence or absence of a specific target sequence.

The analysis of each of the PCR products involves agarose slab-gel electrophoresis and then comparison of the electrophoretic tracks. Of course, for the diagnosis of a large number of samples it would be extremely useful to utilize an automatic technique with high resolving power, such as CZE. CZE is now emerging as a competitive analytical tool for separating a variety of charged and uncharged molecules, including proteins and nucleic acids *(10,11,12)*. Its advantage over slab-gels include, minute sample requirements (the sample zone is just a few nL), extremely high sensitivity (of the order of yocto moles with laser-induced fluorescence detection, LIF), and "on line" peak detection and integrated data collection. The applications in the field of nucleic acid analysis range from DNA sequencing *(13,14)*, to pulse fields for large DNAs *(15)*. Some reviews have already described its use in analysis of mutations in PCR-amplified fragments and for the detection of genetic defects *(16–18)*.

2. Materials

2.1. Polymerase Chain Reaction

1. 20 pmol/μL of a normal ARMS primer (store at –20°C).
2. 20 pmol/μL of a mutant ARMS primer (store at –20°C).
3. 20 pmol/μL of a common primer (store at –20°C).
4. 20 pmol/μL of each control primer. Control primers to coamplify a fragment from a different gene or from a genome region as an internal control of PCR proceeding (store at –20°C).
5. DNA: 250 ng/μL of human genomic DNA, or 5 ng/μL of genomic DNA fragment cloned into plasmid vectors (store at 4°C or –20°C, respectively).
6. 5 U/μL *Taq* DNA polymerase (store at –20°C).
7. 10X PCR amplification buffer containing 15 mM MgCl$_2$ (store at –20°C).
8. 25 mM solution of each deoxynucleotide (store at –20°C).
9. Bidistilled sterile water (store at –20°C in small aliquots).

2.2. Agarose Gel

2.2.1. Agarose

1. 50X TAE buffer: 242 g Tris-base, 57.1 mL glacial acetic acid, 37.2 g Na$_2$EDTA · 2H$_2$O, add ddH$_2$O to 1 L. Working solution: 1X (Store at room temperature. Stable for up to 3 mo).
2. 10 mg/mL ethidium bromide (EtBr). Working concentration: 0.5 mg/mL (**CAUTION: EtBr is a very hazardous, mutagenic agent intercalating DNA.**) It is also light sensitive and solutions should be stored in a darkened vessel at +4°C).
3. Loading buffer: 70% of 1X TAE, 30% glycerol, 0.1% bromophenol blue, 0.1% xylene cyanol (store at –20°C in small aliquots).
4. DNA molecular size standard (store at +4°C).

2.3. Purification of PCRs

1. Microcon 30 (Amicon).
2. Microcentrifuge (Eppendorf type).

2.4. Capillary Zone Electrophoresis

1. 20 mM Tris-acetate (1.211 g Tris-base, 341 μL acetic acid/500 mL, pH 8.0). Stable at room temperature for up to 3 mo.
2. 100-μm id capillaries (Polymicro Technologies) (*see* **Note 1**).
3. 89 mM TBE buffer, pH 8.3 1.078 g Tris-base, 550 mg boric acid, 58.44 mg EDTA, distilled water to 100 mL. After preparation, the buffer is passed through a 0.22-μm filter-membrane (Millipore) and is stable at room temperature for a week.
4. Hydroxyethyl cellulose (HEC), 27,000 Mr is dissolved in filtered TBE buffer at 3 g HEC per 100 mL.
5. SYBR Green I (Molecular Probes) is diluted in running buffer at 1:30,000.

3. Methods

3.1. CZE Equipment Preparation

3.1.1. Capillary Coating

1. The capillaries are coated with a Silane coating procedure. Reagents are prepared in 1-mL disposable vials provided with Swegelok tee which is held in place with the aid of copious amount of Teflon tape. The side of the tee is connected with Teflon tubing to a nitrogen cylinder and the top is connected to a 1/16 in. adapter, which is connected to the capillary. All the reagents are flushed through the capillary under nitrogen pressure at 20 psi.
2. A 4–5-m length capillary is first flushed with 0.1 M NaOH for 30 min and then flushed with water for another 30 min. A 4% sol of γ-methacryloxypropyltrimethoxysilane in a 1:1 mixture of glacial acetic acid and water is prepared, and the capillary is flushed with this solution for 20 min. The reaction is allowed to proceed for 1 h.
3. Next, the capillary is flushed with water for 10 min. 1 mL of 3% T sol of acryloyl-aminopropanol (AAP) monomer is carefully degassed and 4 μL of freshly prepared 10% ammonium persulphate and 1 μL of TEMED are added. The capillary is filled with this solution flushing for 2–3 min, and then allowed to react for 3 h. Water is then flushed through the capillary to replace the polymer, and finally the coated capillary is washed for 10 min with distilled water and then 10 min with running buffer.

3.1.2. Electroosmotic Flow (EOF) Check (see **Note 2**)

1. To evaluate the quality of the coating on the capillary, the EOF is determined. The elution time of a solution of 20 mM acrylamide in 20 mM Tris-acetate, pH 8.0 running buffer at an applied intensity of electric field 400 V/cm is measured by monitoring at 214 nm. The EOF is calculated using the formula: $(L \times Lw)/(V \times T)$, where L = total capillary length, Lw = length to the window, V = total voltage, T = elution time of acrylamide peak. An EOF of at least 7×10^{-5} cm$^2 \times$ V^{-1}s^{-1} is required for a successful coating.

3.1.3. Liquid Sieving Polymer Preparation

1. The sieving liquid polymer is prepared by dissolving 3 g of HEC in 100 mL of 89 mM TBE buffer, pH 8.3 (*see* **Note 3**).

2. After the HEC is completely dissolved into the buffer, SYBR Green I fluorophore is added in the dark to a final dilution of 1:30.000. The SYBR Green I, HEC solution must be freshly prepared every day due to the light sensitivity of the SYBR Green I.

3. The solution is sonicated to eliminate the air bubbles formed during the HEC preparation and the polymer solution is carefully stored in the dark until use.

3.1.4. CZE Run Conditions

1. CZE analyses are performed on a Beckman P/ACE® System 5000 equipped with a LIF detector providing 488-nm excitation with a 520-nm bandpass emission filter.

2. A 100 µm id × 27 cm (20 cm to the detector window) coated capillary is purged with a degassed solution of 1.5% HEC, SYBR Green I in TBE for 5 min at high pressure.

3. The DNA sample is loaded electrokinetically by applying, 200 V/cm for 10 s, and the separation run is made at 100 V/cm.

3.2. PCR Conditions

3.2.1. General ARMS Procedure (see **Note 4**)

1. This procedure is common to the various ARMS protocols. Two ARMS reactions are set up for each DNA sample, each complementary ARMS reaction differs only in the specific primers for the normal or the mutant sequence (*see* **Note 5**).

2. Prepare two reaction premixes for the ARMS reactions with the normal and the mutant primers, containing sufficient reagents to exceed the total number of reactions. The volume of the premix for a single reaction is 40 µL.

3. The premix (N) for the normal ARMS reaction contains the following reagents: the normal ARMS primer, the common primer, the dNTPs mix, 10X PCR buffer, $MgCl_2$ and sterile water to the final volume. The premix (M) for the mutant ARMS reaction contains all the reagents above listed except the mutant ARMS primer instead of the normal one.

4. If an internal control PCR is included in the protocol, add to both the premixes N and M, the pair of specific primers. Dispense 40 µL of N and M premixes in PCR tubes. Add to each tube of N and M premixes, DNA to be amplified in 5 µL.

5. Usually a set of control ARMS reactions is run in separate, parallel reactions, using DNA from known normal homozygous, mutant homozygous, and heterozygous individuals. In addition for each premix N and M, a reaction without DNA (blank) is performed, to check for the absence of contaminant DNA.

6. Add mineral oil and cap. Microcentrifuge the contents to mix them and then place the tubes in a thermal cycler. Proceed with a "hot start PCR" by adding the *Taq* DNA polymerase after 5 min at 95°C. Then the 10X PCR buffer diluted in water and 5 µL of enzyme at a final concentration of 1 U/µL are added to each tube, below the mineral oil layer. During the set up of ARMS program it is advisable to perform reaction previously defined with control DNAs and the blank reaction.

7. Purification of PCR products is performed in order to desalt samples and to partially remove unincorporated oligonucleotides. An Amicon Microcon 30 device is used following the manufacturer's instructions.

3.2.2. Amplification Refractory Mutation System

1. PCR with ARMS oligonucleotides results in the selective amplification of a PCR product of the active *P450c21-B* genes, whereas avoiding the amplification of the inactive *P450c21-A* pseudogenes (*see* **Note 6**).

Table 1
Primers Used for PCR and for Nested ARMS* Assay

A) Primers for first round PCR	Product in bp
606 - aggtcaggccctcagctgccttca 2197R - ctcgggctttcctcactcatc	1590
-280 - cctgcacagtgatgtggaacc 1394R - gctgcatctccacgatgtga	1683
B *gene specificities are bold-faced*	

B) Primers for the second round of PCR

			Ann/ext	ARMS products in bp	
Primers for ARMS	Mutation	ntd	(°C)	WT	Mutant
655C-R: cttagacaccagcttgtctgcaggaggaag	Intron2	C	75	243	
441: tatgttgcccaggctggtcttaaattccta 655A-R: cttagacaccagcttgtctgcaggaggaat 655G-R: cttagacaccagcttgtctgcaggaggaac		A G*	75 77	252	252
433: gttcttgctatgttgcccaggctggtctta					
3wt685: aactacccggacctgtccttggg<u>agactac</u> 3del679: tctaagaactacccggacctgtccttggtc	8bpDel	wt DEL*	77 76.5	295	293
979-R: agagaattcctcctcaatggccacaggggt					
999T: ggaattctctctcctcacctgcagcatcgt 999A: ggaattctctctcctcacctgcagcatcga	I172N	T A*	76.5 75	172	172
1141-R: taggcaggcattaagttgtcgtcctgccag					
1683G-R: gtccactgcagccatgtgcac	V281LG		75	139	
1557: ttcaccctctgcaggagagc 1683T-R: gtccactgcagccatgtgcaa		T*	73		146
1560: cgctcctttcaccctctgcag					
1993C: cgatcattcccagattcagcagcgactcc 1993T: cgatcattcccagattcagcagcgactct	Q318X	C T*	78 77	180	180
2144R: tggtgcggtggggcaaggctaagggcacaa					

*Mismatched positions are bold-faced. The region of the 8-bp deletion is underlined. *Mutated allele.*

*The allele specific primers are named with the nucleotide position and with the base present in the corresponding allele. For the mutation of intron 2, and with the missense at codon 281, the allele specific primers are designed based on the reverse strand and each is used with a different forward primer. R suffix denotes the reverse primers.

2. ARMS reactions for the *P450c21B* genes are based on the gene sequence surrounding those loci associated with mutations. **Table 1B** lists the primer sequences, the mutations detected, the annealing temperatures, and the length of PCR-amplified products. Some of the ARMS primers are designed with a mismatch in their penultimate position, following the rule previously described *(21)* (*see* **Note 5**).

3. To minimize allele "drop out" (a failure to amplify one allele), the *B* genes are selectively amplified in two overlapping fragments by use of the forward B-specific primer 606 with

the reverse primer 2197, and the forward primer –280 with the reverse B-specific primer 1394R (**Table 1A**).

4. The PCR reactions are performed with 2.5 U of *Taq* DNA polymerase in a buffer of 50 μL final volume containing 10 mM Tris-HCl, pH 8.3, 50 mM KCl, 1.5 mM MgCl$_2$, and 0.2 mM of each deoxynucleotide, and 20 pmol of each oligonucleotide primer. The DNA template is either 250 ng of genomic DNA, or 5 ng of constructs obtained by cloning amplified B genes sequences from patients into TA vectors (Invitrogen, San Diego, CA).

5. The first round of amplification is carried out for 30 cycles (denaturation: 30 s at 94°C, annealing for 30 s at 60°C, extension for 2 min at 72°C). The PCR reaction is ended by incubating for 5 min at 72°C.

6. It is not possible to internally control the amplification step in the present protocol, as it is a nested ARMS amplification reaction. DNA samples from previously genotyped subjects, identified as homozygous normal, or as homozygous mutant, or as heterozygous are included as positive and negative controls for each set of reactions. In addition, a double-blind protocol is established using samples from Italian families with 21-hydroxylase deficiency previously screened for gross rearrangements by restriction mapping analysis and for point mutations by PCR and either hybridization with allele specific oligonucleotides (ASO), or by direct gene sequencing *(19,20)* (*see* **Note 7**).

3.2.3. PCR Conditions

1. In the second round of PCR, 0.5 μL of the amplified *B* gene fragments, 1:20 diluted, are reamplified with primers specific for each allele. The amplification reactions are identical to that described above, except that 1 U of *Taq* DNA polymerase is used.

2. The amplifications are performed using a two-step protocol. The protocol comprises 30 cycles of denaturation at 94°C for 30 s, and annealing/extension steps at the temperatures indicated in **Table 1B**. The annealing/extension step is for 1 min initially, with an increase of 1 s in the extension time per cycle, resulting in a final annealing/extension step of 5 min. Amplification is followed by 15 min at 37°C to facilitate the reannealing of the complementary chains.

3. The five mutations tested are located in a region spanning 1400 bp of the gene. For this reason, it was not possible to multiplex the amplification reactions. Therefore, an ARMS reaction is set up individually for each pair of primers.

3.3. Results of ARMS Analysis

3.3.1. Agarose Gel Electrophoresis of PCR Products

1. The amplified products are analyzed on 2–3% agarose minigels (7–10 cm length) and stained with EtBr. 10–20-μL reactions are loaded onto the gels in 1X TAE buffer for 45 min at 50–75 V.

2. **Figure 1** shows an example of agarose analysis of the ARMS for the mutation 1683G-T of the *P450c21B* gene. The inferred genotype of tested samples is described in the figure legend.

3.3.2. CZE of PCR Products

1. PCR products with primers specific for the normal and mutated sequences (*see* **Table 1B**) are pooled into the wild-type and mutant set, respectively before running CZE (*see* **Note 8**).

panel M - ARMS 1683T

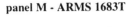

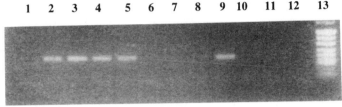

panel N - ARMS 1683G

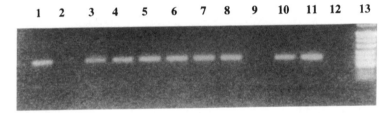

Fig. 1. ARMS analysis of the mutation 1683G-T that causes the missense substitution V281L. Agarose gel analysis of the N and M ARMS reactions is shown. In the upper gel, the reactions with the mutant ARMS (*panel M*), and in the lower gel the reactions are with the normal ARMS (*panel N*). Lanes 1–11 of both panels: Reactions amplified with different DNA samples. The same lane in the two panels corresponds to the same DNA sample (example lane 1, panel N and lane 1 panel M: same DNA sample tested with the N and M ARMS reactions). Lane 12: No DNA added in reactions (blank). Lane 13: DNA molecular weight size standard. The DNA samples are as follows: lane 1: homozygous normal control DNA; lane 2: homozygous mutant control DNA; lane 3: heterozygous control DNA; Lanes 4–11: DNA samples to be tested with the ARMS. In lanes 4 and 5, a fragment was amplified in both the ARMS reactions, thus indicating heterozygosity for the 1683 G-T mutation. In lanes 6, 7, 8, 10, and 11 a fragment is obtained only with the normal ARMS reaction, thus indicating a homozygous normal genotype. In contrast, in lane 9, a homozygous mutant is present since an amplified product is only obtained with the mutant ARMS.

2. **Figure 2** shows all the possible peaks that we can obtain in the wild-type and mutant sets, after resolution in a 27-cm coated capillary with 3% HEC as the sieving liquid polymer in TBE buffer. We can also see that the polymorphism A/C in the same 655 position of intron 2 is well resolved.

3. A series of examples, showing how CZE can identify carrier parents and affected children in families with steroid 21-hydroxylase deficiency are shown in **Figs. 3–6**. The results of CZE of nested-ARMS with the wild-type specific wild-type and with the mutant specific primers are shown in parallel for each subject.

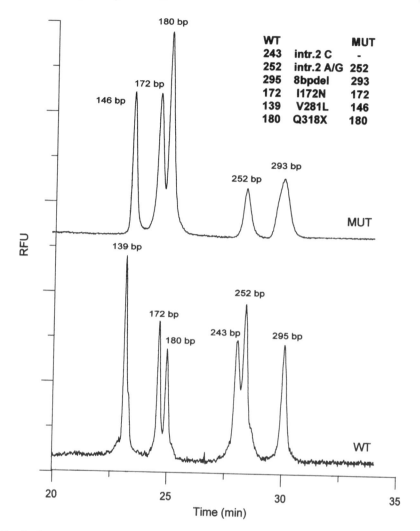

Fig. 2. CZE detection of nested ARMS products. CZE of the set of five mutant (*upper tracing*), and six wild-type (*lower profile*) ARMS-amplified DNA fragments for the detection of 21-hydroxylase deficiency. The corresponding mutations are listed in **Table 1B**. CZE conditions are as described in materials and methods. RFU: relative fluorescence units.

4. In **Figs. 3** and **4**, two families with the classic form of the disease in their offspring and the cryptic form in parents are analyzed. Analyses from two additional families with the salt wasting form of the disease are shown in **Figs. 5** and **6**.
5. A differential analysis of the spectrum of bands produced by amplifying the wild-type and mutant regions of the gene from each member of the family allows a precise diagnosis.

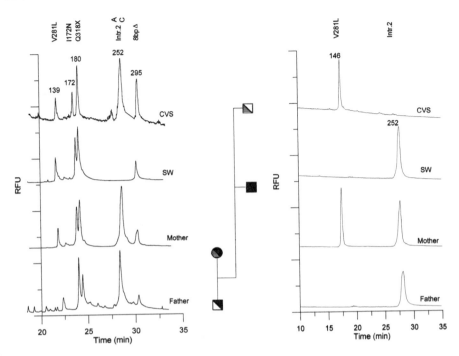

Fig. 3. Family profiling by CZE for detection of 21-hydroxylase deficiency. In this family, the affected child suffers from the salt wasting (SW) form and is homozygous for the mutation in intron 2. The father is heterozygous for the intron 2 mutation, showing the presence of the amplified product corresponding to that mutation in both electropherograms. The mother being "double heterozygous" for the severe mutation in intron 2 and the milder V281L missense and completely asymptomatic, is diagnosed as a cryptic form. The early prenatal diagnosis of her second pregnancy, revealed by direct detection of mutations, that she transmitted the mutation V281L to the carrier fetus which is not inherited by her previous affected child. *Left panel*: wild-type; *right panel*: mutant type.

4. Notes

1. The length of the capillary is not so peculiar, in fact, we have utilized the shortest capillary compatible with our machine. Of course, even shorter or longer capillaries can be adopted, but the running time will change with length.
2. The EOF must be carefully checked, even if a commercially prepared coated capillary is used. Often the EOF is not the same as claimed by the manufacturer. An elevated EOF compromises the separation, especially the reproducibility of elution time, and thus influences the identification of the peaks.
3. The running buffer can be modified adopting other cellulose polymers and different concentrations of polymers, as required by the fragment to be analyzed. As an alternative to cellulose, short chain polyacrylamide can be used at a concentration of between 5% and 7% to improve the sieving properties of the matrix. Higher concentrations of polyacrylamide are difficult to handle because of the high viscosity, and capillary filling is slow.

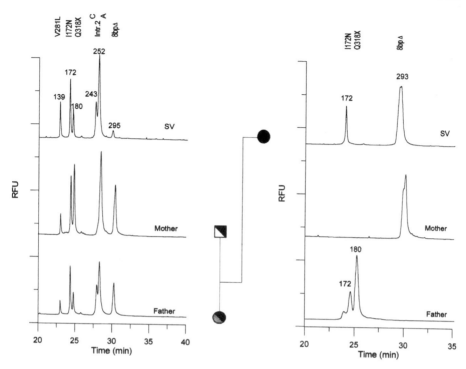

Fig. 4. Family profiling by CZE. In this family, the father has the cryptic form and is double heterozygous for both the I172N and the Q318X mutations, both are associated to the classic form of the disease. The affected daughter suffers from the simple virilizing (SV) form of the disease, inherited the I172N mutation from her father and an 8-bp deletion in exon 3 from the mother.

4. The ARMS protocol requires the design of specific primers, each capable of generating a unique PCR product from the target allele in standardized reaction conditions. The specificity and reproducibility of the protocol should be established by using DNA samples that have already been genotyped. The final definition of an ARMS protocol requires the standardization of parameters critical for reproducibility and stringency, such as thermal cycling profile, enzyme concentration, and the sequence and concentration of the ARMS primers. It is advisable to use a pre-PCR room (for PCR reaction setup) and a post-PCR room (for PCR analysis) to prevent contamination with other PCR products.

5. When 20-mer sequence specific primers were used for all mutations only the V281L substitution oligonucleotide directed specific amplification of the target alleles. Both longer 30-mer primers and a mismatched residue *(21)* at the penultimate position of primers are now used to increase the specificity of the allele targeting, and good reproducibility is obtained. The optimal primer length is around 30 bases and falls in a 28–60-base range *(21)*. With longer primers, it is crucial to perform a double-blind testing of the system on previously genotyped samples to evaluate the selectivity of the ARMS assay.

Carrera et al.

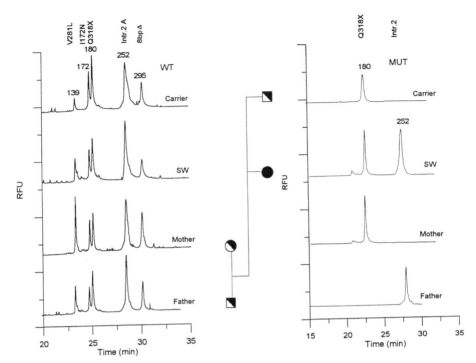

Fig. 5. Family profiling by CZE. In this family the SW patient was double heterozygous for Q318X and intron 2 mutations, transmitted from the mother and the father, respectively. The carrier brother inherited the mutation Q318X from the mother and the healthy gene allele from the father.

6. A nested ARMS assay is set up to detect five common mutations in steroid 21-hydroxy-lase deficiency in order to only amplify mutated/wild-type alleles from the active *P450c21-B* gene and to avoid amplification from the 98% homologous pseudogene. This situation is quite complex, because it is not possible to include an internal control to differentiate between a failed PCR reaction and homozygosity when using such a nested protocol. Therefore, we strongly recommend the inclusion of internal controls if using an "unnested" protocol, with control primers chosen from other gene regions with similar amplification profiles, but with a size distinguishable from the ARMS fragment.

7. ARMS systems have been applied to the detection of mutations in simple and multi-plexed *(22)* tests and also for the detection of mutated alleles representing a small fraction of the DNA, such as in cancer *(23)*.

8. It is not always required for CZE separation, but it is recommended that it should be used in all multiplexed or pooled CZE separations.

References

1. White, P. C., New, M. I., and Dupont, B. (1987) Congenital adrenal hyperplasia. *New Eng. J. Med.* **316,** 1519–1524 and 1580–1586.

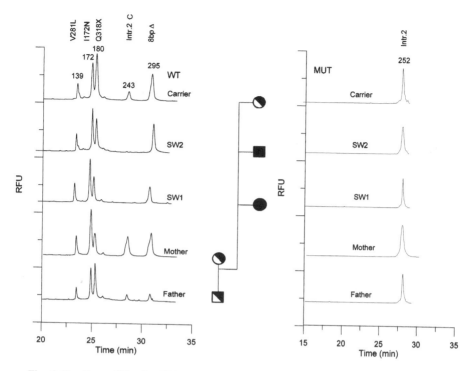

Fig. 6. Family profiling by CZE. In this family, the affected children (SW₁, SW₂) are both homozygous for the mutation in intron 2. The daughter and the parents are heterozygous for the same mutation, having the C polymorphism on the normal allele.

2. Donohoue, P. A., van Dop, C., McLean, R. H., Bias, W., and Migeon, C. J. (1986) Gene conversion in salt-losing congenital adrenal hyperplasia with absent complement C4B protein. *J. Clin. Endocrinol. Metab.* **62,** 995–1002.
3. White, P. C., New, M. I., and Dupont, B. (1986) Structure of the human 21-hydroxylase genes. *Proc. Natl. Acad. Sci. USA* **83,** 5111–5115.
4. White, P. C., Tusie-Luna, M. T., New, M. I., and Speiser, P. W. (1994) Mutations in steroid 21-Hydroxylase (CYP21). *Hum. Mutat.* **3,** 373–378.
5. Wilson, R. C., Ji-Qing, W., Cheng, K. C., Mercado, A. B., and New, M. I. (1995) Rapid deoxyribonucleic acid analysis by allele-specific polymerase chain reaction for detection of mutations on the steroid 21-hydroxylase gene. *J. Clin. Endocrin. Metab.* **80,** 1635–1640.
6. Day, D. J., Speiser, P. W., White, P. C., and Barany, F. (1995) Detection of steroid 21-hydroxylase alleles using gene-specific PCR and a multiplexed ligation detection reaction. *Genomics* **29,** 152–162.
7. Carrera, P., Barbieri, A. M., Ferrari, M., Righetti, P. G., Perego, M., and Gelfi, C. (1997) Rapid detection of 21-hydroxylase deficiency mutations by allele-specific in vitro amplification and capillary zone electrophoresis. *Clin. Chem.* **43,** 2121–2127.
8. Newton, C. R., Graham, A., and Hiptinstall, L. E. (1989) Analysis of any point mutation in DNA. The amplification refractory mutation system (ARMS). *Nucleic Acid Res.* **17,** 2503–2516.

9. Ballabio, A., Gibbs, R. A., and Caskey, C. T. (1990) PCR test for cystic fibrosis deletion. *Nature* **343**, 220.

10. Conti, M., Gelfi, C., and Righetti, P. G. (1995) Screening of umbilical cord blood hemoglobins by isoelectric focusing in capillaries. *Electrophoresis* **16**, 1485–1491.

11. Gelfi. C., Orsi, A., Leoncini, F., Righetti, P. G., Spiga, I., Carrera, P., and Ferrari, M. (1995) Amplification of 18 dystrophin gene exons in DMD/BMD patients: Simultaneous resolution by capillary electrophoresis in sieving liquid polymers. *Biotechniques* **19**, 254–263.

12. Gelfi, C., Gelfi, C., Cremonesi, L., Ferrari, M., and Righetti, P. G. (1996) Temperature programmed capillary electrophoresis for detection of DNA point mutations. *Biotechniques* **19**, 926–932.

13. Huang, X. C., Quesada, M. A., and Mathies, R. A. (1992) Capillary array electrophoresis using laser-excited confocal fluorescence detection. *Anal. Chem.* **64**, 2149–2154.

14. Ueno, K., and Yeung, E. S. (1994) Simultaneous monitoring of DNA fragments separated by electrophoresis in a multiple array of 100 capillaries. *Anal. Chem.* **66**, 1424–1429.

15. Mitnik, L., Heller, C., Prost, J., and Viovy, J.-L. (1995) Segregation of DNA solutions induced by field electric field. *Science* **267**, 219–222.

16. Baba, Y. (1996) Analysis of disease-causing genes and DNA-based drugs by capillary electrophoresis. Towards DNA diagnosis and gene therapy for human diseases. *J. Chromatogr. B* **687**, 271–302.

17. Landers, J. P. (1995) Clinical capillary electrophoresis. *Clin. Chem.* **41**, 495–509.

18. Righetti, P. G. and Gelfi, C. (1997) Recent advances in Capillary electrophoresis of DNA fragments and PCR products in poly(N-substituted acrylamides). *Anal. Biochem.* **244**, 195–207.

19. Carrera, P., Ferrari, M., Beccaro, F., Spiga, I., Zanussi, M., Rigon, F., et al. (1993) Molecular characterization of 21-hydroxylase deficiency in 70 Italian patients. *Hum. Hered.* **43**, 190–196.

20. Carrera, P., Bordone, L., Azzani, T., Brunelli, V., Garancini, M. P., Chiumello, G., and Ferrari, M. (1996) Point mutations in Italian patients with classic, non-classic, and cryptic forms of steroid 21-hydroxylase deficiency. *Hum. Genet.* **98**, 662–665.

21. Dracopoli, N. C., Haimes, J. L., Korf, B. G., Moir, D. T., Morton, C. C., Seidman, C. E., et al. (1994) Clinical molecular genetics in *Current Protocols in Human Genetics* (Boyle, A. L., ed.), Wiley & Sons, New York, pp. 9.8.1–9.8.9.

22. Ferrie, R. M., Schwarz, M. J., Robertson, N. H., Vaudin, S., Super, M., Malone, G., and Little, S. (1992) Development, multiplexing and application of ARMS tests for common mutations in the CFTR gene. *Am. J. Hum. Genet.* **51**, 251–262.

23. Sidransky, D., Tokino, T., Hamilton, S. R., Kinzler, K. W., Levin, B., Frost, P., and Vogelstein, B. (1992) Identification of ras oncogene mutations in the stool of patients with curable colorectal tumors. *Science* **256**, 102–105.

9

SSCP Analysis of Point Mutations by Multicolor Capillary Electrophoresis

Kenshi Hayashi, H. Michael Wenz, Masakazu Inazuka, Tomoko Tahira, Tomonari Sasaki, and Donald H. Atha

1. Introduction

Virtually all methods for the detection of mutations (polymorphism or variant) rely on polymerase chain reaction (PCR). Direct sequence determination of a PCR product is the gold standard for identifying mutations. However, the vast majority of the signal in the sequencing data is derived from nonvariant sequence, and can be a source of noise. Thus, a somewhat high false positive rate is inevitable when rare mutations are searched for in a large genomic region or in a region of many genomes. Techniques to detect variants as positive signals have the advantage of intrinsically low false positive rate, and are suitable methods to preselect fragments that carry mutations among an excess of nonmutated fragments. Such techniques are especially useful, for example in surveying for possible mutations in genes suspected to be responsible for genetic diseases, or finding single-nucleotide polymorphisms (SNPs) in blindly amplified genomic segments.

Single-strand conformation polymorphism (SSCP) analysis of PCR products is the most popular technique to positively detect mutations/polymorphisms in genomic or cDNA sequence *(1)*. In this analysis method, a PCR product is denatured to become single-stranded, and separated by gel electrophoresis using nondenaturing conditions. The mobility of a mutant fragment through a gel matrix is generally shifted from that of a reference fragment, because it has a folded structure that is different from that of the reference fragment. PCR-SSCP analysis is widely used in clinical or basic medical science, because it is less skill-demanding than other mutation detection methods, yet can detect mutations at a high sensitivity *(2)*. Though relatively easy and simple to set up, the method still requires certain skill such as gel preparation and sample loading. Also, the judgment of a mobility shift often requires experience, and tends to be subjective. Capillary electrophoresis SSCP (CE-SSCP) has been developed to overcome

From: *Methods in Molecular Biology, Vol. 163:*
Capillary Electrophoresis of Nucleic Acids, Vol. 2: Practical Applications of Capillary Electrophoresis
Edited by: K. R. Mitchelson and J. Cheng © Humana Press Inc., Totowa, NJ

these disadvantages, and also to adapt the technique to the emerging demand of high-throughput analysis in the age of SNP collection and typing.

The first part of this chapter includes optimization of CE-SSCP analysis of PCR products using fluorescently labeled primers, and efforts of identifying several different mutations from their characteristic mobility shifts. These protocols are developed through collaborative work between PE Applied Biosystems and the National Institute of Standards and Technology. The second part covers post-PCR fluorescent labeling of amplification products and their use in a high-precision calibration of mobility that allows a simplified statistical discrimination between a reference fragment and variant fragments at a high sensitivity. The latter system is developed and is extensively used by Kyushu University group.

2. Optimizing CE SSCP

In this subheading, we will discuss how several features of CE influence the separation of individual strands of a putative mutant, relative to a wild-type sample. Although changes in the polymer concentration, capillary length, and to a lesser degree the applied electric field, have a predictable effect on the outcome of an SSCP experiment, the influence of the applied temperature is not predictable. We hope these results can provide the researcher with guidelines that help to find conditions for optimal separation efficiency. Because the mobility profiles of individual strands of a given sample are very reproducible and often unique to the particular sample, some mutations can be identified by comparing the profiles of samples with unknown genotypes, to the profiles of known reference samples. We will show in this subheading an example of how an automated CE system can be set up to identify single point mutations in the *p53* gene.

2.1. Materials

2.1.1. Apparatus

1. Electrophoresis is performed on an ABI PRISM® 310 CE system that allows the multiplexing and simultaneous detection of up to four different fluorophores (PE Applied Biosystems).
2. The instrument controls the temperature between ambient and 60°C with an accuracy of ±1°C.
3. The electric field can be controlled between 0 and 15 kV.
4. Uncoated fused silica capillaries with a total length varying from 61 to 41 cm (= effective separation length of 50–30 cm) and an id of 50 µm are used. The capillary is typically stable for at least 100 sample injections.
5. A syringe pump is used to automatically replenish the separation polymer after each run.
6. A Perkin-Elmer 9600 thermal cycler and a *GeneAmp* PCR Core Reagents Kit (PE Applied Biosystems) are used for all PCR amplifications.

2.1.2. CE Reagents

1. GeneScan polymer (PE Applied Biosystems) is a low viscosity and replenishable polymer for CE. It is used here in the presence of 10% glycerol (w/w) in 1X TBE buffer, pH 8.3. **NOTE:** We choose to assemble the solutions by weighing the individual components, since it is difficult to pipet the viscous reagents reliably by volume.

2. To prepare the separation polymer, a stock solution in electrophoresis buffer is prepared. The stock polymer is then diluted to the desired concentration with electrophoresis buffer. To prepare 50 g of a 5% GeneScan Polymer stock solution in 1X TBE and 10% glycerol combine:

   35.7 g of GeneScan® Polymer (7%, w/w)
   5 g of glycerol
   5 g of 10X TBE
   distilled, deionized H_2O to a total mass of 50 g
   Mix by inverting several times, then vortex thoroughly for 30 s.

3. To prepare 250 g of electrophoresis buffer combine 1X TBE with 10% glycerol:
   25 g of 10X TBE
   25 g of glycerol
   distilled, deionized H_2O to a total mass of 250 g
   Mix by stirring, then filter through a 0.2-μm cellulose nitrate filter.

4. Both the polymer solution and electrophoresis buffer are stable for at least 3 mo at room temperature.

2.2. Laser-Induced Fluoresence (LIF)

1. One of the advantages of using an automated CE system with an LIF detector is that the PCR fragment can be specifically labeled and detected by fluorescence.

2. If one uses SSCP to screen samples with the goal to detect a large percentage of all known and unknown mutations in a given DNA fragment, it is recommended that each strand of the PCR fragment is labeled with a different fluorescent dye.

3. In many cases, only one of the two strands shows a significant response to the employed separation conditions, whereas the other strand remains relatively stationary. We observe that under certain separation conditions, the relative position of the strands can change. This can only be tracked by LIF detection using different dye labels for each of the two strands (*see* **Subheading 2.3.**).

4. On the other hand, if one is interested in using SSCP to detect or identify a few well characterized mutations in a given gene, conditions can be established that require the labeling of only one of the two DNA strands *(3)*.

2.3. Data Analysis

1. Raw data are analyzed using GeneScan® software version 2.1 (PE Applied Biosystems). Run-to-run variations in retention time are normalized for each run using a size marker as an internal standard.

2. Fragments are selected from the electropherogram of the internal standard that brackets the peaks of the sample fragments in order to determine the "size" of the sample strands. The migration times of these size marker fragments, expressed as data points, are used to produce a sizing curve.

3. The samples from different runs are then analyzed based on this sizing curve using the Local Southern algorithm contained in the GeneScan software.

4. For comparison, electropherograms of wild-type and mutants or reference and blind (unknown) samples are overlaid.

5. To better visually trace each respective sample, the colors of the specific peaks are changed to distinguishing colors, and the peak heights are adjusted using the GeneScan software.

6. The migration times for the two main peaks of each sample are recorded as data points (220 ms/data point).

Table 1
Mutation Reference Samples
of the *p53* Gene Exon 7[a]

Sample name	Mutation	Position[a]
H596	G → T	14060
Colo320	C → T	14069
Namalwa	G → A	14070
Wt	Wild-type	NA

[a]The particular mutation and its position in exon 7 of the
p53 gene are indicated for each reference sample. Nucleotide
position is according to Genbank locus HSP53G, Accession
#X54156.

7. Differences in the migration times of the peaks of the respective samples are compared to
wild-type and are plotted using Microsoft Excel. Negative data points indicate the peaks
that migrate slower than the wild-type reference peak, and positive peaks indicate the
peaks that migrate faster than the reference.

2.4. Samples

1. To characterize the influence of the separation factors we examine used three plasmid
clones of mutant genes derived from tumor cells, the Colo 320, the H596 and the Namalwa
(4~6) mutations, and one wild-type clone harboring exon 7 of the human *p53* gene. The
types and positions of mutations in each clone are shown in **Table 1**.

2.4.1. Generation of Labeled Samples

1. The fluorescently labeled primers are 5' end-labeled, with 5-FAM (5-carboxyfluorescein)
for the sense primer:
Sense primer: (5'-GTGTTGTCTCCTAGGTTGGCTCTG-3') and JOE (2',7'-dimethoxy-
4',5'-dichloro-6-carboxyfluorescein) for the antisense primer:
Antisense primer: (5'-CAAGTGGCTCCTGACCTGGAGTC-3').
These primers are used to amplify a 139-bp segment which corresponded to nucleotide
#13986-14124 in Genbank locus HSP53G, Accession #X54156.
2. All PCR amplification reactions contain the following: 1X PCR buffer II, 200 nM of each
primer, 1.5 mM MgCl$_2$, 0.5 U of Ampli*Taq* DNA polymerase (PE Applied Biosystems),
and 200 µM of each dNTP.
3. The thermal cycling conditions are as follows, 1 cycle of preamplification denaturation at
94°C for 3 min; then 35 amplification cycles of denaturation at 94°C for 30 s, annealing at
66°C for 30 s, then elongation at 72°C for 40 s; and finally 1 elongation cycle at 72°C for
7 min.
4. All amplifications use *GeneAmp* PCR core reagents on a Perkin-Elmer 9600 thermal
cycler (PE Applied Biosystems). Fluorescent dye-labeled PCR samples are used without
prior purification and are diluted 1:10 into deionized water.
5. For analysis, 1 µL of the diluted sample is combined with 10.5-µL deionized formamide,
0.5 µL 0.3 N NaOH and 0.5 µL of a TAMRA labeled internal size standard (GS 350; PE
Applied Biosystems).

6. Samples are denatured for 5 min at 95°C and then subsequently allowed to cool to 4°C.
7. Samples are injected electrokinetically, typically for 10 s at 15 kV.

2.5. Methods (see Note 1)

2.5.1. Effect of Polymer Concentration on Fragment Mobility

1. Separations are performed in GeneScan polymer (PE Applied Biosystems), a low viscosity and replenishable polymer, in the presence of 10% glycerol (w/w) in 1X TBE *(7)*.
2. We use three different concentrations of polymer for the separation of the four p53 reference samples, 1%, 3%, and 5%. At each polymer concentration a significant shift between the main peaks of each mutation and the corresponding peaks of the wild-type is observed (**Fig. 1**). As expected, the separation of the peaks, and therefore the ability to detect differences in the mobility between corresponding strands, increases with increasing polymer concentration. A characteristic pattern is observed for both strands of each mutation that is unique both in the extent of the shift and in the direction of the shift, but which is not affected by the polymer concentration. Changing the polymer concentration influences the electrophoresis time of wild-type from approx 11 min at 1% polymer to 20 min at 5% polymer.
3. When changing either the polymer concentration or the capillary length, the pump time required to replace the polymer within the capillary also has to be adjusted. **Table 2** provides a guideline for adjusting the pump time for different polymer concentrations and capillary lengths for solutions at 30°C. The fill times are adjusted through the manual control window of the system-specific collection software *(8)*.

2.5.2. Effect of Electric Field

1. One of the advantages of CE over slab-gel systems is that the capillaries can be run at permission of very high field strengths, due to their efficient dissipation of heat.
2. We have tested the effect of varying the applied electric field on the separation of the four different p53 reference samples. Whereas the electrophoresis system allows fields between 0 and 15,000 kV, the effects at runs between 9 and 15 kV are examined.
3. The effect of electric field on the separation of three of the p53 mutation is not as predictable as when varying the polymer concentration, or the effective capillary length. Generally for fields up to 12 kV, the effects on separation are less pronounced.
4. Increasing the field decreases the efficiency by which each of the strands of a mutation are separated in mobility, relative to the mobility of each of the corresponding wild-type strands (*see* **Fig. 2**).
5. The sample-dependent degree to which the separation is effected could be a reflection of the sensitivity of the particular strand (sequence) to an increase of temperature within the capillary, in response to the increase in voltage.

2.5.3. Effect of Separation Temperature

1. Since the ABI 310 instrument does not support active cooling, it is necessary to demonstrate that mutations can be detected during runs at or above the ambient room temperature (25–45°C).
2. As an example, we examine the effect of various temperatures on the separation of four of the p53 reference samples. Since the conformation of single-stranded DNA is dependent on the surrounding temperature, and therefore affects the migration through a separation media, unpredictable shifts in mobility of individual strands were observed (*see* **Fig. 3**).

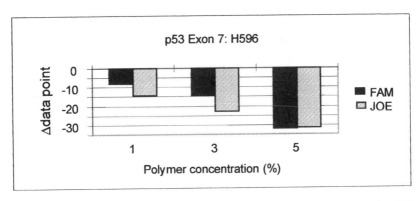

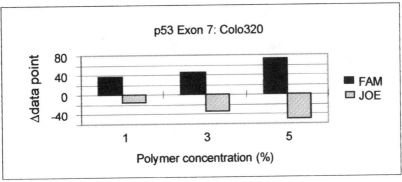

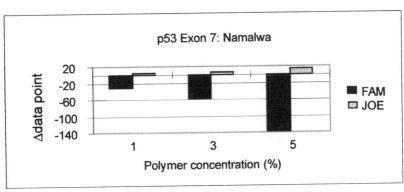

Fig. 1. Effect of separation polymer concentration on detection of *p53*-exon 7 reference mutations. Samples from *p53*-exon 7: H596, Colo320, Namalwa, and wild-type are separated in capillaries in separation buffer with 10% glycerol and at the indicated polymer concentrations. Total capillary length is at 47 cm, the temperature is 30°C and the voltage is 13 kV. The migration time, expressed in data point (~4.5 data point/s), for the two major peaks of each sample (sense and antisense strands) are corrected for minor run-to-run variability using size markers. The difference (in data points) between the mutation samples, relative to the wild-type (baseline), is plotted for each polymer concentration. Negative bars represent SSCP peaks which migrate slower than the corresponding wild-type strand, and positive bars represent SSCP peaks which migrate faster than the corresponding wild-type strand. The fluorescence label of each the two strands (Fam and Joe) are indicated.

Table 2
Capillary Fill Times for GeneScan Polymer Concentrations for Particular Capillary Lengths[a]

Length (cm)	Capillary Fill Time (s)		
	1% GSP (w/w)	3% GSP (w/w)	5% GSP (w/w)
47	25	30	60
61	30	40	100

[a]Capillary fill times in seconds are empirically determined for the indicated GeneScan polymer concentrations and for the indicated total capillary lengths (50 μm id) in a buffer containing 10% glycerol and 1X TBE at 30°C.

3. Although for the Namalwa mutation, the mobility shift in both strands is highest at the lower temperature, the Fam-labeled strand of the Colo320 mutation show a maximum shift in mobility at 40°C, whereas the shift in mobility of the Joe-labeled strand is highest at 25–30°C.

4. The shifts in mobilities of both strands of the H596 mutation, relative to each other, remain similar between 25°C and 35 °C.

5. The two strands of Colo320 mutation shift their position between 25°C and 35°C. This change in position is only detectable when each strand of a sample is labeled with a different fluorophor.

6. The sensitivity with which individual strands of a sample react to the applied column temperature requires that the temperature within a laboratory be controlled as precisely as possible. This is necessary to permit high repeatability/sensitivity when comparing different samples.

2.5.4. Reproducibility of Analysis (see **Note 2**)

1. Since the shifts in mobility between a wild-type and a putative mutant are often small, care must be taken to ensure that these shifts are a result of differences in the DNA conformation. The effect of variation in the performance of a particular instrument or the analysis system must be minimal and its contribution to any mobility changes must be accounted.

2. To test the performance of the ABI PRISM 310 CE system instrument, we subjected the four *p53* reference samples to repeated analyses at different separation temperatures (**Table 3**). For each of the two DNA strands of each sample the reproducibility of mobility at each temperature is very high, with relative standard deviations (RSD) ranging from 0.01% to 0.04%.

2.6. Typing of Unknown Mutations (see **Note 3**)

1. Since the reproducibility from run-to-run is very high using the ABI 310 instrument, we observe that the four *p53* reference samples exhibit unique mobility profiles at different temperatures (*see* **Fig. 3**).

2. Here we test to see if these features can be used to actually identify the mutation in each specific sample mutation. We used a total of 10 blind sample tests and compare their mobility profiles after separation to the four *p53* reference samples: namely to the H596, the Namalwa, and the Colo320 mutations and to wild-type. The 10 blind sample peaks are either identical in mobility to one of the four reference samples, or have closely or were similar mobility to one of the reference samples.

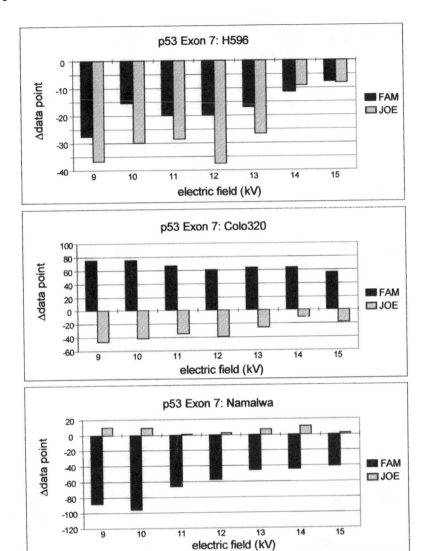

Fig. 2. Effect of electric field on the detection of *p53*-exon 7 reference mutations. Samples from *p53*-exon 7: H596, Colo320, Namalwa and wild-type are separated at the indicated electric fields at 30°C. The polymer concentration is constant at 3% (with 10% glycerol). The total capillary length is at 47 cm. *See* **Fig. 1** for further information.

3. In contrast, mutations artificially generated by site directed mutagenesis contain lesions that are not present in the reference samples *(6)*. All genotypes were called correctly using this approach (data not shown).
4. We cannot exclude that mobility profiles of unrelated mutations in samples can be identical, especially when examining DNA fragments that contain a multitude of different mutations.

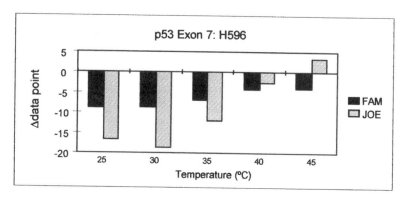

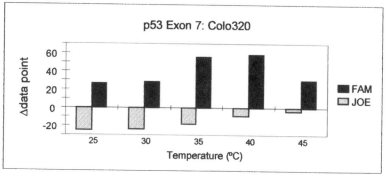

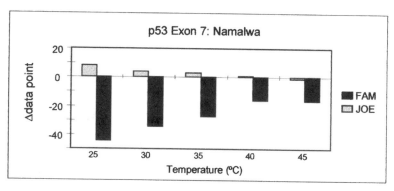

Fig. 3. Effect of separation temperature on the detection of *p53*-exon 7 reference mutations. Samples from *p53*-exon 7: H596, Colo320, Namalwa and wild-type are separated at the indicated temperatures at 13 kV. The polymer concentration is constant at 3% (with 10% glycerol). The total capillary length is 47 cm. *See* **Fig. 1** for further information.

Therefore, we recommend that the mobility shift assay described in this chapter is used only for the analysis of genes or gene fragments that have a small number of well-characterized mutations. We have applied this approach recently to detect the two common mutations in the *HFE* gene that are implicated in the genetic disorder of hereditary hemochromatosis *(3)*.

Table 3
Reproducibility for the Major Peaks of the *p53* Exon 7 Reference Samples[a]

		Migration time (data points)							
		25°C		30°C		35°C		40°C	
Sample		Joe	Fam	Joe	Fam	Joe	Fam	Joe	Fam
H596	Aver.	4512.7	4624.5	4076.2	4168.8	3966.8	4024.0	3723.6	3770.9
	RSD	0.02	0.03	0.03	0.02	0.02	0.03	0.02	0.02
Colo320	Aver.	4529.8	4588.5	4085.6	4109.0	3973.9	3942.2	3729.8	3680.7
	RSD	0.04	0.01	0.02	0.01	0.02	0.03	0.02	0.02
Nam	Aver.	4494.1	4701.3	4066.4	4242.7	3944.5	4056.3	3718.1	3790.9
	RSD	0.02	0.03	0.02	0.01	0.02	0.04	0.03	0.03
WT	Aver.	4505.5	4609.0	4072.1	4170.7	3949.0	4011.7	3720.3	3765.2
	RSD	0.02	0.03	0.02	0.03	0.02	0.04	0.02	0.01

[a]H596, Colo320, Namalwa mutations, and the wild-type of exon 7 of *p53* gene are separated at 25°C, 30°C, 35°C, and 40°C in 3% GeneScan polymer in a buffer comprising 10% glycerol and 1X TBE. Total capillary length was kept at 47 cm and separation voltage was 13 kV. Each sample is injected eight times at the indicated temperature. The average DNA fragment size, expressed in data points (220 ms/data point) and the calculated relative standard deviation (RSD) in size are shown as a percent variation.

2.7. Automated Mutation Calling Using Genotyper Software

1. Because we can correctly identify each of the *p53* samples using the high precision of the ABI 310 instrument, we show that the genotype of the samples can be called automatically using ABI PRISM Genotyper® 2.1 software (PE Applied Biosystems).
2. From three independent sample injections performed at both 25°C and 35°C, the average size for each of the two major peaks of each reference sample is calculated and is then imported into Genotyper.
3. Two groups of categories are set up, one for the Forward Strand labeled in Blue (Fam) and the other for the Reverse Strand labeled in Green (Joe).
4. For each group, four category members are defined that correspond to the four reference samples, Colo320, H596, Namalwa, and Wild-Type.
5. Each category member is defined as the highest peak at the midpoint of the expected peak size (calculated from mobility), with a tolerance of 2 bp.
6. A fifth category member called 'UNKNOWN' is defined to span the entire range of expected peak sizes. The reference category members are further defined as being exclusive size definitions.
7. The initial SSCP call for each of the blind samples is made as follows: Using the "Label and Filter" peaks macro-program, a size labeled peak for each strand (forward or reverse) from a blind sample is compared to each of the four reference category members.
8. The blind sample is called by the name of the respective reference sample if the mobility of its peak(s) fall within the reference peak size range(s).
9. If the mobility of the blind sample strand does not match any of the reference samples, then it is called UNKNOWN.

10. Once the initial calls are made and are put into a "Genotyper" table using the "Build Table" macro, then, a comparison is made between both the forward and reverse strand "calls" for each sample using the "SSCP Calls" macro. This macro utilizes the "Analyze in Table" feature of Genotyper 2.1 to make the comparisons between the strands.

11. If both of the calls are matched, a final "SSCP Call" is made, indicating that the sample in question is matched with a particular reference sample mutation.

12. If there was a mismatch between the "SSCP Call" of the strands of the same sample, then a different "SSCP Call" is made to indicate that the sample belongs to the "UNKNOWN" mutation group.

13. A separate macro has been written to change the column headers from the default in the Genotyper table to allow User-defined names.

14. In addition, a fifth macro called "Run All Macros" has been created to automatically run each of the other four macros sequentially, and then display a table of the results of this analysis in a single step (*see* **Fig. 4**).

15. We find that the Genotyper software "calls" the genotype of the blind samples similarly to the genotype "calls" made when visually comparing the mobility profiles. Further, upon breaking the code it is revealed that all the blind samples have been identified correctly using this automated technique.

3. PLACE-SSCP: A Streamlined, Sensitive Mutation Detection System

A cost-effective, robust, and sensitive mutation detection method is described in this section, called "post-PCR fluorescent labeling and analysis by automated capillary electrophoresis under SSCP conditions" (PLACE-SSCP) (*see* **Note 4**).

The salient features of this method are:

   a. A one-tube reaction to fluorescently label PCR products after amplification using unmodified primers,
   b. A strategy of CE that allows precise calibration of run-to-run variation of reference DNA, and
   c. A simple statistical evaluation of presence or absence of mutation in sample DNA.

3.1. Post-PCR Fluorescence Labeling

Many CE systems suitable for DNA analysis such as the ABI PRISM 310 are developed primarily for use in sequencing, and have the capability of multicolor fluorescence detection.

A simple way to fluorescently label DNA fragments is to amplify the target using PCR amplification primers carrying fluorescence at their 5' ends as described in **Subheading 2.4.1.** Initial attempts of fluorescent detection of bands in slab-gel based SSCP also use this approach *(9)*, but the method has not gained popularity perhaps because of high cost associated with the synthesis of fluorescent PCR primers. Presently, the cost of primer synthesis has been dramatically reduced, but the cost of synthesis of fluorescently modified primers is still considerable.

Recently, a simple method for fluorescent labeling of PCR product after amplification using unmodified primers has been developed *(10)*. **Figure 5** illustrates that the 3'-end nucleotides of a PCR product are replaced with fluorescently labeled nucleotides by the 3'-exchange activity of Klenow fragment of DNA polymerase I (i.e., a combined activity of polymerization and exonucleolysis at the 3'-end).

A

```
• Forward Strand (Blue)
•  UNKNOWN      Highest peak from 4000.00 to 4200.00 bp in blue with scaled ht >= 100
•  COL0320      (X) Highest peak at 4087.70 ±    2.00 bp in blue with scaled ht >= 100
•  H596         (X) Highest peak at 4166.99 ±    2.00 bp in blue with scaled ht >= 100
•  NAM          (X) Highest peak at 4205.90 ±    2.00 bp in blue with scaled ht >= 100
•  WT           (X) Highest peak at 4157.06 ±    2.00 bp in blue with scaled ht >= 100
- Reverse Strand (Green)
•  UNKNOWN      Highest peak from 4000.00 to 4200.00 bp in green with scaled ht >= 100
•  COL0320      (X) Highest peak at 4114.01 ±    2.00 bp in green with scaled ht >= 100
•  H596         (X) Highest peak at 4109.75 ±    2.00 bp in green with scaled ht >= 100
•  NAM          (X) Highest peak at 4090.87 ±    2.00 bp in green with scaled ht >= 100
   WT           (X) Highest peak at 4095.38 ±    2.50 bp in green with scaled ht >= 100
```

B

Sample ID	Forw. Call	Rev. Call	Forw. Size Call	Rev. Size Call	Sample Call
S1/1	NAM	NAM	4205.75	4089.41	NAM
S1/2	H596	H596	4168.08	4110.85	H596
S1/3	COL0320	COL0320	4087.89	4113.93	COL0320
S1/4	WT	WT	4156.04	4093.90	Wild Type
S2/1	WT	UNKNOWN	4158.03	4088.09	UNKNOWN
S2/2	WT	NAM	4157.00	4091.00	UNKNOWN
S2/3	NAM	NAM	4205.94	4090.19	NAM
S2/4	COL0320	COL0320	4088.14	4115.07	COL0320
S2/5	COL0320	COL0320	4087.70	4113.82	COL0320
S2/6	WT	WT	4156.03	4092.95	Wild Type

Fig. 4. Automatic mutation calling with Genotyper software. Categories, including a bin, for each of the major peaks of each reference sample are defined, based upon the calculation of the average size from 3 separate injections (**A**). The tables shown in (**B**) automatically assign ("call") a sample (User Comment 1) as either being identical to a reference DNA, or an "unknown," based upon the matching of the size of the blind sample against the reference. Being "called" as "similar" to a reference DNA requires that both strands are "called" identical to both corresponding strands of the same reference DNA (Blue 1/Green 1). The size information of both peaks of the blind sample (Blue 1/Green 1), next allow comparison to the several preset categories of the various reference DNA standards.

Many fluorescent nucleotides are efficient substrates for the 3'-exchange polymerizing activity of the enzyme, but are poor substrates of the exonuclease activity once they are incorporated to the 3' end of DNA. Consequently, fluorescently labeled products accumulate even in the presence of a high concentration of competing, unlabeled nucleotides.

PCR product can be efficiently labeled just by adding fluorescent nucleotides, Klenow fragment and Mg^{2+} to the tube in which the amplification reaction has been carried out, where unincorporated nucleotides still remain at high concentrations (*see* **Fig. 5**).

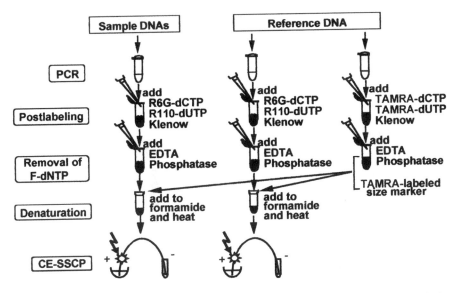

Fig. 5. Schematic diagram of PLACE-SSCP. *See* **Subheading 3.4.** for a detailed description.

Subsequent addition of EDTA and phosphatase enzyme inactivates the Klenow fragment activities and destroys all remaining nucleotides, including fluorescent nucleotides which may otherwise disturb subsequent SSCP analysis as the nucleotides would be detected as a huge peak in the electropherogram.

The Klenow fragment also degrades the remaining unused amplification primers during the fluorescent exchange-labeling reaction. If present at high concentrations, free primers can bind to the single-stranded PCR products at complementary loci. The action of the Klenow enzyme is fortuitous, as unincorporated primers would result in the appearance of anomalous peaks in an SSCP analysis profile.

Importantly, the 5'-ends of PCR primers must be designed to be compatible with the fluorescent-exchange reaction. Primers should have a 5'-G when fluorescent-dCTP is used, and 5'-A when fluorescent-dUTP is used for the labeling. In addition, two Ts adjacent to the 5'-terminal residue seem to assure efficient labeling and to prevent internal incorporation of the fluorescent nucleotides. The 5'-three residues, GTT or ATT can be noncomplementary to the target sequence. The two strands of a product may be independently labeled with different fluorophores. Strands of products amplified with one of the primers having 5'-GTT and the other having 5'-ATT can be differentially labeled with two colors using, for example, [R110]-dUTP and [R6G]-dCTP as nucleotide substrates.

The DNA products resulting from post-PCR fluorescence labeling are exclusively blunt-ended. This is an additional advantage of the postlabeling method over primer labeling method in which the amplified product is often a mixture of blunt-ended and 3'-overhanged fragments, because *Taq* DNA polymerase tends to add a nontemplated extra-nucleotide to the 3'-end.

3.2. Mobility Calibration Using Reference Fragments

Whether a mutant fragment shows a shift in mobility during SSCP analysis is unpredictable, because of the lack of full theoretical understanding of the factors affecting how ssDNA is three dimensionally folded under particular physical conditions. In addition, the manner in which differences in the folded structure of ssDNA affects the extent of retardation during electrophoresis through a matrix is not known. However, it is our experience that the sensitivity of CE-SSCP is much higher than the conventional method owing to the high resolving power (up to 10^7 in theoretical plates) of CE (*see* **ref. 11** for detailed discussion).

In SSCP, a mutation is detected if the mobility of a fragment carrying a mutation is significantly different from the mobility of a reference fragment, which carries a normal or wild-type sequence *(1)*. In other words, a fragment is judged to carry a mutation if it shows a mobility shift, which is unlikely to be explained by any experimental variation of mobility of the reference sequence. Thus, a high reproducibility in mobility of a reference fragment is especially important for sensitive and reliable detection of mutations.

In CE-SSCP, the mobility of any fragment can vary from run-to-run because of possible changes of microenvironments, such as a slight change of the column temperature which is often a critical factor that affects the mobility of individual runs. Thus, it is essential to precisely correct for "run-to-run" variations during the calibration of electropherograms, to allow real comparison of the mobility of ssDNA observed in different runs.

How, then, can the mobility of the reference fragment be accurately calibrated? In fluorescent SSCP analysis using capillary-based automated sequencers, one of the colors is devoted to label an internal standard and the remaining colors are assigned to label the samples. After several electrophoresis runs, the "run-to-run variation" is corrected using the internal standard, that is, one of the runs are taken as a "template" run, and other electropherograms are fitted to the template by referring to the peaks of the internal standard of each sample run *(12)*, *see* **Subheading 3.3.**

3.2.1. Local Southern Analysis

1. Several methods are available for fitting the data points between the peaks to the template peak, among which a method called "local Southern" has been most successfully applied in SSCP.
2. In this method, the calibration of regions in the electropherogram nearer to peaks of internal standard are more likely to be accurate. Therefore, to precisely calibrate the mobility of a *reference* fragment, it is important that the internal standard contains a fragment that shows a peak at a position close to that of the reference.
3. Using commercially available size markers as internal standard in CE-SSCP, we have experienced moderately good calibration. However, it should be noted that some mutant DNA fragments also show mobilities that are within the variation of the mobility of the corresponding reference fragment.
4. One reason for a rather wide variation of mobility of the reference fragment is that in some cases, none of the peaks of size markers are close enough to the peak of reference for accurate calibration.

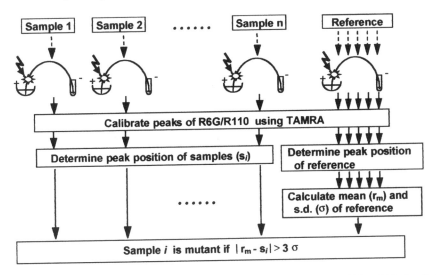

Fig. 6. Data processing for run-to-run calibration and evaluation of mobility shift. *See* **Sub-heading 3.5.** for a detailed description. These results are also applicable to the bundled capillary systems such as the ABI 3700 which are now commercially available.

5. In addition, the size marker fragments have a completely different sequence context from the sequence of the reference, and may behave differently due to changes in the microenvironment between different runs.

6. Using such a fragment as an internal standard can lead to erroneous calibration. A simple answer to always solve these problems is to include the reference fragment itself as a part of the internal standard. In this way, one of the internal standard fragments is always found close to the reference, and the fragment has the same sequence context as that of the reference.

7. In the following experiments, both strands of the reference fragment is labeled with TAMRA and mixed with the "size marker" which is also labeled with TAMRA. This mixture is then used as an internal standard.

8. Samples with strands that are differentially labeled with R6G and R110 are combined with the internal standard, and applied to CE (*see* **Fig. 5**).

9. Examining the reference fragment as a sample, and repeating the analysis five times, the variability of calibrated mobility of reference fragment are evaluated (*see* **Fig. 6**).

10. **Table 4** demonstrates that the standard deviation of the estimated mobilities after calibration using this mixture is much smaller than the estimated mobilities determined using the "size marker" alone. That is, the mobility of the reference fragment is highly reproducible after calibration using an internal standard, which includes the reference fragment itself.

11. The variation the mobility of the reference fragment calibrated in this way is usually within one data point, and the relative standard deviation is less than 0.005%. This has been found to be true for more than 50 fragments of different sequence contexts, which are 200–600 bp, and have mean values of data points of peak positions between 15,000 and 20,000 (data not shown).

Table 4
Calibration of the Peak Positions of Reference DNA

	Sense strand			Antisense strand	
	Calibrated using				Calibrated using	
Raw[a]	M[b]	M + N[c]	Raw[a]	M[b]	M + N[c]
6393d	6393.00	6393.00	6557[d]	6557.00	6557.00
6474	6389.01	6393.03	6638	6556.32	6557.02
6457	6392.16	6392.98	6623	6560.09	6557.02
6476	6391.79	6393.01	6641	6560.12	6557.02
6469	6389.53	6392.01	6635	6557.35	6557.01
Average	6391.10	6392.81		6558.18	6557.01
S.D.[e]	1.55	0.40		1.61	0.01

Data points of peaks of R6G-labeled sense strand and R110-labeled antisense strand of PCR product of wild-type *gyrA* examined over five successive runs.

[a]Raw data without calibration. Data collection started 5 min after the start of electrophoresis at a rate of 70 ms/data point.

[b]Calibrated using size marker alone.

[c]Calibrated using size marker and TAMRA-labeled wild-type PCR products.

[d]Used as a template for calibration.

[e]Standard deviation.

3.3. Statistical Evaluation of Mobility Shift

In conventional SSCP analysis, the shift in ssDNA mobility is judged by visual examination. However, in PLACE-SSCP, mobility is digitally obtained in data points. Thus, whether or not a particular DNA sample shows a mobility shift relative to a reference DNA can be statistically evaluated (*see* **Fig. 6**).

To evaluate PLACE-SSCP analysis, we examine the mobility shift of 34 known mutant fragments of 600–700 bp in three sequence contexts relative to the mobility the reference *(12,13)*. The mobility shift is judged to be significant if it is more than $3 \times \sigma$, where σ is the standard deviation of mobility of the reference. Out of the 34 mutants, 33 (97%) showed a significant shift in the mobility of at least one strand, and so, are judged to be mutant sequences.

Assuming that a (calibrated) mobility fluctuates following Gaussian distribution, probability that at least one strand of reference fragment of showing a mobility more than $3 \times \sigma$ away from the mean is 0.6% (0.3% for each strand). Thus, the false negative rate of this series of experiments is 3%, and the false positive rate is 0.6%.

The estimation of sensitivity is based on the examination of mutations of a bacterial gene, or a gene on X chromosome of male subjects and is valid only for the detection of mutations in a homozygous/hemizygous state, or in a haploid genome. However, in many cases with diploid genomes, we need to find mutations, which are heterozygous with the wild-type allele also present. Such heterozygotes are often electrophoretically unresolved, but are detected as a combined single peak.

Theoretically, the false negative rate of detection of mutations in heterozygotes with wild-type alleles should be twofold the rate of that of homozygous/hemizygous mutation. In general, perhaps it is safe to say that at the above criteria, 95% and 90% of mutations in homozygous and heterozygous states, respectively, are detected at a false positive rate of 0.6%, by PLACE-SSCP method using only one electrophoresis condition.

4. Notes

1. We demonstrate how several factors influence sensitivity in detecting single point mutations by SSCP. These factors include the polymer concentration, the capillary length, and the applied electric field. We also show how capillary temperature can be particularly important in resolving mutations that are ordinarily difficult to resolve.
2. In addition, we show that capillary systems with multicolor detection capabilities such as the ABI PRISM 310 are necessary to accurately record changes in the electrophoretic mobility which occur in multicomponent systems such as SSCP of DNA. The multicolor detection system of the ABI PRISM 310 also allows the simultaneous monitoring of an internal standard for the very high precision and reproducibility that these measurements require. The analysis of SSCP mobility profiles using GeneScan and Genotyper software affords an automated method for identifying DNA point mutations.
3. A sensitive, robust, and cost-effective mutation detection system is needed to satisfy a growing demand for a high-throughput mutation/polymorphism detection method. CE-SSCP, though it offers many advantages over conventional gel-based SSCP analysis, still requires optimization of experimental conditions to enhance the mobility shift and to attain a high rate of accurate detection.
4. The running costs associated with SSCP analysis of mutations is rather high, due mainly to the use of fluorescence-labeled primers for amplification of target sequences. In contrast, PLACE-SSCP does not require fluorescence-labeled primers. Also in PLACE-SSCP, the calibration of runs are dramatically improved so that even a small shift in the mobility of a mutant fragment can be distinguished from fluctuation of experimental data, thus allowing high detectability without meticulous optimization. The whole system including fluorescence labeling of PCR products and data interpretation is streamlined and does not require special expertise. Because of the high rate of detection even using one experimental electrophoresis condition, the algorithm of judgment of presence of mutation is simple, and should be easily adaptable to high-throughput analyses such as single-nucleotide polymorphism finding.

Disclaimer

References

1. Orita, M., Suzuki, Y., Sekiya, T., and Hayashi, K. (1989) Rapid and sensitive detection of point mutations and DNA polymorphisms using the polymerase chain reaction. *Genomics* **4,** 874–879.

2. Hayashi, K. (1999) Recent enhancements in SSCP. *Gen. Anal. Biomol. Eng.* **14**, 193–196.

3. Wenz, H. M., Baumhueter, S., Ramachandra, S. and Worwood, M. (1999) A rapid automated SSCP multiplex capillary electrophoresis protocol that detects the two common mutations implicated in hereditary hemochromatosis (HH). *Human Genet.* **104**, 29–35.

4. O'Connell, C. D., Tiang, J., Juhasz, A., Wenz, H. M., and Atha, D. H. (1998) Development of standard reference materials for diagnosis of p53 mutations: Analysis by slab gel-SSCP. *Electrophoresis* **19**, 164–171.

5. Atha, D. H., Wenz, H. M., Morehead, H., Tian, J., and O'Connell, C. D. (1998) Detection of p53 point mutations by single strand conformation polymorphism (SSCP): Analysis by capillary electrophoresis. *Electrophoresis* **19**, 172–179.

6. Wenz, H. M., Ramachandra, S., O'Connell, C. D., and Atha, D. H. (1998) Identification of known p53 point mutations by capillary electrophoresis using unique mobility profiles in a blinded study. *Mutat. Res.* **382**, 121–132.

7. Kukita, Y., Tahira, T., Sommer, S.S., and Hayasi, K. (1997) SSCP analysis of long DNA fragments in low pH gel. *Human Mutat.* **10**, 400–407.

8. ABI Prism 310 Genetic Analyzer (1997) http://www2.perkin-elmer.com:80/ga/pdf/310GSRG.pdf

9. Makino, R. Yazyu, H., Kishimoto, Y., Sekiya, T., and Hayashi, K. (1992) F-SSCP: A fluorescent polymerase chain reaction-single strand conformation polymorphism (PCR-SSCP) analysis. *PCR Meth. Appl.* **2**, 10–13.

10. Inazuka, M., Tahira, T., and Hayashi, K. (1996) One-tube post-PCR fluorescent labeling of DNA fragments. *Genome Res.* **6**, 551–557.

11. Guttman, A., Cohen, A. S., Heiger, D. N., and Karger, B. L. (1990) Analytical and micropreparative ultrahigh resolution of oligonucleotides by polyacrylamide gel high-performance capillary electrophoresis. *Anal. Chem.* **62**, 137–141.

12. Inazuka, M., Wenz, H. M., Sakabe, M., Tahira, T., and Hayashi, K. (1997) A stream-lined mutation detection system: Multicolor post- PCR fluorescence-labeling and SSCP analysis by capillary electrophoresis. *Genome Res.* **7**, 1094–1103.

13. Hayashi, K., Kukita, Y., Inazuka, M., and Tahira, T. (1998) Single-strand conformation polymorphism analysis, in *Mutation Detection: A Practical Approach* (Cotton, R. G. H., Edkins, D., and Forrest, S. eds.), IRL Press, Oxford, UK, pp. 7–24.

10

SSCP Analysis by Capillary Electrophoresis with Laser-Induced Fluorescence Detector

Jicun Ren

1. Introduction

Among the various techniques developed for mutation detection, single-strand conformation polymorphism (SSCP) analysis has become the most popular method for the screening of unknown mutation in small stretches of DNA. Its widespread use is related to its simplicity and low cost. SSCP analysis is based on the principle that ssDNA with a single base substitution (or containing a small sequence deletion or insertion) often assumes an altered folded conformation to an equivalent wild-type sequence lacking the base substitution. The single base substituted and wild-type ssDNA molecules therefore may migrate with different mobility from each other in a nondenaturing electrophoresis gel (1). This is demonstrated in **Fig. 1**, in which the conventional SSCP procedure is shown. Here, the gene DNA fragment of interest is amplified by use of the polymerase chain reaction (PCR). The DNA fragments are [^{32}P]-radiolabeled (or using radiolabeled primers). Then following nondenaturing gel electrophoresis, the mobility (position) of the ssDNA is detected by autoradiography (2–4). In spite of the simplicity of the technique, conventional SSCP analysis remains both labor-intensive and time-consuming, involving the preparation of large-format polyacrylamide gels for multiple sample handling, the long run time of the electrophoresis gels, and the manipulation of radioactive compounds and radioactive wastes.

Capillary electrophoresis (CE) in entangled polymers is an attractive alternative to slab-gel electrophoresis techniques for DNA analysis (5–7). CE can be automated and the technique is characterized by short analysis time, small sample and small reagent requirements, high resolution and separation efficiency, and when coupled to laser-induced fluorescence (LIF) detector has unsurpassed detection sensitivity. Notably, when used in a multiple injection mode, the average analysis time per sample is reduced to about 4 min (8). An additional important advantage of CE is that the running temperature can be precisely controlled to allow SSCP analysis with high resolution and good reproducibility between experiments. This is because the change of the

From: *Methods in Molecular Biology, Vol. 163:*
Capillary Electrophoresis of Nucleic Acids, Vol. 2: Practical Applications of Capillary Electrophoresis
Edited by: K. R. Mitchelson and J. Cheng © Humana Press Inc., Totowa, NJ

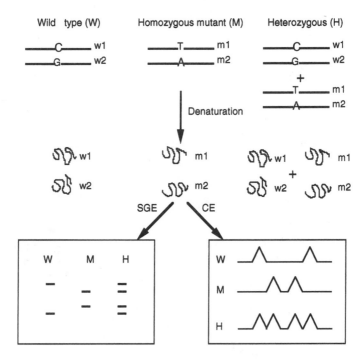

Fig. 1. The principle of SSCP analysis. Slab-gel electrophoresis (SGE). Capillary electrophoresis (CE).

temperature in the capillary markedly affects the sensitivity and the reproducibility of SSCP analysis. CE-LIF has successfully been used instead of slab-gel electrophoresis for the separation of DNA fragments as part of SSCP analysis *(8–18)*. The technique is simple and rapid and is well suited to the analysis of a large number of clinical samples with high throughput.

2. Materials
2.1. DNA Extraction
1. Whole human blood samples (*see* **Note 1**).
2. The "DNA Direct" Preparation Kit (Dynal, Oslo, Norway).
3. The "QIAquik Blood Kit" (HT Biotechnology Ltd., UK).

2.2. Polymerase Chain Reaction
1. Reaction tubes, thin-walled: (Gene Amp).
2. Fluorescein-labeled primers: 20 µM stock solution of primers is stored at –20°C until use. Examples of primers are:
   a. The amplification of C677T mutation of *methylenetetrahydofolate reductase* gene includes the following oligonucleotide primers *(8)*:
   5'-fluorescein-GGAGCTTTGAGGCTGACCTGAA-3'(forward primer),
   5'-fluorescein-GACGATGGGGCAAGTGAT-3' (reverse primer);

b. The amplification of G1691A mutation of *factor V* gene uses the following primers *(15)*:
5'-fluorescein-CTTTGGGGAGCTGAAGGACTACTAC (forward primer),
5'-fluorescein-TCAAGGACAAAATACCTGTATTC (reverse primer).

3. PCR buffer, 10X (HT Biotechnology Ltd.): 100 mM Tris-HCl, pH 9.0, 500 mM KCl, 15 mM MgCl$_2$, 0.1% (w/v) gelatin and 1% Triton X-100, stored at –20°C until use.

4. dNTPs solutions with 1.25 mM of each dNTP (dATP, dGTP, dCTP, and dTTP), stored at –20°C.

5. Super *Taq* DNA Polymerase: (HT Biotechnology Ltd.) stored at –20°C.

6. dd-Water, double-distilled and purified on a Milli-Q Plus water purification system (Millipore, Bedford, MA).

7. PCR thermocycler (Perkin-Elmer).

2.3. Purification of PCR Products

1. The "QIAquick PCR Purification Kit" (QIAGEN Co. Germany).
2. 100% Ethanol (store at –20°C).
3. 70% Ethanol (store at –20°C).
4. Tris-EDTA buffer, 1X: 10 mM Tris-HCl, 1 mM EDTA, pH 8.0.
5. Eppendorf Centrifuge with refrigeration.

2.4. Synthesis of Short-Chain Linear Polyacrylamide (SLPA)

1. Acrylamide (Sigma).
2. N,N,N',N'-tetramethylenediamine (TEMED).
3. Ammonium peroxydisulfate (APS).
4. Isopropanol (A.R. grade).
5. Dialysis membrane tubing, 12,000 mol wt cutoff (Thomas Scientific, Philadelphia).
6. Ultrapurified helium gas.
7. A thermostated water bath.

2.5. Capillary Coating

1. Fused silica capillaries with 50 µm or 75 µm id from Polymicro Technologies Inc. (Phoenix, AZ).
2. 3-Methacryloxypropyltrimethoxysilane (Pharmacia LKB Biotechnology AB, Uppsala, Sweden).
3. 50% Acetic acid solution.
4. 1 M HCl.
5. 0.1 M NaOH.
6. 3% Acrylamide solution.

2.6. Capillary Electrophoresis

1. CE Instrument with LIF detector.
2. Argon-ion laser with 488-nm emission (Uniphase Ltd., Herts, UK).
3. TBE buffer, 1X: 89 mM Tris-base, 89 mM boric acid, 1 mM EDTA, pH 8.3.

3. Methods

3.1. DNA Extraction

1. Collect a whole blood sample (at least 20 µL), (*see* **Note 1**).
2. Extract the genomic DNA from whole blood using the "DNA Direct Kit" or the "QIAquik Blood Kit" according to the instructions from the manufacturer.
3. Store the purified DNA at 4°C until use.

3.2. PCR Reaction

1. A 100-µL PCR mixture contains: 10 µL of 10X PCR buffer, 10 µL of 1.25 mM dNTP mixtures, 2 µL of 10 µM each of the forward and reverse primers, 1 µL of *Taq* DNA polymerase (0.5 U), 5 µL of genomic DNA (about 100 ng) and 72 µL water.
2. Amplify by PCR using the following cycle profile (*see* **Note 2**): Initial denaturation: 94°C for 4 min. Then, 32–36X thermal cycles of: denaturation: 94°C, 30 s; annealing: 60°C, 30 s; extension: 72°C, 10 s. Final extension: 72°C, 7 min.

3.3. Purification of PCR Products

1. Transfer 100 µL of PCR product into an 1-mL Eppendorf tube, add 200 µL of 100% ethanol (cold) and then mix gently (*see* **Notes 3** and **4**).
2. Place the mixture into refrigerator and stand it for 1 h at –20°C.
3. Recover the PCR product by centrifugation at 8000g for 10 min at 4°C.
4. Rinse the DNA pellet twice with 300 µL of 70% cold ethanol.
5. Resuspend the PCR product in 50 µL of 1X TE buffer and store at –20°C until use.

3.4. Synthesis of Short-Chain Linear Polyacrylamide

1. Short-chain linear polyacrylamide (SLPA) is synthesized according to a slight modification *(8)* of the procedure as described by Grossman *(19)* (*see* **Note 5**).
2. The reaction mixture contains 111 mL of dd-water, 12.5 g of acrylamide, 3.75 mL of isopropanol.
3. The mixture is degassed with helium for 0.5–1 h and then heated to 60°C in water bath for about 15 min.
4. Add 0.625 mL of 10% (v/v) TEMED, 0.625 mL of 10% (w/v) APS to the mixture and allow the polymerization to take place at 60°C for 2 h.
5. Dialyze the product against water for 2 d using the 12,000-mol wt cutoff dialysis membrane tubing to remove short chain molecules, then lyophilize the product.

3.5. Capillary Coating

1. The capillary is coated with linear polyacrylamide (LPA) according to a slight modification *(8)* of the method described by Hjertén *(20)*.
2. Wash an uncoated capillary (90 cm length) with 1.0 M HCl for 40 min.
3. Wash the capillary with 0.1 M NaOH for 10 min.
4. Flush the capillary extensively with water.
5. Fill the capillary with 50% acetic acid solution containing 2% of 3-methacryloxy-propyltrimethoxysilan, and allow it to stand for 2 h at room temperature.
6. Flush capillary extensively with water.
7. Fill up capillary with 3% acrylamide solution (5 mL of 3% acrylamide, 2 µL TEMED, and 20 µL 10% APS), and allow it to stand for 2 h at room temperature.
8. Wash the LPA-coated capillary with water, and store it at 4–8°C with the two ends of the capillary inserted into water to prevent the coating drying out (*see* **Note 6**).

3.6. Capillary Electrophoresis Procedure

1. Prepare a 6% SLPA solution with 1X TBE buffer as the sieving medium (*see* **Note 7**).
2. Use the 1X TBE as the electrophoresis running buffer (*see* **Note 8**).
3. Rinse the new PA-coated capillary with TBE buffer for 5 min.
4. Fill the capillary with the 6% SLPA sieving medium.
5. Introduce the ssDNA samples by electrokinetic injection at –2 kV for 6 s.

6. Run capillary electrophoresis at reverse polarity mode at constant electric field, typically at –20 kV for 15–20 min. The column temperature is 25°C. (*see* **Note 9**).
7. Between each run, wash the capillary with TBE buffer for about 2 min and refill the capillary with fresh sieving medium.

3.7. SSCP Analysis

1. Take 5 µL the purified PCR product into a 0.5-mL Eppendorf tube and dilute by 10-fold with double-distilled water, (*see* **Notes 10** and **11**).
2. Heat the PCR product at 94°C for 4 min to denature the DNA strands and then cool the mixture in ice water for 10 min.
3. Introduce the ssDNA samples by electrokinetic injection at –2 kV for 6 s.
4. Run CE at reverse polarity mode at constant electric field, typically at –20 kV for 15–20 min, (*see* **Note 12**). The column temperature is 25°C.
5. Identify the genotype of the sample according to the migration time and number of peaks in electropherogram (*see* **Note 13**).
6. **Figure 2** shows the C677T polymorphism of the *methylenetetrahydrofolate reductase* gene by SSCP analysis *(8)*. The homozygous normal (––), homozygous mutant (++) and heterozygous (+–) genotypes were clearly distinguished by CE-LIF.

3.8. Poly(Dimethyl Acrylamide) (PDMA)

1. PDMA *(22)* can form a dynamic coating on inner surface of silica capillary and suppress the electroosmotic flow and the interaction of DNA and the inner surface of the capillary, (*see* **Note 14**).
2. Using PDMA as sieving medium *(16)* SSCP analysis successfully detected point mutations/polymorphisms of *methylenetetrahydrofolate reductase* gene, *factor V* gene and *cystathionine β-synthase* gene in uncoated capillary (not shown).

4. Notes

1. Template DNA used in PCR reaction can be conveniently extracted from whole blood samples using the "DNA Direct Kit" or the "QIAquik Blood Kit." Conveniently, when the LIF detector with high sensitivity is used in SSCP analysis sufficient ssDNA fragments can be amplified by PCR using 1 µL of unpurified whole blood as the template DNA.
2. The conditions of the PCR reaction described above can be only applied to the amplification of the C677T mutation of the *methylenetetrahydrofolate reductase* gene using the primer pairs indicated. The cycle number and thermal cycle profile should be optimized the PCR reactions to amplify other gene fragments.
3. The PCR products are purified by precipitation with 70% ethanol in order to remove salts from the DNA fragments and thereby increase the resolution of the SSCP analysis.
4. Occasionally, the unincorporated fluorescein-labeled primers interfered with SSCP analysis. We found that the fluorescent materials and fluorescein-labeled primers could be efficiently removed using the "QIAquick PCR Purification Kit" *(21)*.
5. The SLPA medium is well suited for SSCP analysis on a commercial CE instrument. Commercial CE instruments can easily carry out capillary filling and medium replacement with the low viscosity SLPA medium. The resolution of ssDNA fragments is markedly dependent on the concentration of SLPA. Our studies showed that the best resolution is obtained at 6% SLPA for analysis of the C677T mutation *(8)*. We have also successfully detected eight of nine different point mutations from the *methylenetetrahydrofolate reductase* gene, the *factor V* gene, and the *cystathionine β-synthase* gene using 6% SLPA as the sieving medium *(15)*.

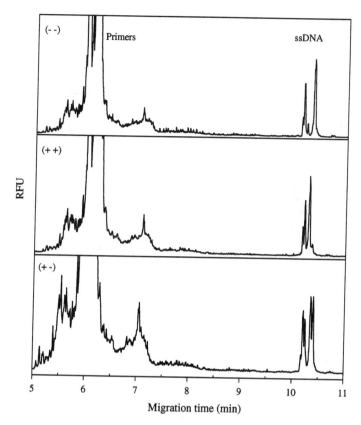

Fig. 2. SSCP analysis by CE-LIF. The C677T polymorphism of the *methylenetetra-hydrofolate reductase* gene is determined by SSCP analysis. The homozygous normal (--), homozygous mutant (++), and heterozygous (+-) genotypes are indicated in upper left corners of the panels. Relative fluorescence units (RFU). The single-strand DNA fragments are separated in the PA-coated capillary (39.5 cm length and 50 μm id) filled with 6% SCLP in 1X TBE buffer, pH 8.3. Electrokinetic injection at –2 kV for 6 s is used. The run temperature is 25°C and the applied voltage is –20 kV.

6. The new LPA-coated capillary should be stood for overnight after coating. This is observed empirically to prolong the lifetime of the LPA-coated capillary. After finishing the analysis each day, the capillary should be stored filled with water to avoid air entering into capillary. Surface drying will decrease the efficiency of the coated capillary markedly. The commercial DB-17 coated capillary (J & W Scientific) also has high resolution, long lifetime and is suitable for SSCP analysis *(15)*.

7. The sieving medium is an important factor that may affect SSCP analysis using CE. SLPA sieving medium has low viscosity and high ssDNA resolution compared to both hydroxypropyl methyl cellulose and poly(ethylene oxide) media.

8. TBE buffer (pH 8.3) is a widely used running buffer in SSCP analysis based on slab-gel electrophoresis or CE. Most of mutations can be detected under this pH value. Recently,

our results have demonstrated that the pH value of the running buffer strongly affects the pattern of SSCP and the resolution of ssDNA conformations *(15)*. We suggest that the pH of the running buffer should be systematically optimized in SSCP analysis based on CE.

9. Temperature significantly affects the conformation of ssDNA and thereby alters the sensitivity of the SSCP analysis. The increase in temperature will remarkably reduce the resolution of ssDNA fragments, which attain a less folded structure at high temperature. Some studies have shown that a low temperature (15–20°C) is beneficial to maintenance of the sensitivity of SSCP analysis by CE.

10. The use of a thin capillary (e. g., 50 μm id) is suggested if the sensitivity of the detector is compatible with the concentration of PCR products. The thin capillary can more efficiently dissipate Joule heat and the high electric field can be applied in SSCP analysis compared to thicker capillaries, which can shorten the analysis time without loss of resolution.

11. The sensitivity of the UV-detector is much (about 1000 times) lower than that of the LIF detector. However, the concentration of PCR product can dramatically increase by adding more template DNA, increasing the cycle number of PCR, optimizing the thermocycle profile of PCR reaction or using nested PCR technique. The UV-detector can be used in SSCP analysis of some mutations *(9,11)*.

12. The appropriate electric field strength should be applied according to the dimensions of the capillary and the electrical conductivity of the running buffer. An extremely high electric field will produce serious Joule heating. This increases the temperature in center of the capillary, which may alter the conformation of ssDNA, increase analyte diffusion, and thereby cause peak broadening and a reduction in resolution. However, an insufficiently strong electric field will prolong the migration time of ssDNA fragments, which will also increase the sample zone diffusion and also lead to loss of resolution. We suggest that an electric field should be chosen to give an optimal current of about 10–20 μA, in order to avoid Joule heating.

13. The effects of glycerol on SSCP analysis are dependent on the sequence context of DNA fragments. A similar effect on SSCP analysis has been observed by a reduction of the pH value of sieving medium (*see* **Note 8**). We have found that glycerol reduces the resolution in SSCP analysis of the C677T mutation of the *methylenetetrahydrofolate reductase* gene *(8)*. The results of SSCP analysis should therefore be compared with and without glycerol present.

14. PDMA *(22,23)* can form a dynamic coating on inner surface of silica capillary and suppress the electroosmotic flow and the interaction of DNA and the inner surface of the capillary. Our experiments *(16)* have shown that PDMA may become a very useful sieving medium in SSCP analysis in uncoated capillaries.

References

1. Orita, M., Iwahana, H., Kanazawa, H., Hayashi, K., and Sekiya, T. (1989) Detection of polymorphisms of human DNA by gel electrophoresis as single-strand conformation polymorphisms. *Proc. Natl. Acad. Sci. USA* **86,** 2766–2770.

2. Glavac, D. and Dean, M. (1993) Optimization of the single-strand conformation polymorphism (SSCP) technique for detection of point mutations. *Hum. Mutat.* **2,** 404–414.

3. Hayashi, K. (1993) PCR-SSCP: A method for detection of mutations. *Genet. Anal. Tech. Appl.* **9,** 73–79.

4. Sekiya, T. (1993) Detection of mutant sequences by single-strand conformation polymorphism analysis. *Mutation Res.* **288,** 79–83.

5. Grossman, P. D. and Soane, D. S. (1991) Capillary electrophoresis of DNA in entangled polymer solutions. *J. Chromatogr.* **559,** 257–266.

6. Ren, J., Deng, X., Cao, Y., and Yao, K. (1996) Analysis of DNA fragments and polymerase chain reaction products from Tx gene by capillary electrophoresis with a laser-induced fluorescence detector using non-gel sieving media. *Anal. Biochem.* **233**, 246–249.

7. Ulvik, A., Ren, J., Refsum, H., and Ueland, P. M. (1997) Simultaneous determination of methylenetetrahydrofolate reductase C677T and factor V G1691G genotypes by mutagenically separated PCR and multiple-injection capillary electrophoresis. *Clin. Chem.* **44**, 264–269.

8. Ren, J., Ulvik, A., Ueland, P. M., and Refsum, H. (1997) Analysis of single-strand conformation polymorphism by capillary electrophoresis with laser-induced fluorescence detection using short-chain polyacrylamide as sieving medium. *Anal. Biochem.* **245**, 79–84.

9. Kuypers, A. W. H. M., Willems, P. M. W., van der Schans, M. J., Linssen, P. C. M., Wessels, H. M. C., de Bruijn, C. H. M. M, et al. (1993) Detection of point mutation in DNA using capillary electrophoresis in a polymer network. *J. Chromatogr.* **621**, 149–156.

10. Hebenbrock, K., Williams, P. M., and Karger, B. L. (1995) Single strand conformational polymorphism using capillary electrophoresis with two-dye laser-induced fluorescence detection. *Electrophoresis* **16**, 1429–1436.

11. Arakawa, H., Nakashiro, S., Maeda, M., and Tsuji, A. (1996) Analysis of single-strand DNA conformation polymorphism by capillary electrophoresis. *J. Chromatogr. A* **722**, 359–368.

12. Katsuragi, K., Kitagishi, K., Chiba, W., Ikeda, S., and Kinoshita, M. (1996) Fluorescence-based polymerase chain reaction-single-strand conformation polymorphism of p53 gene by capillary electrophoresis. *J. Chromatogr. A* **744**, 311–320.

13. Inazuka, M., Wenz, H.-M., Sakabe, M., Tahira, T., and Hayashi, K. (1997) A streamlined mutation detection system: multicolour post-PCR fluorescence labeling and single-strand conformational polymorphism analysis by capillary electrophoresis. *Genome Res.* **7**, 1094–1103.

14. Atha, D. H., Wenz, H.-M., Morehead, H., Tian, J., and O'Connell, C. D. (1998) Detection of p53 point mutations by single strand conformation polymorphism: Analysis by capillary electrophoresis. *Electrophoresis* **19**, 172–179.

15. Ren, J. and Ueland, P. M. (1999) Temperature and pH effects on single strand conformation polymorphism analysis by capillary electrophoresis. *Hum. Mutat.* **13**, 458–463.

16. Ren, J., Ulvik, A., Refsum. H., and Ueland, P. M. (1999) Application of short chain polydimethylacrylamide as a sieving medium for the electrophoretic separation of DNA fragments and mutation analysis in an uncoated capillary, *Anal. Biochem.* **276**, 188–194.

17. Iwahana, H., Yoshimoto, K., Mizusawa, N., Kudo, E., and Itakura M. (1994) Multiple fluorescence-based PCR-SSCP analysis. *BioTechniques* **16**, 296–297, 300–305.

18. Kuypers, A. W. H. M., Linssen, P. C. M., Willems, P. M. W., and Mensink, E. J. B. M. (1996) On-line melting double strand DNA for analysis of single-stranded DNA using capillary electrophoresis. *J. Chromatogr. B* **675**, 205–211.

19. Grossman, P. (1994) Electrophoretic separation of DNA sequencing extension products using low-viscosity entangled polymer network. *J. Chromatogr.* **663**, 219–227.

20. Hjertén S. (1985) High-performance electrophoresis elimination of electroendosmosis and solute adsorption. *J. Chromatogr.* **347**, 191–198.

21. Ren, J, Ulvik, A., Refsum, H., and Ueland, P. M. (1998) Chemical mismatch cleavage combined with capillary electrophoresis: detection of mutations in exon 8 of the cystathionine b-synthase gene. *Clin. Chem.* **44**, 2108–2114.

22. Madabhushi, R. S. (1998) Separation of 4-colour DNA sequencing extension products in noncovalently coated capillaries using low viscosity polymer solutions. *Electrophoresis* **19**, 224–230.

23. Madabhushi, R. S. (2001) DNA sequencing in noncovalently coated capillaries using low viscosity polymer solutions, in *Capillary Electrophoresis of Nucleic Acids*, Vol. 2 (Mitchelson, K. R. and Cheng, J., eds.), Humana Press, Totowa, NJ, pp. 309–315.

11

Magnetic Bead-Isolated Single-Strand DNA for SSCP Analysis

Takao Kasuga, Jing Cheng, and Keith R. Mitchelson

1. Introduction

1.1. SSCP Definition

Single-strand DNA conformational polymorphism (SSCP) makes use of sequence-dependent folding and structural conformation assumed by ssDNA for detection of small sequence changes or point mutations in DNA fragments. The technique was developed using polyacrylamide gel electrophoretic (PAGE) fractionation of distinctive conformational isomers (conformers) of ssDNAs. These are formed by denaturation of short polymerase chain reaction (PCR) products (typically <200 bp) *(1–7)*. Conventional understanding of SSCP suggests that conformational distortion created by the folding (loops and bends) stabilized by the formation (short) regions of intramolecular duplex between complementary regions within each single-stranded molecule alters the three dimensional shape of the molecule. Crudely, the migrating DNA strand collides with the sieving polymer media and through the temporary interaction of DNA and sieving media molecules, the mobility of the DNA is reduced *(8)*. Differences in the size or conformation of DNA molecules affect the duration of these sieving interactions. Since the two complementary DNA strands differ in sequence, each strand assumes an independent conformation, which has a characteristic mobility. Thus, minor differences in the sequence between DNAs from two sources, such as point mutations, may affect the conformation assumed by each of the strands and identify the sequence differences as a shift in mobility. High-performance capillary electrophoresis (HPCE) is increasingly used for SSCP analysis, with detection by UV absorption of DNA *(9,10)*, or by LIF of ssDNA tagged with fluorescent dye-labeled primers *(11–14)*.

1.2. SSCP of PCR Products

The direct use of unlabeled PCR products for SSCP analysis is attractive because of the low cost and potential convenience, however, the complementary DNA strands

From: *Methods in Molecular Biology, Vol. 163:*
Capillary Electrophoresis of Nucleic Acids, Vol. 2: Practical Applications of Capillary Electrophoresis
Edited by: K. R. Mitchelson and J. Cheng © Humana Press Inc., Totowa, NJ

may easily reanneal following denaturation and rapid handling is required to prevent reannealing occurring *(1–3,6,7)*. Unincorporated PCR primers may also modify SSCP profiles of PCR products, as the primers can reanneal with the single-strand termini and alter the electromigration of the major strands *(2–4)*.

1.2.1. SSCP Conformation Profiles

The interpretation of SSCP profiles obtained using PCR products can be complex because although only two ssDNAs are anticipated during SSCP, more than two DNA bands may be detected *(1,2)*. Conventional PAGE profiles of SSCP usually have three major bands, representing the two single-strand molecules and some reannealed duplex DNA *(1)*. However, the number of different conformational isomers (conformers) that may be assumed by each ssDNA depends both upon the stability of duplex formed by (small) complementary regions and the DNA sequence, presumably because of strand reannealing and partial reannealing phenomena may create several different conformers *(2)*. When multiple bands are seen, the major species are usually the two complementary single DNA strands, whereas other minor species may represent different stable conformational isomers of the single DNA strands, partially annealed or fully annealed duplex DNA, or single-strands annealed to residual PCR primer *(2–4)*. The presence of self-complementary sequence regions within the single-strands also suggests that regions of short duplex could also form between two or more copies of the same strand, and could contribute to the slower migrating isomers which are occasionally detected in a SSCP mixture *(2,5)* (*see* **Notes 1** and **2**). Many of these problems may be overcome by techniques such as magnetic-bead purified single DNA strands *(2,15–18)* in which unique ssDNA strands are isolated, free of complement strand and unincorporated PCR primers, or by asymmetric-PCR to selectively amplify one strand *(5)*.

1.2.2. Development of SSCP Analysis by Capillary Electrophoresis

Several groups have described the application of (CE) for conventional SSCP analysis (SSCP-CE) *(9–14)*. The various DNA species observed during conventional SSCP-PAGE analysis may also influence the profile of SSCP-CE, and may depend upon the method of purification of the test DNA, the manner of preparing the ssDNA, and the manner of loading the ssDNA into the capillary column *(10,13)*. The time taken to load heat denatured DNA on to the capillary column is critical as ssDNA has the tendency to reanneal with complementary sequence, and significant renaturation of duplex DNA may occur at permissive temperatures *(12,14)*. Conditions that provide for contact between strands and thermal conditions that allow the destabilization of the ssDNA conformation to occur may allow for the formation of these conformations.

1.3. Recent Developments in SSCP-CE Analysis

One approach that can overcome strand reannealing is the use of "On-line in-capillary melting of PCR strands" in which strand melting and conformation formation are rapidly achieved during the electrophoretic analysis, and strands are electrophoretically separated before significant reannealing can occur *(19)*. Atha et al. and Wenz et al. *(12,14)* observed that ssDNA assumes different characteristic mobilities (determined by ssDNA folding) at different temperatures. They developed a panel of DNA fragments representing the 10 most common mutations found in the human *p53* tumor suppressor gene and created a database of conformation polymorphism (mobility shifts in the temperature range 25–45°C) which are characteristic of each mutation. Notably, different mobilities corresponding to different conformational isomers could be detected for single-strands electrophoresed under different thermal conditions. The combination of detailed thermal profiling of each mutation with "on-line in-capillary melting of DNA" could provide a rapid method to identify both known mutations and novel mutations within the target gene region. Interestingly, SSCP-PAGE analysis of the same *p53* gene mutations suggest that subambient temperature (20°C) was preferable for discrimination of these mutations as a group, which is probably indicative of the necessity for good thermal control to ensure the stable assumption of particular conformers *(6,7)*.

Using a different approach, Ren et al. *(20)* applied CE to mutation detection using a combination of SSCP and specific chemical mismatch cleavage (CMC) at the mutation sites. The cleavage products were analyzed under denaturing conditions as ssDNA, affording relatively high resolution of ssDNA (down to 1 nt). The precise size assessment of CMC products gave additional information about the location of the mutation. The method allows screening of known mutations that give expected CMC products, but also detects unknown mutation locations which are indicated by novel CMC fragment sizes and novel SSCP profiles.

1.3.1. Magnetic Bead Capture of ssDNA

Biotin-labeled DNA is used to specifically bind to immobilized avidin for DNA subtraction *(21)*. PCR products can be made using one biotinylated primer and one nonmodified primer to allow selective purification of the unmodified strand by removal of biotinylated strand using Dynal magnetic Dynabeads®. Following biotin-avidin capture of isolated single-strand DNA, the detection of SSCPs may be compared to conventional SSCP profiles observed either by CE analysis *(15)* or by PAGE analysis *(2,16–18)*. The clarity of profiles and interpretation of characteristic mobility shifts is usually improved over the profiles seen with duplex DNA which is rapidly melted before loading on to the electrophoretic medium

1.3.2. Pretreatment of Bead-Purified ssDNAs

1.3.2.1. STANDARD TREATMENT

The pretreatment of the bead-purified ssDNA (before electrophoresis) influences the information content of the SSCP profiles. When bead-purified ssDNA are thermally melted and snap cooled prior to loading, the SSCP profiles of each of the strands are similar to the profiles of individual ssDNA seen when duplex DNA is treated in the same manner *(2,16–18)*.

1.3.2.2. NO THERMAL TREATMENT

Since the complementary strands are not present with bead-purified ssDNA, the heat denaturation and snap cooling procedures to melt strands prior to analysis are not necessary *(2)*. The ssDNA is released from the duplex DNA bead-complex using alkali then ethanol precipitated and dissolved in water as a single-strand. A completely different SSCP profile is seen if these ssDNAs are analyzed directly without further treatment *(2)* and show two closely migrating major (doublet) bands, as well as a number of slow migrating metastable species (**Fig. 1**). The entangled solution capillary electrophoresis (ESCE)-SSCP profile of rapidly melted DNA *(9)* is directly comparable to the PAGE-SSCP profile of the same DNA strands as bead-isolated ssDNAs. The doublet bands represent stable conformations assumed by single DNA strand which are resolved by the sieving gel, although some conformation isomers result from the annealing of residual PCR-primers and the ssDNA. Notably, neither of the pairs of doublet ssDNA bands is prominent in the conventional SSCP profiles thermally melted duplex DNA, even when most residual PCR-primers are removed by agarose gel-purification. Importantly, the true ssDNA conformers seen in bead-purified single-strand SSCP are not usually abundant in a conventional SSCP assay, even for gel-purified PCR products *(2)*.

1.3.2.3. METASTABLE SPECIES

The low electrophoretic mobility of the mSSCP suggests large molecular species, such as multimers resulting from bi- or tri-molecular self-annealing interactions between single-strands at regions of sequence complementary (**Fig. 2**). Studies suggests that duplex and triplex structures between both folded and linear oligonucleotides can form in solution which are very energetically stable *(22,23)*. Since conventionally, SSCP are thought to result from intramolecular base pairing between complementary regions within a single DNA strand, additional intermolecular duplex regions are likely to occur between complementary regions of two or more molecules. The multimers would be partially single-stranded, because the strands are not fully self-complementary and could form several different conformations. These multimer species form upon storage of the bead-purified ssDNAs and are observed only when such structures are not disrupted by heating (*see* **Notes 3–6**).

The highly regulated and uniform thermal environment of the capillary contributes to the stability of the metastable single-stranded isomers during the electrophoretic separation. Konrad and Pentoney *(23)* also showed that formation of DNA secondary structure and short duplex regions can effect the separation of ssDNA oligomers (12 mers) by HPCE even in 7 *M* urea at 35°C. It is likely that small DNA regions of 4–6 bp may form intermolecular homoduplex with high structural stability between the two oligomers.

The ssDNAs were stable for subsequent electrophoretic analysis even after several weeks of storage at 4°C. This stability of single-strand conformers is advantageous for

---

Fig. 1. *(opposite page)* Electropherogram of SSCP of purified PCR products of the same ribosomal gene region: (**A**) P_{GB} strain; (**B**) S_{Fi} strain. Nucleotide differences between S_{Fi} and P_{GB} are 2 Deletion, 4 Substitution, 2 bp length = (2D + 4S/2B). S1 and S2 are the major (stable) SSCP peaks, D is reannealed homoduplex and M indicates metastable SSCP peaks. Electropherograms are aligned from the reannealed peak D. The sample was electrokinetically loaded

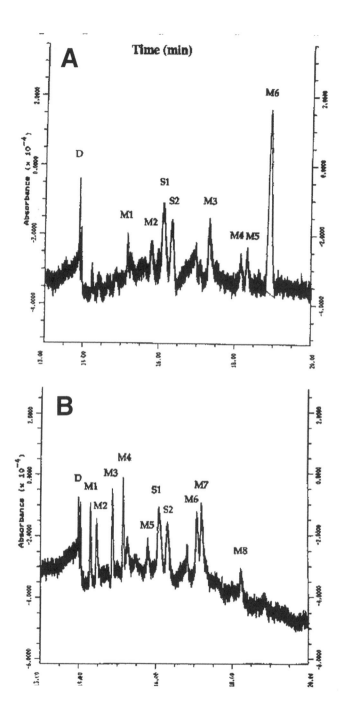

into the capillary at 175 V/cm for 13 s. The separation was carried out at 228 V/cm for 20 min. The UV absorbance was monitored at 260 nm. The temperature of the external surface of the capillary was maintained at 20 ± 0.3°C. Reprinted with permission from *(9)* *Journal of Capillary Electrophoresis* **2,** 24–29.

A   Single strand DNA

5'-----zzz----nnn----xxx----uuu-

B   Bi-molecular interactions
                        5'----zzz----nnn----xxx----uuu-
                               | | | | | |
                        -uuu----xxx----nnn----zzz----5'

                        5'-----zzz----nnn----xxx----uuu-
                                 | | | | | |
                          -uuu----xxx----nnn----zzz----5'

C   Tri-molecular interactions
                     -uuu----xxx----nnn----zzz----5'
                            | | | | | |
                     5'-----zzz----nnn----xxx----uuu-
                            | | | | | |
                        -uuu----xxx----nnn----zzz----5'

Fig. 2. Model of low mobility metastable conformational conformers. Regions of complementarity (x::z; n::u) allow duplex formation between two or more ssDNA molecules: (**A**) illustrates a single strand of DNA. Interactions between two or more strands can create partially duplex, (**B**) bi-molecular and (**C**) tri-molecular complexes. Reprinted with permission from *(9)* *Journal of Capillary Electrophoresis* **2**, 24–29.

the automated handling of samples, or for automated HPCE. Glycerol may help containing to maintain the stability of the single-stranded isomers in sieving buffers *(1,6,7,24)*. During HPCE using borate buffer with different sugar polymer sieving matrices, the matrix may be crosslinked and present selective separation to DNAs of particular sizes or physical conformations *(24)*, although a recent study of PAGE-SSCP suggests that some of the beneficial effect of glycerol may be mediated by a decrease in pH *(25)* which would decrease electrorepulsion between backbone phosphates and allow closer compaction and assumption of more stable conformations. A wide range of sugar-polymers used commonly for ESCE may interact with borate to form dynamic crosslinked matrices, which provide mild environmental conditions that are favorable for the stability of short duplex regions, and hence, may allow separation of different conformation isomers of ssDNA molecules *(26)*.

1.4. Novel Applications of Magnetic Bead Capture

Rashkovetsky et al. *(27)* have described a novel microanalytical method using magnetic beads and CE instrumentation. A short plug of magnetic beads can be localized into the capillary and held fixed by a magnet, allowing biological assays to be performed using enzymes or specific ligands attached to the beads. Such capillary-located magnetic beads linked to avidin could be used to capture biotin-tagged PCR products and would allow rapid, selective SSCP analysis of the untagged strand in combination with an "On-line in-capillary melting of PCR strands" procedure *(19)*.

Biotin-labeled DNA primers specific for gene regions can amplify particular cDNA fragments from cellular mRNA using reverse transcriptase combined with PCR (RT-PCR), and these cDNA can be collected specifically by binding to immobilized avidin *(28)*. Both the quantitative level of gene expression, or putative sequence variation within such cDNA fragments could be examined by SSCP analysis following release of a single DNA strand from the amplified dsDNA *(29)*. One application is the genotyping of bacteria *(30)* in which length and sequence polymorphism analysis of rRNA-intracistronic and tRNA-intergenic regions could be performed. The use of SSCP-CE analysis in combination with magnetic bead separation of one DNA strand from the other would allow detailed CE migration profiles and conformation isoforms to be created for each bacterial isolate. An alternative approach might be a capture and detection system *(31)* using biotin-labeled probes to capture the DNA segments that contain specific regions of the genes by DNA-DNA hybridization following PCR amplification or RT-PCR.

1.5. Interpretation of the SSCP Profile

1.5.1. Direct PCR Products

Capillary SSCP experiments *(9)* of agarose gel purified PCR products show several strong bands which correspond to the reannealed duplex DNA as well as the two stable ssDNA conformers (labeled S1 and S2), as well as a number of other metastable ssDNA peaks (mSSCP) (*see* **Fig. 1**). Each of the peaks had marked different retention times for the two variant DNA strands, and the number of metastable peaks also differed. The reproducibility of the CE conditions was high and base-line resolution of the peaks was observed. The assay of SSCP-CE can be resolved rapidly, in approx 20 min. Reannealing between residual unincorporated primer and its complementary ssDNA may have produced some of the bands that migrate close to each ssDNA band *(2–4)*. We believe that during ESCE, the electrokinetic loading contributes to the rapid fractionation of all DNA forms into separate zones in the capillary sieving media, preventing "zippering up" of complementary DNAs into duplex molecules and limiting the loss of metastable isomers. The effectiveness of "on line melting" for rapid creation of ssDNA conformers and subsequent rapid electrophoretic separation of single DNA strands described by Kuyper et al. *(19)*, supports this view. In addition, the use of glycerol in the sieving buffer may help to maintain the stability of the single-stranded isomers (*see* **Notes 7–9**).

1.5.2. PAGE of Bead-Purified ssDNAs

The adoption of magnetic bead-purified ssDNA provided a marked increase in the information content of the PAGE-SSCP analysis. In each case, runs with SSCP of bead-purified ssDNA which correspond to the two stable ssDNA conformers, as well as a number of other slow migrating metastable ssDNA peaks (mSSCP) *(2)*. Each isolated strand of the PCR product provided several novel SSCP bands, which are stable to temperature. The mobility of these bands are different between the two sets of DNA fragments that differed by (2D + 4S/2B), suggesting that several stable conformations may be assumed by a single DNA strand, which each present different profiles to the sieving gel matrix *(2)*.

1.5.3. Metastable Conformation Isomers of ssDNA in PAGE

The low mobility metastable bands were readily detectable in bead-purified ssDNA preparations, but were consistently observed by PAGE electrophoresis only when DNAs were not thermally treated prior to loading on the gel *(2)*. Metastable isomers of ssDNA are also not generally obvious in conventional PAGE-SSCP analyses, which involve a thermal treatment to melt the duplex DNA *(1,6)*, nor in bead-purified ssDNAs, which are also thermally treated prior to electrophoresis *(16–18)*. However, in both asymmetric PCR-SSCP *(5)*, cDNA-SSCP *(1)* and unheated, bead-purified ssDNA *(2)*, the abundance of one DNA strand permits multiple slow migrating species to also be readily observed by conventional PAGE. It is likely that the slow migrating species seen in asymmetric-PCR-SSCP may also represent metastable isomers of molecules, similar to the metastable isomers observed bead-purified ssDNA. The failure to detect these slow migrating bands during most conventional SSCP protocols that use a duplex DNA source is most likely because of the higher thermodynamic stability of interactions between complementary strands than between two molecules of the same sequence. Cai and Touitou *(3)* noted that annealed molecules could form between residual PCR primers and ssDNAs and which result in altered mobility of bands. Formation of the annealed primer-ssDNA molecules also occurred in SSCP of PCR products despite purification by gel electrophoresis *(2)*. Some of the metastable SSCP peaks therefore may result from the resolution of these additional partially annealed molecules *(4)*.

1.5.4. SSCP Assay of Single-Strand RNA

Danenberg et al. *(32)* transcribed short RNA copies (~250 bp) from regions of PCR amplified target genes by T7 or SP6 RNA polymerases from appropriate promoters in the PCR primers. The RNA molecules could be separated into numerous metastable conformational forms by nondenaturing gel electrophoresis. Pronounced changes in the conformational patterns including mobility shifts of major and minor conformations, the appearance of new conformations and the loss of other conformations occurred with single-base substitutions in the RNA. Both sense and antisense RNA strands of the same gene segment give unique conformational patterns. Presumably, the large repertoire of secondary structure assumed by RNA is the source of such diverse metastable conformations, which are stabilized by the shorter hairpin turns and H-bonding between the 2'-hydroxyl groups and available bases and sugars in the RNA oligonucleotide. Although analysis of metastable RNA conformations by CE has not been reported, the conformational analysis is likely to be similar to the slow and fast migrating RNA species detected during nondenaturing gel electrophoresis *(32)*. The metastable RNA conformational forms are also likely to correspond to some of the metastable species occurring during analysis bead-purified ssDNA *(2)*.

2. Materials

1. DNA is isolated using a Nucleon PhytoPure DNA Isolation Kit (Amersham Life Sciences).
2. Ultrafree-MC filters (cat. no. UFC30HV100) (Millipore).
3. Magnetic M-280 strepavidin Dynabeads are used according to Biomagnetic Techniques in Molecular Biology—Technical Handbook *(21)*.

4. MPE washing buffer: 1 M NaC1, 10 mM Tris-HCl, 1 mM EDTA, pH 8.0.
5. 1X TE: 10 mM Tris-HCl, 1 mM EDTA, pH 8.0.
6. Bovine serum albumin (BSA) is from Sigma.
7. ESCE sieving media: 1X TBE: 90 mM borate, 90 mM Tris-base, pH 8.3, 1 mM EDTA; 0.5% (w/v) hydroxypropyl methyl cellulose (HPMC) (Sigma H7509), 3 μM EtBr and 4.8% glycerol *(8,23)*.
8. A surface-modified fused-silica capillary, DB-17 (57 cm × 100 μm, effective length 50 cm) (J & W Scientific, Folsom, CA).
9. HPCE is performed on a P/ACE® 5050 HPCE system (Beckman Instruments Inc., Fullerton, CA).
10. Post-run data analysis is performed using the Gold Chromatography Data System, version 8.13 (Beckman Instruments Inc.).

3. Methods
3.1. PCR Amplification of DNA Fragments

1. Separate PCR reactions are performed in a total mixture of 25 μL containing 5 ng of template DNA and 225 nM of primer pairs, only one of which is biotin labeled *(9)*.
2. The PCR products are separated on agarose gels and are then purified by Ultrafree-MC filters.

3.2. Purification of ssDNA

1. A biotin-5'-labeled primer is used in combination with nonmodified primers for PCR amplification *(9,21)*. Briefly, 100 μL of magnetic M-280 strepavidin Dynabeads are washed twice in an Eppendorf tube with 200 μL PBS, 0.1% BSA, pH 7.5 and are collected on the magnetic particle concentrator (MPC).
2. Beads are resuspended in 50 μL of MPE washing buffer and 15 μL PCR-amplified DNA and 85 μL water are added and the solution is mixed gently for 30 min at room temperature on a rotator at 20 rpm.
3. When the biotin is bound to the streptavidin, the beads are concentrated on the MPC. The supernatant is removed and residual unbound primers are eliminated by washing the beads three times with washing buffer for 10 min at room temperature on a rotator at 20 rpm (*see* **Note 10**).
4. The unmodified ssDNA is released by the addition of 40 μL of 120 mM NaOH. The supernatant is then collected and neutralized with 3.6 μL of 1.7 M acetic acid.
5. The ssDNA is collected by precipitation at –70°C, by adding 150 μL of ethanol without carrier.
6. Purified ssDNA is dissolved in 15 μL of 1X TE and is stored at 4°C until use. The ssDNA could be directly analyzed for SSCP after up to 4 wk of storage, without detectable change in the electrophoresis profile.

3.2.1. SSCP Assay of PCR Products by CE

1. The capillary column is conditioned daily with 5 vol of ESCE buffer and then subjected to voltage equilibration for 15 min until the baseline is stabilized. ESCE is performed on the P/ACE 5050 system (Beckman) in the reversed-polarity mode (negative potential at the injection end of the capillary).
2. For direct SSCP of PCR products, 2 mL of each DNA sample (in 0.1X TE) is mixed with 5 mL of water and 3 mL of formamide. The mixtures are boiled for 3 min and then snap frozen in a dry ice/acetone bath for 3 min, and stored at –20°C until use (*see* **Notes 7–9**).

3. Frozen mixtures are allowed to thaw on the P/ACE loading tray immediately before elec-trokinetic loading on to the capillary column. Samples are introduced into the capillary by electrokinetic injection at negative polarity of 175 V/cm for 10–13 s (*see* **Notes 11–13**).
4. Separations are performed under constant voltage at 228 V/cm for 30 min. The capillary temperature is set at 22°C and UV absorbance is monitored at 260 nm in the presence of 3 µ*M* EtBr.
5. Fluorescent dyes such as TOTO, TO-PRO, or YO-PRO at 0.2 µ*M* can also be used for the detection of SSCP conformers using laser induced fluorescence (LIF).

3.2.2. SSCP Assay of Bead-Isolated ssDNA by CE

1. For SSCP of bead-isolated single strands, 2 mL of each DNA sample (in 0.1X TE) is diluted with 5 mL of water. The mixtures are placed on the P/ACE loading tray immedi-ately before loading on to the capillary column. DNA is detected by UV absorbance at 260 nm in the presence of 3 µ*M* EtBr. The CE conditions are as described above in **Sub-heading 3.3.1.** (*see* **Note 14**).

4. Notes

1. Atha et al. and Wenz et al. (*12,14*) have analyzed several different common, single point mutations of the *p53* gene by SSCP using CE. The CE-SSCP for all mutations at ambient temperature (25°C) showed characteristic shifts in the direction and migration times of both strands relative to the wild-type, which identified each of these mutations. The direction of the shifts in the migration times for particular ssDNA molecules vary with temperature between 20–45°C, in a discrete pattern for each mutation. These mobility shifts allow a temperature-specific profile for each mutation to be derived. In addition, discrete intra-strand isoforms are observed at different temperatures, which may correspond to ssDNA conformations seen with bead-purified ssDNA (*2*). Detailed profiles of mutations are derivable, which combine both the mutation-specific temperature profiling and an analy-sis of conformational isoforms. These profiles aid the detection of new genetic mutations.
2. Danenberg et al. (*32*) describe the SSCP analysis of RNA transcribed from target gene fragments by nondenaturing gel electrophoresis. Unique conformational patterns such as mobility shifts of conformational species, loss and gain of conformations occur with single-base substitutions in the RNAs. The rate of formation of the equilibrium confor-mations may be accelerated by heating or by mild denaturing conditions. Both sense and antisense RNA strands of the same gene segment give unique conformational patterns. Because large amounts of ssRNA can be specifically transcribed from each strand of the gene template, the ssRNA species may be detected by UV absorption. Biotin-tagged DNA fragments of genes could also be conveniently used to remove the DNA template from the RNA transcription products by magnetic-bead capture.
3. The low mobility metastable SSCP bands (mSSCP) are consistently observed only in SSCP of purified ssDNA, which are not thermally treated (*2*) and are not observed in samples boiled then cooled in the conventional way (*15*).
4. The metastable species represent low stability structures, which require low thermal energy and sufficient time for formation, and may be eliminated from the SSCP profile by thermal treatment.
5. Conventional SSCP profiles typically display two major ssDNA bands, compared to the four or five major ssDNA bands seen with the magnetic-bead purified strands. Thus, the conven-tional SSCP have a reduced informational content compared to the metastable SSCP (mSSCP).

6. The low electrophoretic mobility of the mSSCP suggests large molecular species, such as multimers resulting from bi- or tri-molecular self-annealing interactions between single-strands at complementary regions (*see* **Fig. 2**).

7. A model CE-SSCP experiment analyzed by ESCE and using agarose gel-purified DNA fragments, which differed by two deletions and four substitutions (2D + 4S/2B), is illustrated in **Fig. 1A, B**, respectively.

8. In each case, two peaks were always present (labeled S1 and S2), which represent two stable ssDNA molecules, as well as a number of other metastable ssDNA peaks (mSSCP). Each of the peaks has markedly different retention times for the two strains, and the number of metastable peaks also differed. The reproducibility of the CE conditions is high, and base-line resolution of the peaks is observed. The assay of SSCP by ESCE is rapidly resolved into eight peaks in 20 min.

9. When double-strand PCR products is used as the source DNA for ESCE-SSCP analyses, care must be taken to avoid reannealing of duplex DNA during sample preparation and especially during the loading onto the capillary column.

10. Residual PCR primers are known to interact with ssDNA and to alter the mobility of SSCP conformers *(2–4)*. The use of bead-isolated ssDNA for SSCP analysis should eliminate the spurious bands from the SSCP profiles, and therefore minimizes the potential for false negative and false positive band shifts.

11. The adoption of magnetic-bead purified ssDNA provides a marked increase in the information content of the SSCP analysis *(2)*. When SSCP of magnetic-bead purified strands are separated by nondenaturing gel electrophoresis, two fast migrating (doublet) conformers and several slow migrating bands are seen, distinct from the typical three (2 ssDNA + 1 DsDNA) SSCP bands seen in conventional SSCP preparations.

12. Each isolated strand of the PCR product provides several novel SSCP bands, which are stable to temperature *(2)*. The mobility of these bands is different between the two sets of DNA fragments, suggesting that several stable conformations may be assumed by a single DNA strand, which each present different profiles to the gel matrix.

13. Thermal heating and quenching of bead-isolated ssDNA removes the multimer species and provides a simple SSCP profile. Magnetic bead-isolated ssDNA profiles of sequence variants can be differentiated more readily than conventional SSCP profiles.

14. Glycerol is thought to maintain the stability of the secondary structure of ssDNA during SSCP analysis *(1,6,7)*. Glycerol provides improved resolution of molecular species using a free solution sieving medium with TBE buffer *(24,26)* and aids the resolution of metastable DNA conformers during SSCP analysis *(9)*.

References

1. Humphries, S. E., Gudnason, V., Whittall, R., and Day. I. N. (1997) Single-strand conformation polymorphism analysis with high throughput modifications, and its use in mutation detection in familial hypercholesterolemia. International Federation of Clinical Chemistry Scientific Division: Committee on Molecular Biology Techniques. *Clin. Chem.* **43**, 427–435.

2. Kasuga, T., Cheng, J., and Mitchelson, K. R. (1995) Metastable single strand DNA conformational polymorphism (mSSCP) analysis results in enhanced polymorphism detection. *PCR Methods Applic.* **4**, 227–233.

3. Cai, Q.-Q. and Touitou, I. (1994) Excess PCR primers may dramatically affect SSCP efficiency. *Nucleic Acids Res.* **21**, 3909–3910.

4. Almeida, T. A., Cabrera, V. M., Miranda, J. G. (1998) Improved detection and characterization of mutations by primer addition in nonradioisotopic SSCP and direct PCR sequencing. *Biotechniques* **24,** 220–221.
5. Ainsworth, P. J., Surh, L. C., and Coulter-Mackie, M. B. (1991) Diagnostic single strand conformational polymorphism, (SSCP): a simplified non-radioisotopic method as applied to a Tay-Sachs B1 variant. *Nucleic Acids Res.* **19,** 405–406.
6. O'Connell, C. D., Tian, J., Juhasz, A., Wenz, H.-M., and Atha, D. H. (1998) Development of standard reference materials for diagnosis of p53 mutations: analysis by slab gel single strand conformation polymorphism. *Electrophoresis* **19,** 164–171.
7. Teschauer, W., Mussack, T., Braun, A., Waldner, H., and Fink, E. (1996) Conditions for single strand conformation polymorphism (SSCP) analysis with broad applicability: a study on the effects of acrylamide, buffer and glycerol concentrations in SSCP analysis of exons of the p53 gene. *Eur. J. Clin. Chem. Clin. Biochem.* **34,** 125–131.
8. Slater, G. W., Desruisseaux, C., and Hubert, S. J. (2001) DNA separation mechanisms during capillary electrophoresis, in *Capillary Electrophoresis of Nucleic Acids,* Vol. 1 (Mitchelson, K. R., and Cheng, J., eds.), Humana Press, Totowa, NJ, pp. 27–41.
9. Cheng, J., Kasuga, T., Watson, N. D., and Mitchelson, K. R. (1995) Enhanced single-stranded DNA conformational polymorphism analysis by entangled solution capillary electrophoresis. *J. Capillary Electrophor.* **2,** 24–29.
10. Kuypers, A. W., Willems, P. M., van der Schans, M. J., Linssen, P. C., Wessels, H. M., de Bruijn, C. H., Everaerts, F. M., and Mensink, E. J. (1993) Detection of point mutations in DNA using capillary electrophoresis in a polymer network. *J. Chromatogr. A.* **621,** 149–156.
11. Inazuka, M., Wenz, H.-M., Sakabe, M., Tahira, T., and Hayashi, K. (1997) A streamlined mutation detection system: multicolor post-PCR fluorescence labeling and single-strand conformational polymorphism analysis by capillary electrophoresis. *Genome Res.* **7,** 1094–1103.
12. Atha, D. H., Wenz, H.-M., Morehead, H., Tian, J., and O'Connell, C. D. (1998) Detection of p53 point mutations by single strand conformation polymorphism: analysis by capillary electrophoresis. *Electrophoresis* **19,** 172–179.
13. Arakawa, H., Tsuji, A., Maeda, M., Kamahori, M., and Kambara, H. (1997) Analysis of single-strand conformation polymorphisms by capillary electrophoresis with laser induced fluorescence detection. *J. Pharm. Biomed. Anal.* **15,** 1537–1544.
14. Wenz, H.-M., Ramachandra, S., O'Connell, C. D., and Atha, D. H. (1998). Identification of known p53 point mutations by capillary, electrophoresis using unique mobility profiles in a blinded study. *Mutat. Res.* **382,** 121–132.
15. Debernardi, S., Luzzana, M., and DeBellis, G. (1996) Solid phase purification and SSCP analysis of amplified genomic DNA by capillary electrophoresis. *Front. Biosci.* **1,** c1–c3.
16. Virdi, A. S., Loughlin, J. A., Irven, C. M., Goodship, J., and Sykes, B. C. (1994) Mutation screening by a combination of biotin-SSCP and direct sequencing. *Human Genet.* **93,** 287–290.
17. Weidner, J., Eigel, A., Horst, J., and Kohnlein, W. (1994) Nonisotopic detection of mutations using a modified single-strand conformation polymorphism analysis. *Hum. Mutat.* **4,** 55–56.
18. Selvakumar, N., Ding, B. C., and Wilson, S. M. (1997) Separation of DNA strands facilitates detection of point mutations by PCR-SSCP. *Biotechniques* **22,** 604–606.
19. Kuypers, A. W., Linssen, P. C., Willems, P. M., and Mensink, E. J. (1996) On-line melting of double-stranded DNA for analysis of single-stranded DNA using capillary electrophoresis. *J. Chromatogr. B Biomed. Appl.* **675,** 205–211.

20. Ren, J. (2001) Chemical mismatch cleavage analysis by capillary electrophoresis with laser-induced fluorescence detection, in *Capillary Electrophoresis of Nucleic Acids*, Vol. 2 (Mitchelson, K. R., and Cheng, J., eds.), Humana Press, Totowa, NJ, pp. 231–239.

21. Biomagnetic Techniques in Molecular Biology—Technical Handbook. 2nd ed., (1993). Dynal AS, Oslo, Norway, pp. 153–157.

22. Gryaznov, S. M. and Lloyd, D. H. (1993) Modulation of oligonucleotide duplex and triplex stability via hydrophobic interactions. *Nucleic Acids Res.* 21, 5909–5915.

23. Konrad, K. D. and Pentoney, S. L., Jr. (1993) Contribution of secondary structure to DNA mobility in capillary gels. *Electrophoresis* 14, 502–508.

24. Cheng, J. and Mitchelson, K. R. (1994) Glycerol improves the separation of DNA fragments in entangled solution capillary electrophoresis (ESCE). *Anal. Chem.* 66, 4210–4214.

25. Kukita, Y., Tahira, T., Sommer, S. S., and Hayashi, K. (1997) SSCP analysis of long DNA fragments in low pH gel. *Hum. Mutat.* 10, 400–407.

26. Mitchelson, K. R. and Cheng, J. (2001) Capillary electrophoresis with glycerol as an additive, in *Capillary Electrophoresis of Nucleic Acids*, Vol. 1 (Mitchelson, K. R., and Cheng, J., eds.), Humana Press, Totowa, NJ, pp. 259–277.

27. Rashkovetsky, L. G., Lyubarskaya, Y. V., Foret, F., Hughes, D. E., and Karger, B. L. (1997) Automated microanalysis using magnetic beads with commercial capillary electrophoretic instrumentation. *J. Chromatogr. A* 781, 197–204.

28. Augenstein, S. (1994) Superparamagnetic beads: applications of solid-phase RT-PCR. *Am. Biotechnol. Lab.* 12, 12–14.

29. Maekawa, M. and Sugano, K. (1998) Quantification of relative expression of genes with homologous sequences using fluorescence-based single-strand conformation polymorphism analysis- application to lactate dehydrogenase and cyclooxygenase isozymes. *Clin. Chem. Lab. Med.* 36, 577–582.

30. Vaneechoutte, M., Boerlin, P., Tichy, H. V., Bannerman, E., Jager, B., and Bille, J. (1998) Comparison of PCR-based DNA fingerprinting techniques for the identification of *Listeria* species and their use for atypical *Listeria* isolates. *Int. J. Syst. Bacteriol.* 48, 127–139.

31. Chen, J., Johnson, R., and Griffiths, M. (1998) Detection of verotoxigenic *Escherichia coli* by magnetic capture-hybridization PCR. *Appl. Environ. Microbiol.* 64, 147–152.

32. Danenberg, P. V., Horikoshi, T., Volkenandt, M., Danenberg, K., Lenz, H.J., Shea, L. C., Dicker, A. P., Simoneau, A., Jones, P. A., and Bertino, J. R. (1992) Detection of point mutations in human DNA by analysis of RNA conformation polymorphism(s). *Nucleic Acids Res.* 20, 573–579.

III

PRACTICAL APPLICATIONS FOR GENETIC ANALYSIS

12

Analysis of Short Tandem Repeat Markers by Capillary Array Electrophoresis

Elaine S. Mansfield, Robert B. Wilson, and Paolo Fortina

1. Introduction

The chapter describes the use of capillary array electrophoresis (CAE) for the detection of triplet repeat expansion at the *huntingtin* locus associated with autosomal dominant Huntington disease (HD), an adult-onset neuro-degenerative disorder. The region of this gene that expands to disease-causing mutations consists of two adjacent tandemly repeated polymorphic triplet repeats: a diagnostic (CAG)n repeat immediately followed by a (CCG)n region. Early polymerase chain reaction (PCR)-based methods used to genotype disease-causing mutations amplified through both of the repeat regions. More accurate diagnostic risk assessment for expansion is obtained when the (CAG)n and the (CCG)n repeats are amplified separately, as well as together in a single reaction. However, these tests require three separate PCR amplifications to be run using conventional methods. The use of multicolor fluorescence detection and CAE can simplify this improved diagnostic test.

The CAE method described here makes use of reverse-strand primers labeled with two different fluorescent dyes that are strategically positioned to amplify the (CAG)n region separately from the (CCG)n region. In this study, DNA samples previously analyzed by conventional slab-gel electrophoresis with autoradiographic detection are analyzed by CAE with multicolor fluorescence detection in a blinded manner. Results from the two methods are compared and a consistent diagnosis of all disease-causing mutations is obtained. The sizing precision of CAE permits the length of certain elongated HD alleles to be more accurately measured, since each capillary contains a coelectrophoresing DNA sizing standard and the intensities of each fragment is reported. We generated an allelic ladder standard from a mixture of both patient and control DNA samples to further improve the diagnostic identification of intermediate length alleles. The successful resolution of the normal, intermediate, and disease-causing alleles demonstrates the feasibility of detection of HD using CAE methods. Simi-

From: *Methods in Molecular Biology, Vol. 163:*
Capillary Electrophoresis of Nucleic Acids, Vol. 2: Practical Applications of Capillary Electrophoresis
Edited by: K. R. Mitchelson and J. Cheng © Humana Press Inc., Totowa, NJ

lar CAE protocols have been applied to the genotyping CA-repeat markers commonly used in linkage analysis studies.

1.1. Huntington Disease

At present, twelve neuro-degenerative diseases including HD have been proven to be caused by an expanded trinucleotide repeat (CAG)n, that is located within a specific gene responsible for each disease. HD, an autosomal dominant disorder, results in progressive, selective neuronal cell death associated with dementia and choreic movements. The severity of the disease correlates directly, and "age of onset" of the symptoms correlates inversely, with the length of the (CAG)n triplet repeat region present in the first exon of the *"huntingtin"* gene located on chromosome 4p16.3 *(1)*. This repeat encodes a polyglutamine track of different lengths in the N-terminal portion of the protein beginning 18 codons downstream of the first ATG codon. The polyglutamines encoded by the (CAG)n repeats are involved in the formation of intranuclear and cytoplasm inclusions that contain both the polyglutamine region of the protein and ubiquitin. These inclusions are thought to cause dysfunction and eventual degeneration of specific populations of neuronal cells (*see* **ref.** *2* for a recent review of insights gained from transgenic models of HD and other neuro-degenerative disorders caused triplet repeat expansions). In addition, the (CAG)n repeat length correlates positively with the degree of apoptosis-induced DNA fragmentation observed in the striatum of HD patients *(3)*. The huntingtin protein from normal individuals contains 9–35 glutamines, whereas huntingtin protein from HD patients contains 36–86 glutamines *(3)*. All mutations for Huntington disease arise from expansion of intermediate sized alleles that contain between 27 and 35 (CAG)n repeats (also referred to as "mutable normal alleles"). In disease-causing alleles, the CAG repeats expand upon transmission through the paternal germline to a size of 36 repeats, or more. Intermediate size alleles are present on approx 1% of normal chromosomes of Caucasian descent.

1.2. Diagnosis of Huntington Disease

Accurate diagnosis of HD is complicated by the molecular organization of the repeat region (**Fig. 1**). The unstable (CAG)n repeat lies immediately upstream from a moderately polymorphic polyproline-encoding (CCG)n repeat. The (CCG)n repeat may vary in size, from between 7 and 12 repeats in both affected and normal individuals. In addition, a CAA trinucleotide deletion was found between the CCG and CAG repeats in HD chromosomes in two different affected families *(4)*. The CAA deletion is positioned within the reverse-strand primer traditionally used in the PCR-based HD test (*see* **Fig. 1**), hampering the detection of the HD mutation when only the (CAG)n tract is amplified. Therefore, three separate diagnostic PCR reactions are now recommended: one amplifying the (CAG)n repeat, one amplifying the (CCG)n repeat, and one covering the whole region, encompassing both repeats *(3)*.

Prior to the discovery of polymorphism in the (CCG)n repeat, a worldwide study was undertaken in 1994 by Kremer et al. *(5)* to assess the sensitivity and specificity of the CAG expansion as a diagnostic test for HD. The study covered 1007 patients in 565 families from 43 national and ethnic groups. Of these, 995 had an expanded CAG repeat that contain between 36 to 121 repeats (sensitivity = 98.8%). However, 12

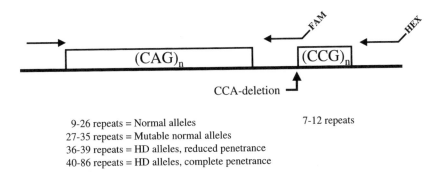

9-26 repeats = Normal alleles   7-12 repeats
27-35 repeats = Mutable normal alleles
36-39 repeats = HD alleles, reduced penetrance
40-86 repeats = HD alleles, complete penetrance

Fig. 1. Organization of the repeat region in the *HD* locus. Genomic mutations responsible for HD have been shown to be due to an unstable expansion of a (CAG)n triplet repeat in the coding region of the *IT15* gene. The highly polymorphic (CAG)n repeat is followed 12 bases downstream by a modestly polymorphic (CCG)n repeat. The relative positions of PCR primers used to amplify both the (CAG)n and (CCG)n repeats are shown. A common forward primer beginning at base 335 of the gene is used to amplify through the regions covering both repeats in separate PCR reactions. The reverse-strand primer used to amplify the (CAG)n region was labeled at the 5'-end with the blue dye FAM whereas the reverse primer for the (CCG)n region with the green dye HEX. The resulting FAM-labeled PCR products contain 43 bases outside of the (CAG)n repeat, whereas the HEX-labeled PCR products contain 82 bases outside the two repeats. The range of normal, intermediate (mutable normal or HD alleles with reduced penetrance) and disease-causing (CAG)n repeats is also indicated.

patients in the study that had been previously diagnosed with HD had alleles in the normal size range. Clinical reevaluation revealed that 11 of these patients had clinical features atypical of HD. Of the 1595 control chromosomes, 1581 contained from 10 to 29 CAG repeats (99.1%). The remaining 14 control chromosomes had 30 or more repeats, including two with expansions of 37 and 39 repeats. In 1998, the American College of Medical/American Society of Human Genetics (ACMG/ASHG) published laboratory guidelines for HD testing, recommending four categories of (CAG)n allele sizes (*see* **Fig. 1**): normal alleles contain fewer than 26 repeats, mutable normal alleles contain 27–35 repeats, HD alleles with reduced penetrance contain 36–39 repeats and fully penetrant HD alleles contain 40 or more repeats *(6)*. In the blinded study of 104 HD alleles *(6)*, a comparison of allele size estimates by CAE and by conventional electrophoresis using radioactive detection identified five allele call differences between the two methods. All calls were within one repeat of the reference. This yielded >95% concordance between the two methods and a ninefold improvement in typing consistency over a recent multilaboratory study.

Rubinsztein et al. *(7)* conducted a blinded multilaboratory study of HD diagnosis using a series of DNA samples previously genotyped in 1996 at several diagnostic centers. Overall, there was only 45% concordance in "called alleles" among the three participating centers, and 25% of the allele "calls" differed by greater than 2 repeats. Not all of the participating laboratories measured the adjacent (CCG)n repeat and its length can alter diagnostic risk assessment of HD. Furthermore, different size standards and analytical methods were used in the participating laboratories, resulting in a 36 repeat allele being called from 34 to 38 repeats by different groups *(7)*. The results of this study point to a need to standardize molecular diagnostic test methods for HD diagnosis.

1.3. Allele Sizing by Capillary Electrophoresis

Patients affected with HD will show a normal allele and enlarged alleles, with ≥ 36 repeats. Frequently, alleles with >40 repeats exhibit a fuzzy signal, or a smear in a conventional polyacrylamide gel separation. The imprecise signal makes the estimation of allele size difficult to interpret precisely using polyacrylamide gel electrophoresis (PAGE). More recently, Le et al. *(8)* reported that accurate HD diagnosis was "potentially problematic using CE" to size the alleles. They concluded that "extra care is required in interpreting CE results" because of an observed systematic bias in sizing results when comparing data from an automated slab-gel electrophoresis system (377 DNA Sequencer, Perkin-Elmer, Applied Biosystems Division, Foster City, CA) to the 310 CE system (Perkin-Elmer). In this study, fragment sizing was determined in 79 HD cases using both the slab-gel system and the 310 CE system. A 3–6-bp difference in sizing estimates was observed when comparing the two electrophoretic separation methods *(8)*.

Most simple sequence loci demonstrate sequence-dependent migration anomalies even when analyzed under denaturing conditions *(9)*. The resulting systematic bias in estimated fragment sizes is generally more pronounced in CAE data than slab-gel electrophoresis. However, by normalizing results to known typing controls, one can obtain locus-averaged accuracy of <0.06 bp and normalized results within 1 bp of actual size *(9)*. One DNA size standard that is effective for normalization of results is an allelic ladder standard. Zhang and Yeung *(10)* used an allelic ladder of the variable number of tamden repeats (VNTR) polymorphism in the human D1S80 locus as the absolute standard in CE separations. Statistical analysis of the data indicates a high level of confidence in matching the bands despite variations in the injection process, or in the CE system. The sizing of over 240 allelic ladder samples yielded an average within-run precision of ±0.13 bp and between-run precision of ±0.21 bp for fragments up to 350 bp on a prototype CAE system *(11)*. Here, an allelic ladder is generated from a mixture of DNA from the HD patient and control DNAs to determine its utility in HD genotyping by CAE using the MegaBACE 1000 capillary DNA sequencing system (**Fig. 2**).

Previous work from our laboratory demonstrated a standard deviation in the precision of sizing of <±0.12 bp *(12)*. In the run conditions used in the study, we observe an average bias in size estimate of under 1 bp from actual size, suggesting that run conditions used in CE-based separations strongly influence the accuracy of the sizing estimate. This is confirmed independently by Yang et al. *(13)*, who demonstrate that sizing measurements are more accurate when energy-transfer dyes are used on both the PCR primers and the DNA sizing standards. We have developed a uniformly migrating, energy-transfer dye-labeled DNA standard (TMR-50-350 and TMR-50-500, Amersham International, Chicago, IL) *(12)* and add it to each capillary to allow conversion of fragment mobility data to DNA size estimates.

In order to determine the diagnostic accuracy of HD diagnosis by CAE, we analyze DNA from 26 individuals "at risk" for HD in a blinded manner. The study set includes

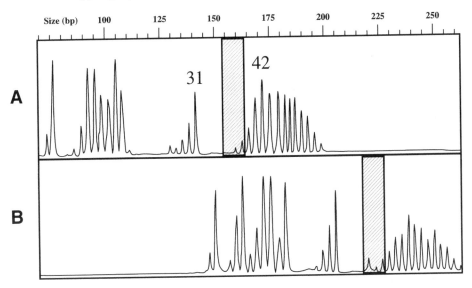

Fig. 2. Allelic ladder of triplet repeats in the *HD* locus. A pool of previously genotyped DNAs from both control patients and HD patients is prepared and appropriate fragments are PCR-amplified. The samples are selected so that all common (CAG)n alleles are represented in the pool. (**A**) Shows the separation of FAM-labeled PCR product amplified across the disease-causing (CAG)n repeat. The peaks corresponding to alleles with 31 and 42 (CAG)n repeats are indicated. The gray box indicates the migration of the 36–39 repeat alleles, which is the range of HD alleles with incomplete penetrance. (**B**) Shows the separation of HEX-labeled PCR product amplified from the same pool of DNA through both the (CAG)n and adjacent (CCG)n repeats. The positions of alleles with incomplete penetrance corresponding to $(CAG)_{36-39}$ and $(CCG)_9$ which are indicated by the gray box in this panel.

a total of 22 elongated HD alleles that were previously diagnosed using P^{32} labeling and slab-gel electrophoresis, followed by autoradiographic detection of labeled PCR fragments. The PCR fragments were separated in denaturing DNA sequencing gels using an M13 T-track in adjacent lanes as the fragment sizing standard. The length of (CCG)n had not been measured in all samples since the original diagnostic testing was performed before 1996, and before the three-PCR test method became standard. The objective of this study was to compare allele sizes, and to study the inferred lengths for both the (CAG)n and (CCG)n repeats in all samples. In addition, the diagnostic status of all samples are determined using the (CAG)n allele size guidelines established by the ACMG/ASHG statement on HD testing (*5*).

When processing samples by CAE, a total of 104 alleles are determined because both the (CAG)n and (CCG)n repeats are measured. For each sample, the (CAG)n and (CCG)n are amplified separately, then pooled and coelectrophoresed with a DNA sizing standard labeled with a separate color (**Fig. 3**). The CAE software reports the fragment length, the peak height and the peak area of all detected PCR product bands. These results are used to determine the primary peak labeled in each color, since the

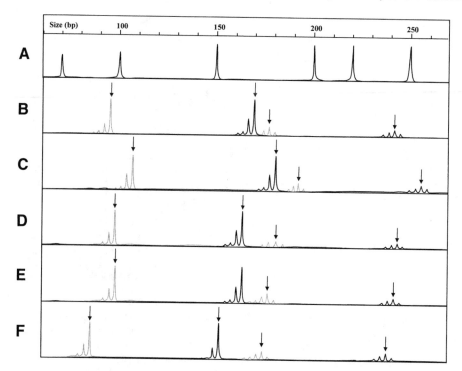

Fig. 3. Electropherogram profiles of the repeat regions amplified from five HD patients. (**A**) Contains the electropherogram plot of the fragment size vs the peak intensity of a DNA sizing standard, which is included in each capillary of a CAE separation. The standard contains 50-, 100-, 150-, 200-, 220-, 250-, 300-, and 350-bp energy-transfer dye-labeled DNA fragments (the region of 40–270 bp is shown). This standard is used to determine the size of the comigrating PCR fragments labeled with different color dyes. The electropherograms in (**B–F**) show the separation of (CAG)n and (CCG)n repeat regions which are aligned by size and are amplified from five different HD patients. In these panels, FAM-labeled PCR products amplified across the (CAG)n repeat are shown by dotted lines, whereas HEX-labeled PCR product amplified through both the (CAG)n and the adjacent (CCG)n repeats are indicated by solid lines. The arrows indicate the positions of the "called allele" peaks in the HD triplet repeat profiles.

triplet repeats frequently contain several "stutter" peaks in addition to the allele peak. The size of the peak with the greatest area is then used to assign the number of repeats in each allele, and these genotyping results are compared with those determined previously by slab-gel electrophoresis.

2. Materials

2.1. Biologicals

1. Human genomic DNA extracted from peripheral whole blood. (The method in **ref. *14*** is appropriate.)
2. Spectrophotometer for measuring the DNA concentration using OD 260/280.

3. TE buffer: 1 mM Tris-HCl, pH 8.0, 100 μM EDTA. Used to adjust the DNA concentration to 20 ng/μL.
4. *Taq* DNA polymerase (5 U/μL).
5. 10X PCR buffer: 100 mM Tricine buffer, pH 8.3, 500 mM KCl, 15 mM MgCl$_2$, 0.1% Triton X-100, 1% gelatin.
6. Deoxynucleotides: 1.25 mM solution are prepared by mixing and diluting 10 mM stock solutions of dATP, dCTP, dGTP, dTTP (Amersham).
7. Oligonucleotide primers: 5-μM solutions are used to amplify the triplet repeat region of the *HD* locus:
   Common forward primer:
   5'-ATG AAG GCC TTC GAG TCC CTC AAG–3'
   Reverse primer for CAG repeat:
   5'–FAM–CGG CGG AGG CGG CTG TTG CTG–3'
   Reverse primer for CAG /CCG repeat:
   5'-HEX–GGC GGC TGA GGA AGC TGA GGA G–3'
8. 96-Well membrane-lined tray for desalting PCR samples (mixed ester-cellulose Multiscreen, VMWP 0.5 μm; P/N SA2M267E1; Millipore Corp., Bedford, MA). Use to desalt and prepare the PCR samples for electrokinetic injection into the capillaries.
9. An 8-in. square Pyrex tray (or similar container), magnetic stir rod, and magnetic stir plate.
10. Hydroxyethyl cellulose (HEC): Stock of 2.0% HEC separation medium (Amersham). Prepare according to the method of Bashkin et al. *(15)*. Store at –20°C.
11. Stock electrophoresis buffer (diluted from 10X TBE): 1.33 M Tris-base, pH 8.2, 0.45 M boric acid, 25 mM EDTA.
12. Deionized formamide.
13. Energy-transfer dye-labeled DNA standard (TMR-50-350, Amersham International, Chicago, IL).
14. DNA thermal cycler. We used either the Perkin-Elmer model 9600 or MJ Research Tetrad model.
15. MegaBACE 1000® DNA Sequencing CAE system manufactured by Molecular Dynamics Division of Amersham.

3. Methods

3.1. PCR Amplification Procedure

1. Add the following reagents to 0.2-mL conical PCR tubes:

Component	Volume	Final concentration
10X PCR buffer	2.5 μL	1X
1.25 mM dNTP mixture	4.0 μL	100 μM
Unlabeled forward primer (5 μM)	2.5 μL	0.5 μM
Labeled reverse primer (5 μM)	2.5 μL	0.5 μM
Template DNA (20 ng/μL)	5.0 μL	100 ng
1:10 dilution *Taq* polymerase	2.0 μL	1 U
Dimethylsulfoxide (DMSO)	2.5 μL	10%
Autoclaved nuclease-free distilled water	4.0 μL	—

1. A master mix of components can be prepared for multiple PCR reactions (*see* **Note 1**). This helps minimize reagent loss and enables more accurate pipeting of reagents.
2. Cap the tubes and microfuge briefly to mix and bring all components to the bottom of the tubes.

3. PCR amplify for 30 cycles run as follows:
   Denature: 95°C for 1 min.
   Anneal: 65°C for 30 s.
   Extend: 72°C for 2 min.
4. A final extension is performed for an additional 8 min at 72°C and the samples are maintained at 4°C. Samples can be stored at –20°C in light-protected containers until analyzed.

3.2. Preparation of Amplified Products for CAE Separation

1. Combine the PCR products for the CAG and CAG /CCG repeat regions for each patient sample (*see* **Note 2**).
2. Transfer the PCR mixture to the wells of 96-well float dialysis tray. Cover securely with a microplate-sealing lid.
3. Prepare approx 2 L of 1X TE buffer and place into an 8-in. square Pyrex tray containing a magnetic stir rod. Place the buffer tray on the top of a magnet stir plate.
4. Carefully remove the plastic tray bottom and transfer the 96-well float dialysis trays to the buffer chamber taking care to keep the PCR liquid on the bottom of the wells.
5. Slowly activate the magnetic stir plate and permit the 96-well dialysis tray to remain on the buffer surface for 10 min.
6. Transfer the desalted PCR products to sterile 0.2-mL tubes adjusting the volume of each mixture to 60 µL.
7. Combine 1 µL of desalted PCR product mixture, 0.5 µL of TMR-50-350 sizing standard (*see* **Note 3**), and 3.5 µL deionized formamide in thin-walled 96-well conical tubes.
8. Immediately prior to electrophoresis, denature the DNA sample by heating for 2 min at 90°C and then snap-chill on ice.

3.3. CAE Separation

1. Thaw 6 tubes of the stock 2.0% HEC DNA separation medium (Amersham).
2. Degas the separation medium by centrifuging the 1.5-mL tubes at 10,000*g* for 5 min.
3. Wet the capillary walls with deionized water and pressure inject the DNA separation medium into the capillaries for 2 min using the programmed high pressure injection methods provided with the MegaBACE 1000.
4. Permit the HEC separation medium to equilibrate to ambient pressure for 5 min before initiating a prerun.
5. Preelectrophorese the separation medium for 5 min at a constant voltage of 185 V/cm.
6. Electrokinetically inject the DNA samples into the capillaries by applying 185 V/cm constant voltage for 5–10 s. Samples can be reinjected up to 10–15 times, if reanalysis of sizes is required.
7. The fluorescently labeled DNA is size-separated for 40 min at a constant voltage of 185 V/cm. **Figure 1** shows the electrophoretic profiles of both normal and HD disease alleles.
8. Both the sizes and the peak intensities of the HD alleles are measured using the prototype Genetic Analysis Software tools for the MegaBACE 1000, (*see* **Notes 4** and **5**).

3.4. Analysis of STR Genotypes by CAE

1. In CAE separations, samples with up to 48 repeats of (CAG)n continue to show discrete "stutter" peaks in the electropherogram. However, the peak intensities of the elongated alleles are consistently weaker than alleles in the normal size range. This is predominantly due to reduced amplification efficiency of the elongated alleles, but it is also due to a bias in electrokinetic sampling (*see* **Note 6**).

2. Short DNA fragments consistently inject more efficiently into the capillary than longer DNA (*see* **Note 7**). The assay could be improved by using the more sensitive FAM dye to label the combined CAG/CCG repeat region. The increased sensitivity relative to the HEX dye would help compensate for the losses in signal intensity due to preferential amplification, as well as the preferential electrokinetic loading of the shorter PCR products.
3. The combination of the four-color fluorescence detection, multicolor-multiplex PCR, and a MegaBACE 1000 CAE system has the capacity to generate up to 5.5 million genotypes per year *(9)*. Furthermore, an average sizing precision of ±0.12 bp for fragments up to 350 bp can be realized in 1-h runs.
4. Marsh et al. *(16)* used a 48-capillary prototype to conduct 1 h DNA sequencing runs (of approx 500 bases/h/capillary). They estimate the throughput to be of the order of 720 templates/d, or 360,000 bases/d *(16)*.
5. The same CAE system can also be used separate a broad size range (80–40 kbp) of DNA fragments, under nondenaturing matrix conditions without the use of pulse-field separation *(17,18)*.

4. Notes

1. The triplet-repeat region of the *HD* locus is very GC-rich. We found that addition of 10% DMSO to the PCR mixture is required to achieve both specific amplification and good product yields. The addition of lesser amounts of DMSO result in the detection of the normal alleles, but not the longer expanded HD alleles.
2. Several features of the triplet repeat region in the *HD* locus makes the accurate diagnosis of the disease by STR genotyping difficult to perform. The diagnostic (CAG)n repeat in the *HD* gene is tightly linked to a noninformative, but moderately polymorphic (CCG)n repeat. Multicolor fluorescence detection permits both of the repeat regions to be analyzed simultaneously, along with a coelectrophoresing DNA sizing ladder. In this manner, the lengths of both repeats are measured directly in a single capillary run and the number of PCR reactions required reduced from 3 to 2.
3. The allelic ladder standards might further improve reliability of HD typing results. These DNA are of identical composition to the "unknown samples" being amplified, thus sequence migration anomalies are eliminated. Furthermore, the MegaBACE analysis software permits unknown sample profiles to be overlaid with the profiles of allele standards produced from previously typed DNA samples. An example HD allele standard is shown in **Fig. 2**.
4. The triplet repeat profiles frequently contain several "stutter" peaks in addition to the allele peak, thus complicating the interpretation of allele calls (*see* **Fig. 3** for an example of such a profile). For this reason, a report of peak intensities provides an objective measure of the most abundant PCR product amplified for each allele, and allows the weaker "stutter" peaks to be identified.
5. CAE peak area/intensities are useful in typing HD alleles or CA-repeat alleles because they provide an objective measurement of relative abundance of amplified peaks.
6. The average allele size estimates made by CAE was +0.55 bp from actual size, thus electrophoretic migration anomalies that are associated with repetitive DNA sequence are minimal under the experimental conditions used.
7. The (CCG)n allele size did not change the diagnostic status of the sample set tested.

Acknowledgments

Molecular Dynamics and SBIR grant #2R44NS34589 awarded to E.S.M. by the National Institutes of Health funded this research. We thank Marina Vainer and Curtis Kautzer for technical assistance with this work.

References

1. MacDonald, M. E., Barnes, G., Srinidhi, J., Duyao, M. P., Ambrose, C. M., Myers, R. H., et al. (1993) Gametic but not somatic instability of CAG repeat length in Huntington's disease. *J. Med. Genet.* **30,** 982–986.
2. Price, D. L., Sisodia, S. S., and Borchelt, D. R. (1998) Genetic neurodegenerative diseases, the human illness and transgenic models. *Science* **282,** 1079–1083.
3. Butterworth, N. J., Williams, L., Bullock, J. Y., Love, D. R., Faull, R. L., Dragunow, M. (1998) Trinucleotide (CAG) repeat length is positively correlated with the degree of DNA fragmentation in Huntington's disease striatum. *Neuroscience* **87,** 49–53.
4. Gellera, C., Meoni, C., Castellotti, B., Zappacosta, B., Girotti, F., Taroni, F., et al. (1996) Errors in Huntington disease diagnostic test caused by trinucleotide deletion in the *IT15* gene. *Am. J. Hum. Genet.* **59,** 475–477.
5. Kremer, B., Goldberg, P., Andrew, S. E., Theilmann, J., Telenius, H., Zeisler, J., et al. (1994) A worldwide study of the Huntington's disease mutation: the sensitivity and specificity of measuring CAG repeats. *New Eng. J. Med.* **330,** 1401–1406.
6. The American College of Medical Genetics/American Society of Human Genetics Huntington Disease genetic testing working group. (1998) ACMG/ASHG Statement: Laboratory guidelines for Huntington disease genetic testing. *Am. J. Hum. Genet.* **62,** 1243–1247.
7. Rubinsztein, D. C., Leggo, J., Coles, R., Almqvist, E., Biancalana, V., Cassiman, J. J., et al. (1996) Phenotypic characterization of individuals with 30-40 CAG repeats in the Huntington disease (HD) gene reveals HD cases with 36 repeats and apparently normal elderly individuals with 36-39 repeats. *Am. J. Hum. Genet.* **59,** 16–22.
8. Le, H., Fung, D., and Trent, R. J. (1997) Applications of capillary electrophoresis in DNA mutation analysis of genetic disorders. *J. Clin. Path. Mol. Path.* **50,** 261–265.
9. Mansfield, E. S., Vainer, M., Enad, S., Barker, D.L., Harris, D., Rappaport, E., et al. (1996) Sensitivity, reproducibility, and accuracy in short tandem repeat genotyping using capillary array electrophoresis. *Genome Res.* **6,** 893–903.
10. Zhang, N. and Yeung, E. S. (1996) Genetic typing by capillary electrophoresis with the allelic ladder as an absolute standard. *Anal. Chem.* **68,** 2927–2931.
11. Mansfield, E. S., Robertson, J. M., Vainer, M., Isenberg, A. R., Frazier, R. R., Ferguson, K., et al. (1998) Analysis of multiplexed short tandem repeat (STR) systems using capillary array electrophoresis. *Electrophoresis* **19,** 101–107.
12. Mansfield, E. S., Vainer, M., Harris, D. W., Gasparini, P., Estivill, X., Surrey, S., et al. (1997) Rapid sizing of polymorphic microsatellite markers by capillary array electrophoresis. *J. Chromatogr. A* **781,** 295–305.
13. Wang, Y., Wallin, J. M., Ju, J., Sensabaugh, G. F., and Mathies, R. A. (1996) High-resolution capillary array electrophoretic sizing of multiplexed short tandem repeat loci using energy-transfer fluorescent primers. *Electrophoresis* **17,** 1485–1490.
14. Sambrook, J., Fritsch, E. F், and Maniatis, T. (1989) Molecular Cloning: A laboratory manual. Cold Spring Harbor Laboratory, Cold Spring Harbor, NY.
15. Bashkin, J., Marsh, M., Barker, D., and Johnston, R. (1996) DNA sequencing by capillary array electrophoresis with a hydroxyethylcellulose sieving buffer. *Appl. Theor. Electrophoresis* **6,** 23–28.
16. Marsh, M., Tu, O., Dolnik, V., Roach, D., Solomon, N., Bechtol, K., Smietana, P., et al. (1997) High-throughput DNA sequencing on a capillary array electrophoresis system. *J. Capillary Electrophor.* **4,** 83–89.

17. Madabhushi, R. S., Vainer, M., Dolnik, V., Enad, S., Barker, D. L., Harris, D. W., et al. (1997) Versatile low-viscosity sieving matrices for nondenaturing DNA separations using capillary array electrophoresis. *Electrophoresis* **18**, 104–111.

18. Madabhushi, R. S. (2001) DNA sequencing in noncovalently coated capillaries using low viscosity polymer solutions, in *Capillary Electrophoresis of Nucleic Acids*, Vol. 2 (Mitchelson, K. R., and Cheng, J., eds.), Humana Press, Totowa, NJ, pp. 309–315.

13

Genotyping by Microdevice Electrophoresis

Dieter Schmalzing, Lance Koutny, Aram Adourian, Dan Chisholm, Paul Matsudaira, and Daniel Ehrlich

1. Introduction

DNA genotyping has traditionally been performed by slab-gel electrophoresis. The method is well established, very effective and reliable, but inherently slow and labor intensive. Capillary gel electrophoresis is now becoming recognized as an important alternative to slabs since it offers higher throughput and automation *(1)*. Nevertheless, the fast growing need for DNA analysis capacity due, for example, to the sequencing of entire genomes and the establishment of complex DNA data banks, seems to demand even more powerful DNA analysis tools. Microdevice electrophoresis is being increasingly explored since it may allow electrophoretic DNA analysis approaching the theoretical performance limits of the method due to unique sample loading characteristics and the employment of very short separation distances *(2)*. It has already been demonstrated that genotyping can be performed 10–100 faster than on capillaries and slabs (*see* **Subheading 3.4.**). In addition, the fabrication and operation of high-density electrophoretic microdevices for high-sample throughput should be straightforward. Microfabrication might also permit the total integration of entire sample processing sequences (e.g., PCR, sample cleanup, separation) on one single device, potentially leading to drastically decreased sample and reagent volumes, significantly less human interference, and increased speed of analysis *(3)*.

Microdevice electrophoresis is performed in microchannel structures embedded in glass, fused silica, or plastic substrates *(4,5)*. The structures can be microfabricated through a variety of processes, e.g., photolithography, chemical wet etching, and wafer bonding or injection-molding, depending on the material. **Figure 1** illustrates a microfabrication process for fused-silica wafers. Microfabrication is extremely versatile enabling virtually the realization of any desired channel geometry regardless of its complexity. The underlying DNA analysis principle is essentially the same as in capillaries. The microchannels are chemically passivated to exclude any perturbation of the separation process through surface effects. The channels are filled with a viscous polymeric buffer solution for the sorting of DNA fragments according to size. High field strengths, typically between 200–400 V/cm,

From: *Methods in Molecular Biology, Vol. 163:*
Capillary Electrophoresis of Nucleic Acids, Vol. 2: Practical Applications of Capillary Electrophoresis
Edited by: K. R. Mitchelson and J. Cheng © Humana Press Inc., Totowa, NJ

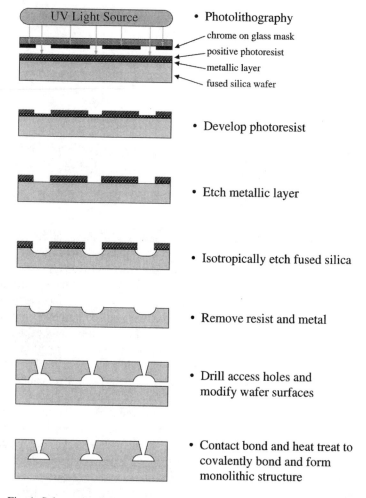

- Photolithography
  - chrome on glass mask
  - positive photoresist
  - metallic layer
  - fused silica wafer

- Develop photoresist

- Etch metallic layer

- Isotropically etch fused silica

- Remove resist and metal

- Drill access holes and modify wafer surfaces

- Contact bond and heat treat to covalently bond and form monolithic structure

Fig. 1. Schematic of the microfabrication process for fused silica wafers.

are used to drive the DNA through the channel structures. DNA fragments are detected on-line by a laser-induced fluorescence (LIF) detector at the end of the separation channel. The DNA is rendered fluorescent either through covalent labeling or intercalation with fluorescent dyes. Genotyping information is achieved either through comparison of DNA migration time data with a preestablished calibration curve or through the addition of a DNA sizing ladder to the DNA genotyping sample prior to microdevice electrophoresis.

2. Materials

2.1. Coating Materials

1. 3-(Trimethoxysilyl)propylmethacrylate (Fluka, Buchs, Switzerland).
2. Acrylamide (Pharmacia, Uppsala, Sweden).

Fig. 2. Schematic diagram of the basis building block of a microfabricated electrophoretic device, depicting the typical cross-structure. Reprinted from **ref. 9**.

3. *N,N,N',N'*-Tetramethylethylenediamine (TEMED) (Sigma, St. Louis, MO).
4. Ammonium persulphate (APS) (Sigma).

2.2. Sieving Matrix Materials

1. Hydroxyethyl cellulose (HEC), mol wt 90,000–105,000 (Polysciences, Warrington, PA).
2. Acrylamide (Pharmacia).
3. TEMED (Sigma).
4. APS (Sigma).

2.3. Buffers, Enzymes, and Dyes

1. Powerplex tandem repeat DNA kit (Promega, Madison, WI).
2. The restriction buffer is 10 m*M* Tris-acetate, 10 m*M* magnesium acetate, and 50 m*M* potassium acetate.
3. The plasmid concentration in the DNA reservoir is 125 ng/μL and the enzyme concentration in the enzyme reservoir is 4 U/μL.
4. 10X TBE (Gibco-BRL, Gaithersburg, MD): 0.89 *M* Tris-borate, pH 8.3, 20 m*M* EDTA.
5. Ethidium bromide (Molecular Probes, Eugene, OR).
6. TOTO-1 (Molecular Probes).

2.4. Sieving Matrices

1. The sieving matrix for short dsDNA was typically 4% (w/v) high molecular weight linear polyacrylamide dissolved in 1X TBE with 3.5 *M* urea and 30% formamide.
2. Another sieving matrix was 0.75% (w/v) HEC as sieving matrix in 1X TBE buffer with 1 μ*M* ethidium bromide as intercalating dye.
3. The separation buffer for on-line restriction analysis consists of 9 m*M* Tris-borate with 0.2 m*M* EDTA and 1% (w/v) HEC as sieving matrix with 1 μ*M* of the fluorescent intercalating dye TOTO-1.

3. Methods
3.1. Microdevice Designs

1. Microdevice electrophoresis is still in the midst of rapid technological development. There is currently no instrument commercially available that uses this powerful method. The subsequent paragraphs will therefore primarily focus on the general features of the technique to familiarize the reader with its main aspects. The researcher interested in building apparatus should refer to the cited literature for specific details.
2. The simple cross-structure shown in **Fig. 2** is the basic building unit of most microfabricated devices tested so far. There are three short side channels, originating from sample, buffer, and waste reservoirs each typically with a length of a few millimeters.

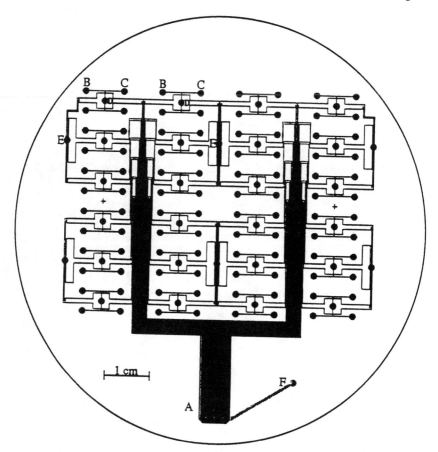

Fig. 3. Mask pattern for a 96-sample capillary array electrophoresis microdevice. Reprinted from **ref. 6**.

3. The separation channel has a length of 1–10 cm depending on the separation requirements. The cross-sectional shape of the channels is usually half-cylindrical with a typical depth of 20–45 μm and a top-width of 60–100 μm, depending on the microfabrication process.
4. The channel junction produces an injection plug typically between 50 μm and 250 μm in length with an injection volume in the subnanoliter range.
5. **Figure 3** shows a highly multiplexed device for the simultaneous analysis of 96 genotyping samples iterating the described cross-structure *(6)*.

3.2. Instrumentation

1. A schematic of a single channel genotyping apparatus is shown in **Fig. 4**. The microfabricated device is mounted on a temperature-controlled heating stage.
2. High voltage is provided to platinum wire electrodes mounted in the four fluid reservoirs by high-voltage powers supply. The voltages in each fluid reservoir are individually addressed by a means of switching circuit.

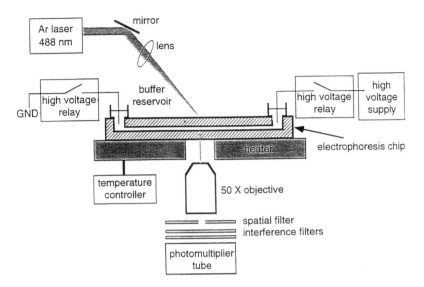

Fig. 4. Schematic of a genotyping apparatus with laser-induced fluorescence detection optics and temperature controlled heating stage (for clarity only a single-color detection setup is shown). Reprinted from **ref. 9**.

3. A laser beam is focused into the separation channel at the desired separation distance. A microscope objective collects the fluorescence emission. Four dichroic mirrors with four photo multiplier tubes allow for the simultaneous detection of up to four different fluorophores.

3.3. Device Operation

1. The general genotyping procedure is illustrated in **Fig. 5**. The entire channel structure is filled with the appropriate polymeric sieving matrix and buffer is placed into the waste and the anodic and cathodic reservoirs.
2. A few microliters of sample are pipeted into the sample reservoir. For loading of the sample into the channel structure, an electric field is applied from the sample to waste reservoir that fills the connecting channel with sample.
3. For subsequent injection of a representative sample plug into the separation channel, the voltages are switched and a high electric field is applied across the perpendicular separation channel.
4. Simultaneously, a much lower electric field is applied to the side channels to prevent excess sample from entering the separation channel during electrophoresis.

3.4. Genotyping Examples

3.4.1. Short Tandem Repeat (STR) Analysis for Human Identification (7–9)

1. STRs are short stretches of repetitive sequences found throughout the genome. The repeat units are 2–7 bases long. A typical STR locus consists of 7–20 repeats.
2. STRs are becoming very important for forensics, clinical diagnostics, and genetic linkage studies since they are highly polymorphic and can easily be amplified even from degraded

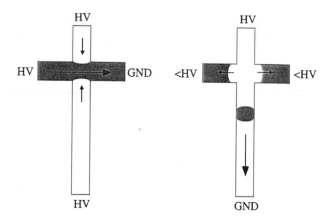

Fig. 5. Illustration of the sample loading and the sample injection/separation process. Reprinted from **ref. 9**.

DNA. In this example, human DNA is extracted from blood stain cards and 8 STR loci are simultaneously PCR-amplified using the PowerPlex kit.

3. The primers for the four loci: D16S539, D7S820, D13S317, D5S818 and for the four loci: CSF1PO, TPOX, TH01, vWA are fluorescently-labeled with fluorescein and tetramethyl-rhodamine, respectively.

4. Prior to microdevice electrophoresis 3 µL of PCR-amplified sample is added to 1 µL of fluorescent allelic standard ladder for allele identification and diluted to a final volume of 10 µL with water.

5. The mixture is briefly vortexed, denatured for 2 min at 95°C and chilled on ice.

6. The microdevice has the cross-structure described above. The effective length of the separation channel is 2 cm. The entire channel structure is filled with 4% (w/v) high molecular weight linear polyacrylamide dissolved in 1X TBE with 3.5 M urea and 30 % formamide.

7. Buffer is loaded into the waste and the two main buffer reservoirs. After pipeting the sample into its reservoir, the sample is electrophoresed at 200 V/cm through the loading channel.

8. After 3 min, the sample is injected and separated applying the same field strength. The microdevice is kept at 50°C during the whole time to help denaturation and to decrease the analysis time.

9. **Figure 6** shows the 8-loci genotyping result for a given individual. All the alleles are unambiguously identified in 2 min. For example, the individual is homozygous (one tall peak) for D5S818 and heterozygous for vWA (two tall peaks). The analysis is 10× to 100× faster than capillary slab-gel electrophoresis without compromising the genotyping information.

3.4.2. Screening for Hereditary Hemochromatosis (HHC) (6)

1. Samples are prepared using PCR amplification and digestion to assay for the C282Y mutation in the *HFE* gene, a candidate for HHC, whose presence creates an *Rsa*I restriction side in this gene. Samples are dialyzed against deionized water on 96-sample dialysis plates prior to loading on the 96-channel device shown in **Fig. 3**.

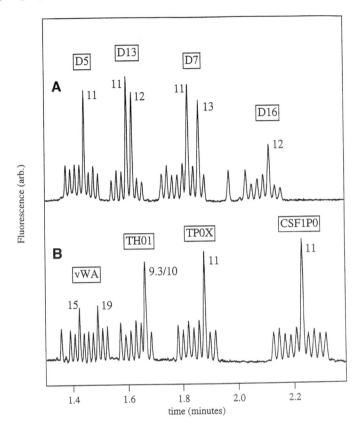

Fig. 6. Microdevice electropherograms of the simultaneous two-color analysis of 8 STR loci **(A)** D16S59, D7S820, D13S317, D5S818 and **(B)** CSF1P0, TPOX, TH01, vWA (CTTv). The allelic standard ladders were mixed with the PCR-amplified sample of an individual before injection. Allele numbers are given above the peaks. Separation conditions: 40-μm deep channel, 150-μm long injector, 2-cm separation distance, 50°C, 200 V/cm, 4% (w/v) linear polyacrylamide in denaturing buffer (1X TBE, 3.5 M urea, 30% formamide. Reprinted from **ref. 8**.

2. The entire microchannel structure is filled with 0.75% (w/v) HEC as sieving matrix in 1X TBE buffer with 1 μM ethidium bromide as intercalating dye for detection. The anode and the cathode reservoirs are filled with 10X TBE buffer to reduce buffer depletion during electrophoresis.

3. Sample reservoirs are rinsed with deionized water, prior to the loading of 3.5 μL of sample per reservoir.

4. The separation distance is 10 cm and the load and run voltages are 300 V/cm. A laser-excited confocal galvanometric scanner is used for the simultaneous detection of all 96 samples seen in **Fig. 7**.

5. The wild-type (845G) shows the 140- and 167-bp fragments, the variant (845A) shows 29-, 111-, and 167-bp fragments, and the heterozygote shows all 4 fragments. The separation is completed in only 220 s. The total analysis time is less than 8 min, which is 50–100× faster than traditional slab-gel systems.

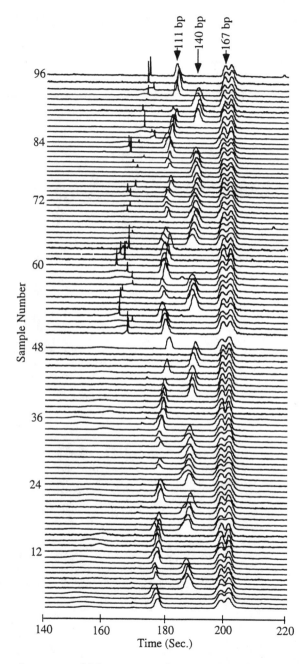

Fig. 7. Electropherograms of 96 *Rsa*I-digested HFE amplicons. Samples are separated on 0.75 % (w/v) hydroxyethyl cellulose in 1X TBE buffer with 1 μ*M* ethidium bromide in the running buffer. The channel length is 10 cm and the separation voltage is 300 V/cm. Reprinted from **ref. 6**.

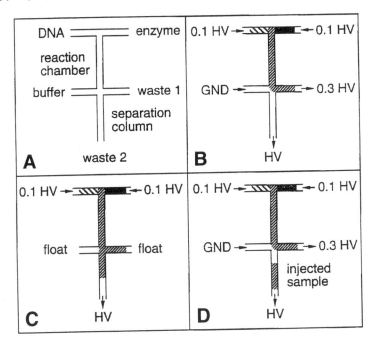

Fig. 8. Schematic of: (**A**) The reaction chamber and injection cross; (**B**) the loading of the reaction chamber with DNA and enzyme for restriction digestion; (**C**) injecting the digestion products onto the separation column; (**D**) separating the product fragments. Arrows depict direction of flow for anions. Reprinted with permission from **ref.** *10*. Copyright (1996) American Chemical Society.

3.4.3. On-Line Restriction Fragment Length Polymorphism (RFLP) Analysis (10)

1. Digestion of the plasmid pBR322 with restriction enzyme *Hin*fI, followed by electrophoretic sizing, is performed sequentially on an integrated microfabricated device depicted in **Fig. 8.**
2. The reaction buffer consisting of 10 m*M* Tris-acetate, 10 m*M* magnesium acetate, and 50 m*M* potassium acetate is placed into the DNA, enzyme and first waste reservoirs.
3. The separation buffer consisting of 9 m*M* Tris-borate with 0.2 m*M* EDTA and 1% (w/v) HEC as sieving matrix is put into the microchannels and the buffer and second waste reservoirs.
4. The plasmid concentration in the DNA reservoir is 125 ng/ μL and the enzyme concentration in the enzyme reservoir is 4 U/μL.
5. DNA and enzyme are then electrophoretically loaded into the reaction chamber. For efficient digestion all electric fields are removed for 2 min after loading.
6. An aliquot of the digestion products is then injected into the 6-cm long separation channel and is separated in 3 min (**Fig. 9**) resulting in a total analysis time of only 5 min.
7. The products are detected by laser-induced fluorescence detection using 1 μ*M* of the fluorescent intercalating dye TOTO-1.

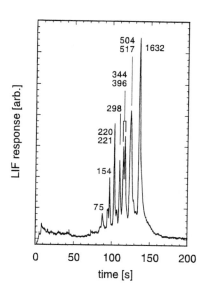

Fig. 9. Electropherograms of products from the digestion of the plasmid pBR322 by the enzyme *Hin*fI. The separation field strength is 380 V/cm, the separation distance is 67 mm. The numbers correspond to the fragment lengths in base pairs. Reprinted with permission from **ref. *10***. Copyright (1996) American Chemical Society.

4. Outlook

1. It can be speculated that within the next few years we will not only see a drastic increase in research on microdevice electrophoresis but also the first practical devices entering the laboratories of biologists, forensic scientists, and so on. The general acceptance in the life science community should be greatly facilitated by the scientists familiarity with the underlying electrophoretic process and by the fact that all the protocols of DNA sample preparation and handling originally developed for slabs can be directly put to use in microdevice electrophoresis *(11)*.
2. The usage of microdevices should not be limited to large research facilities, like genome centers, where there is an urgent need for total automation and ultra-high sample throughput. Smaller laboratories should profit as well from this powerful technology through designated semiautomated apparatus with modest multiplexing capabilities offering ease of operation, almost instant access to results, and savings in reagent costs.

References

1. Mullikin, J. C., and McMurray, A. A. (1999) Sequencing the genome fast. *Science* **283,** 1867–1868.
2. Freemantle, M. (1999) Downsizing chemistry. Chemical analysis and synthesis on microchips promise a variety of potential benefits. *Chem. Eng. News* **22,** 27–36.
3. Service, R. F. (1998) Coming soon: the pocket DNA sequencer. *Science* **282,** 399–401.
4. Harrison, D. J., Fluri, K., Seiler, K., Fan, Z. Effenhauser, C. S., and Manz, A. (1993) Micromachining a miniaturized capillary electrophoresis-based chemical analysis system on a chip. *Science* **261,** 895–897.

5. Simpson, P. C., Wooley, A. T., and Mathies, R. A. (1998) Microfabrication technology for the production of capillary array electrophoresis chips. *Biomed. Microdevices* **1**, 7–25.

6. Simpson, P. C., Roach, D., Woolley, A. T., Thorsen, T., Johnston, R., Sensabaugh, G. F., et al. (1998) High-throughput genetic analysis using microfabricated 96-sample capillary array electrophoresis microplates. *Proc. Natl. Acad. Sci. USA* **95**, 2256–2261.

7. Schmalzing, D., Koutny, L., Adourian, A., Belgrader, P. H., Matsudaira, P., and Ehrlich, D. (1997) DNA typing in thirty seconds with a microfabricated device. *Proc. Natl. Acad. Sci. USA* **94**, 10,273–10,278.

8. Schmalzing, D., Koutny, L., Chisholm, D., Adourian, A., Matsudaira, P., and Ehrlich, D. (1999) Two-color multiplexed analysis of eight short tandem repeat loci with an electrophoretic microdevice. *Anal. Biochem.* **270**, 148–152.

9. Schmalzing, D., Koutny, L., Adourian, A., Matsudaira, P., and Ehrlich, D. (1997) Ultrafast STR analysis by microchip gel electrophoresis. In: *Proceedings from the 8th International Symposium on Human Identification*. Promega Corp., pp. 112–118.

10. Jacobson, S. C., and Ramsey, J. M. (1996) Integrated microdevice for DNA restriction fragment analysis. *Anal. Chem.* **68**, 720–723.

11. Xie, W., Yang, R., Xu, J., Zhang, L., Xing, W., and Cheng, J. (2001) Microchip-based capillary electrophoresis systems, in *Capillary Electrophoresis of Nucleic Acids*, Vol. 1 (Mitchelson, K. R., and Cheng, J., eds.), Humana Press, Totowa, NJ, pp. 67–83.

14

Applications of Constant Denaturant Capillary Electrophoresis and Complementary Procedures

Measurement of Point Mutational Spectra

Andrea S. Kim, Xiao-Cheng Li-Sucholeiki, and William G. Thilly

1. Introduction

Constant denaturant capillary electrophoresis (CDCE) separates macromolecules based on differences in their melting temperatures. The specific apparatus and operating conditions have been described previously that allow CDCE to separate point mutants among 100–150-bp iso-melting DNA sequences [1,2]. CDCE coupled with high-fidelity DNA polymerase chain reaction (PCR) has been applied to the measurement of point mutational spectra in human cell and tissue samples [3–6]. For reviews of this field *see* **refs.** [7,8]. This chapter describes additional techniques which when combined with CDCE and high-fidelity PCR allow point mutation detection at fractions as low as 10^{-6}.

The specific protocol for measuring point mutations depends on the desired degree of sensitivity as outlined in **Fig. 1**. To detect mutations at fractions (mf) down to 5×10^{-4}, one isolates genomic DNA (**Fig. 1**, step 1), amplifies the desired sequence with a high-fidelity DNA polymerase (**Fig. 1**, step 6), and enriches mutant sequences relative to the wild-type sequences (**Fig. 1**, step 7). Mutants in the enriched mixture are then separated from each other and measured by CDCE (**Fig. 1**, step 8). At this stage, the individual mutants are purified and sequenced for their identification (**Fig. 1**, step 9). This method has found three important applications: detection of point mutations in phenotypically altered human cells after chemical treatment, detection of cancer cells in normal tissues [9], and identification of single-nucleotide polymorphisms (SNPs) in pooled human blood samples [10].

The analysis of point mutations at fractions down to 10^{-6} requires three additional steps prior to high-fidelity PCR (**Fig. 1**, step 6). These steps are liberation of a desired sequence from genomic DNA by restriction digestion (**Fig. 1**, step 2), the enrichment of the desired sequence from restriction-digested DNA (**Fig. 1**, step 3), and pre-PCR

From: *Methods in Molecular Biology, Vol. 163:*
Capillary Electrophoresis of Nucleic Acids, Vol. 2: Practical Applications of Capillary Electrophoresis
Edited by: K. R. Mitchelson and J. Cheng © Humana Press Inc., Totowa, NJ

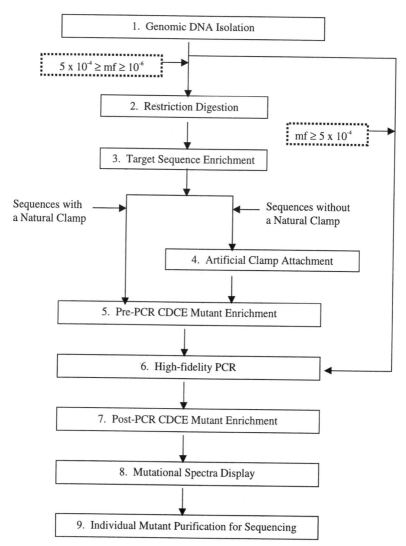

Fig. 1. Flow diagram of the sample handling steps necessary to detect point mutations at different fraction levels. This procedure is for all DNA sequences including ones without a neighboring natural clamp. An additional procedure, step 4, is introduced for DNA sequences without a natural clamp. For mutation detection at fractions down to 5×10^{-4}, include the steps 1, 6–9. For mutation detection at fractions as low as 10^{-6}, include the steps 1–3, 5, prior to performance of the steps necessary for mutation detection at fractions at or above 5×10^{-4}.

mutant enrichment by CDCE (**Fig. 1, step 5**). Even *Pfu*, the highest fidelity DNA polymerase available *(11,12)*, creates mutations at a level which interferes with the initial mutant fractions of approx 5×10^{-4} after about 20 doublings. Thus, the pre-PCR mutant enrichment is essential for detecting low fraction mutations as low as 10^{-6} *(13)*. The sensitivity of this method has been demonstrated in chemically-treated human cells without reference to phenotypic selection *(6)*.

A sequence suitable for CDCE analysis is a 100-bp iso-melting DNA domain juxtaposed to a domain of a higher melting temperature. This higher melting region works as a "clamp," that allows the separation of point mutations in the target domain from the wild-type under appropriate denaturing conditions *(2)*. Approximately 9% of the human genome is comprised of 100-bp sequence elements with such a natural neighboring clamp. If a sequence of interest does not have a natural clamp, an artificially-created clamp can be attached by PCR *(14)* to detect point mutations at fractions down to 5×10^{-4}. However, this PCR-based clamp attachment method also introduces polymerase-created mutations, which prevents mutation detection at fractions below 5×10^{-4}. For this reason, we have been developing a ligation-based artificial clamp attachment method (*see* **Fig. 1**, step 4) as a means to extend our ability to measure point mutations at fractions as low as 10^{-6} in the entire human genome.

A part of the human adenomatous polyposis coli (*APC*) gene (*APC* cDNA-bp 8543–8683) is used as a target sequence in the development of the point mutation detection methods described in this chapter. This *APC* sequence is an iso-melting DNA domain juxtaposed to a natural clamp (*APC* cDNA-bp 8441–8542), suitable for CDCE analysis (**Fig. 2**). If the same target sequence did not have the natural clamp, an artificial clamp can be attached to the sequence.

2. Materials

Unless otherwise specified, all materials and solutions should be stored at room temperature.

2.1. Genomic DNA Isolation

Genomic DNA can be isolated from either blood or from solid tissues. **Items 1–4** are necessary for the tissue samples, and **items 5** and **6** are needed for the blood samples.

1. Dimethyl sulfoxide (DMSO).
2. Surgical scalpels.
3. Liquid nitrogen.
4. Mortar and pestle.
5. ACD Solution B: 0.48% citric acid, 1.32% sodium citrate, 1.47% glucose.
6. Phosphate-buffered saline (PBS): Dissolve 8 g of NaCl, 0.2 g of KCl, 1.44 g of Na_2HPO_4, and 0.24 g of KH_2PO_4 in 800 mL of distilled H_2O (dH_2O). Adjust the pH to 7.4 with HCl and add dH_2O to 1 L.
7. 1X TE buffer: 50 mM Tris-HCl, pH 8.0, 10 mM EDTA; 0.1X TE (dilute from 1X).
8. Proteinase K (Boehringer Mannheim, Indianapolis, IN). Make a fresh solution of proteinase K in dH_2O at 20 mg/mL on the day of DNA isolation. Store at −20°C until use to limit autodigestion.
9. 10% Sodium dodecyl sulfate (SDS): Dissolve 100 g of SDS in 900 mL of dH_2O (heat to 68°C to assist dissolution). Adjust the pH to 7.2 with HCl and add dH_2O to 1L.
10. 10 mg/mL RNaseA (Boehringer Mannheim). Store at −20°C.

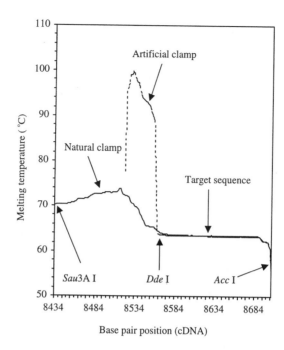

Fig. 2. Melting profiles of the wild-type *APC* sequence (cDNA-bp 8543–8683). The solid line represents the melting profile of the wild-type sequence with a natural clamp (*APC* cDNA-bp 8441–8542), whereas the dotted line represents the melting profile of the same sequence with an added artificial clamp. Restriction digestion of genomic DNA with *Acc*I and *Sau*3AI liberates the target *APC* sequence with the natural clamp, suitable for CDCE analysis. If the same target sequence did not have the natural clamp, an artificial clamp can be attached to the *Dde*I restriction end of the sequence (*APC* cDNA-bp 8570–8572).

11. 5 *M* NaCl: Dissolve 292.2 g of NaCl in 800 mL of dH$_2$O and add dH$_2$O to 1 L.
12. 100% Ethanol. Chill to –20°C before using.
13. 70% Ethanol. Chill to –20°C before using.

2.2. Restriction Digestion and Internal Standard Introduction

1. Restriction endonucleases: *Hae*III and *Xba*I (New England Biolabs, Beverly, MA). Store at –20°C.
2. 10X NEBuffer #2: 10 m*M* Tris-HCl, pH 7.9 at 25°C, 10 m*M* MgCl$_2$, 50 m*M* NaCl, 1 m*M* dithiothreitol (New England Biolabs). Store at –20°C.
3. Items required for PCR (*see* **Subheading 2.6.**) and CDCE (*see* **ref. 2**).
4. Primers for mutant internal standard construction (polyacrylamide gel electrophoresis [PAGE]-purified). Primers for the target *APC* sequence are as follows: Primer 1, 5'-CCATCTCAGA TCCCAACTCC-3' (*APC* cDNA-bp 8422–8441; the fluorescein-labeled thymine in bold). Primer 2, 5'-AACAAAAACC CTCTAACAAG AATCA AACCT A**C**TTAC-3' (complementary to *APC* cDNA-bp 8648–8683; underlined C forms a A:C mismatch at bp 8562), and Primer 3, 5'-TATAATCTAGAAATGATTGA (complementary to *APC* cDNA-bp 8894–8913) (Synthetic Genetics, San Diego, CA). Dilute each primer with dH$_2$O to a concentration of 1.2×10^{13} mol/μL and store at –20°C.
5. Prepare 2X PCR master mix #1 with Primers 1 and 2; 2X PCR master mix #2 with Primers 1 and 3. For the preparation of 2X PCR master mix, *see* **Subheading 2.6.**, **item 5**.

2.3. Target Sequence Enrichment

1. Two oligonucleotide probes that are complementary to the Watson and Crick strands of the target sequence. The probes must meet the three requirements as follows: biotinylated at the 5' end through a 12–18 carbon spacer arm, purified by high-performance liquid chromatography (HPLC) or PAGE, and have similar melting temperatures. Probes for the target *APC* sequence are Probe 1, 5'-CAAAACTGAC AGCACAGAAT CCAGTGGAAC-3' (*APC* cDNA-bp 8472–8501), and Probe 2, 5'-AAGACCCAGA ATGGCGCTTA GGACTTTGGG-3' (complementary to *APC* cDNA-bp 8501–8530) (Synthetic Genetics). Dilute each probe with dH_2O to a concentration of 1.2×10^{13} mol/µL and store at –20°C.
2. 20X SSPE: Dissolve 17.53 g of NaCl, 2.76 g of $NaH_2PO_4 \cdot H_2O$, and 0.74 g of EDTA in 90 mL of dH_2O. Adjust the pH to 7.4 with NaOH and add dH_2O to 100 mL.
3. 10 mg/mL streptavidin-coated paramagnetic beads (MPG®) (CPG, Lincoln Park, NJ). Store at 4°C and prewash with the washing buffer (**item 5**) before using.
4. Neodymium magnet.
5. Washing buffer: 1 *M* NaCl, 10 m*M* Tris-HCl, pH 7.6, 2 m*M* EDTA.
6. 10X Reannealing buffer: 2 *M* NaCl, 100 m*M* Tris-HCl, pH 7.6, 20 m*M* EDTA.
7. 0.025-µm membrane filters (Millipore, Marlborough, MA).

2.4. Artificial Clamp Attachment

1. Restriction endonuclease: *Dde*I (New England Biolabs). Store at –20°C.
2. 10X NEBuffer #3: 50 m*M* Tris-HCl, pH 7.9 at 25°C, 10 m*M* $MgCl_2$, 100 m*M* NaCl, 1 m*M* dithiothreitol (New England Biolabs). Store at –20°C.
3. Pair of complementary oligonucleotides rich in GC bases (PAGE-purified). The pair of oligonucleotides for the target *APC* sequence is: Clamp 1, 5'FITC-CGCCCGCCGC GCCCCGCGCC CGTCCCGCCG CCCCCGCCCG ATAATAAC-3' (fluorescein-labeled at the 5' end [5'FITC]), and Clamp 2, 5'P-TTAGTTATTA TCGGGCGGGG GCGGCGGGAC GGGCGCGGGG CGCGGCGGGC G-3' (phosphate group attached at the 5' end [5'P]), (Synthetic Genetics). Store at –20°C.
4. T4 DNA ligase (Boehringer Mannheim). Store at –20°C.
5. 10X Reaction buffer for T4 DNA ligase: 660 m*M* Tris-HCl, pH 7.5 at 20°C, 50 m*M* $MgCl_2$, 10 m*M* dithioerythritol, 10 m*M* ATP (Boehringer Mannheim). Store at –20°C.

2.5. Pre-PCR Mutant Enrichment

1. Restriction endonucleases: *Acc*I and *Sau*3AI (New England Biolabs). Store at –20°C.
2. 10X NEBuffer #4: 20 m*M* Tris-acetate, pH 7.9 at 25°C, 10 m*M* magnesium acetate, 50 m*M* potassium acetate, 1 m*M* dithiothreitol (New England Biolabs). Store at –20°C.
3. 100X bovine serum albumin (BSA): 10 mg/mL BSA (New England Biolabs). Store at –20°C.
4. Dialysis buffer: 0.1X TBE (dilute from 5X); 5X TBE: dissolve 54 g of Tris-base, 27.5 g of boric acid, and 10 mL of 0.5 *M* EDTA, pH 8.0, in 1 L of dH_2O.
5. Items required for CDCE (*see* **ref. 2**).
6. 542-µm inner diameter (id) and 665-µm outer diameter (od) fused-silica capillaries (Polymicro Technologies, Phoenix, AZ). The DNA loading capacity of 540-µm id capillaries is approx 10 µg *(15)*.
7. Stainless steel tubing (6 cm or longer, 0.042" id, 0.027" od, and 19-in. gauge) (Small Parts Inc., Miami, FL).
8. 0.5-mm id and 1.5-mm od Teflon tubing (Varian Associates, Inc., Walnut Creek, CA).
9. 0.8X TBEB elution buffer: 0.8X TBE, 0.24 mg/mL BSA. Additional dilutions are 0.4X TBEB and 0.2X TBEB. Store at –20°C.

2.6. High-Fidelity PCR

1. Set of primers (PAGE-purified). Primers for the target *APC* sequence with a natural clamp are Primer 4, 5'FITC-GAATAACAAC ACAAAGAAGC-3' (*APC* cDNA-bp 8441–8460), and Primer 5, 5'-AACAAAAACC CTCTAACAAG-3' (complementary to *APC* cDNA-bp 8664–8683) (Synthetic Genetics). Primer 4 is replaced with Clamp 1 (*see* **Subheading 2.4., item 3**) for the same target sequence attached to an artificial clamp. One primer is fluorescein-labeled at the 5' end (5'FITC) (Primer 4 or Clamp 1, the primer in the clamp region). Dilute each primer with dH$_2$O to a concentration of 1.2×10^{13} mol/µL and store at –20°C.

2. dNTP mix: 25 m*M* mixture of four dNTPs (Pharmacia, Piscataway, NJ)—dATP, dCTP, dGTP, and dTTP—each 100 m*M* in equal volume and store at –20°C.

3. Native *Pfu* DNA polymerase (2.5 U/µL) (Stratagene, La Jolla, CA). Store at –20°C.

4. 10X reaction buffer for native *Pfu* DNA polymerase: 200 m*M* Tris-HCl, pH 8.0, 20 m*M* MgCl$_2$, 100 m*M* KCl, 60 m*M* (NH$_4$)$_2$SO$_4$, 1% Triton X-100, 100 µg/mL nuclease-free BSA (Stratagene). Store at –20°C.

5. 2X PCR master mix with Primers 4 and 5 (**item 1**): 100 µL of 2X master mix contains 20 µL of 10X native *Pfu* DNA polymerase reaction buffer, 2.0 µL of each primer (1.2×10^{13} mol/µL), 0.8 µL of 25 m*M* dNTPs, 2.0 µL of 100X BSA, and 73.2 µL of dH$_2$O. Store the mixture at –20°C.

6. 10- and 50-µL glass capillary tubes (Idaho Technology, Idaho Falls, ID).

7. Air Thermo-Cycler® (Idaho Technology).

8. Glass-cutter.

9. In order to perform post-PCR mutant enrichment (**Subheading 3.7.**), mutational spectra display (**Subheading 3.8.**), and individual mutant purification (**Subheading 3.9.**), *see* the materials required for both PCR (**Subheading 2.6.**) and for CDCE (*2*).

3. Methods

1. To minimize potential contamination of samples by artificially-created or PCR-generated DNA mutants, the procedures performed before high-fidelity PCR (*see* **Subheading 3.6.**) must be carried out in a separate laboratory equipped with high-throughput HEPA air filters.

2. No PCR products can be permitted in such a laboratory except for very diluted mutant internal standard stocks (<10^5 copies/µL).

3.1. Genomic DNA Isolation

1. For tissue samples: Cut a tissue sample (stored in 20% DMSO at –70°C or –20°C) into small pieces with a scalpel. Deep-freeze the pieces of the tissue in liquid nitrogen and grind them into a fine powder using a mortar and pestle. Place the powdered tissue in a centrifuge tube and suspend it in 1 mL of 1X TE buffer per 50 mg of tissue. Proceed to **step 3**.

2. For blood samples: Add one volume of PBS to a blood sample (stored in ACD solution B (1 mL of ACD solution B/6 mL of blood) at –70°C or –20°C) and after mixing, centrifuge at 3500*g* for 15 min. Carefully discard as much of the supernatant as possible without disturbing the pellet at the bottom of the tube. Resuspend the pellet in 1 mL of 1X TE buffer for each 3 mL of blood.

3. Add 20 mg/mL Proteinase K and 10% SDS to the final concentrations of 1 mg/mL and 0.5%, respectively. Incubate the solution while continuously mixing the contents thoroughly in a water-bath shaker (100–200 rpm) at 50°C for 3 h.

4. Add 10 mg/mL RNaseA to a final concentration of 20 µg/mL. Incubate the suspension in a water-bath shaker and mix thoroughly (100–200 rpm) at 37°C for 1 h.

5. Centrifuge at 10,000g (13,000 rpm) for 15 min while keeping the centrifuge temperature at 4°C. Transfer the central portion of the supernatant into a new tube by careful pipeting. Repeat this step two or three times and combine all of the transferred supernatant (*see* **Note 1**).

6. Add 5 M NaCl to the transferred supernatant tube to a final concentration of 250 mM and add two volumes of chilled 100% ethanol. Mix by inverting the tube gently several times (a DNA spool should start to form upon mixing).

7. Transfer the DNA spool into a microcentrifuge tube and wash with 1 mL of chilled 70% ethanol. Repeat this washing and discard as much of 70% ethanol as possible.

8. Air-dry the DNA spool by leaving it in the tube with the cap open at room temperature for approx 30 min (*see* **Note 2**). Add 0.1X TE buffer to a DNA concentration of 2–4 mg/mL. Pipet the DNA sample mixture up and down several times upon dissolving the spool at room temperature for complete mixing.

9. Add 5 µL of the DNA sample mixture into 495 µL of dH$_2$O and measure A$_{260}$ nm and A$_{280}$ nm with a UV spectrophotometer. A typical DNA yield and ratio of A$_{260}$ to A$_{280}$ are over 90% and 1.4–1.6, respectively with the DNA suitable for both restriction digestion and PCR amplification. For the detection of point mutations at fractions down to 5×10^{-4}, proceed to **Subheading 3.6.**, **step 2**. For the detection of point mutations at fractions down to 10^{-6}, proceed to **Subheading 3.2.1.**, **step 1**.

3.2. Restriction Digestion and Internal Standard Introduction

3.2.1. Restriction Digestion

1. Add 10X NEBuffer #2 to a final concentration of 1X and *Hae*III and *Xba*I to an enzyme/DNA ratio of 1 U/µg to the DNA sample with a DNA concentration of 2–3 mg/mL.

2. Mix thoroughly and incubate at 37°C overnight (*see* **Note 3**).

3.2.2. Internal Standard Preparation

1. A 492-bp artificial mutant (*APC* cDNA-bp 8422–8913) with an AT → GC transition at *APC* cDNA-bp 8652 is constructed to serve as an internal standard. This internal standard mutant is compatible with the 482-bp *APC* fragment liberated from genomic DNA by *Hae*III and *Xba*I (*see* **Subheading 3.2.1.**).

2. Add 1 µL of the restriction-digested DNA sample into 9 µL of dH$_2$O (10-fold dilution).

3. Preparation of a 262-bp *APC* mutant fragment (*APC* cDNA-bp 8422–8683): mix 1 µL of the diluted sample with 5 µL of 2X PCR master mix #1, 3.6 µL of dH$_2$O, and 0.4 µL of *Pfu* DNA polymerase. Amplify the fragment (*see* **Subheading 3.6.**, **steps 4–7**).

4. Preparation of a 492-bp *APC* wild-type fragment (*APC* cDNA-bp 8422–8913): Amplify the fragment as described in **step 3** except use 2X PCR master mix #2.

5. The preparation of a 492-bp *APC* mutant fragment (internal standard) mixture involves: mix together 1 µL of the PCR product from **step 3**, 1 µL of the PCR product from **step 4** (diluted 10-fold in dH$_2$O), 5 µL of 2X PCR master mix #2, 2.6 µL of dH$_2$O, and 0.4 µL of *Pfu* DNA polymerase. Amplify the fragment with an appropriate number of cycles to convert all the primers into product (*see* **Notes 4** and **5**).

6. Make subsequent stock dilutions of the amplified mutant internal standard (492 bp) with dH$_2$O, 10-fold each time at concentrations down to 10^2 copies/µL (*see* **Notes 4** and **5**).

3.2.3. Internal Standard Introduction

1. Mix together 1 µL of the 10-fold diluted restriction digested DNA sample, 1 µL of the internal standard stock of 10^4 copies/µL (*see* **Subheading 3.2.2.**, **step 6**) 5 µL of 2X PCR master mix (*see* **Subheading 2.6.**, **item 5**), 2.6 µL of dH$_2$O, and 0.4 µL of *Pfu* DNA polymerase. Amplify the target sequence (*see* **Subheading 3.6.**, **steps 4–7**).

2. Separate the PCR product by CDCE and measure the copy number of the target sequence in the restriction-digested DNA sample (*see* **Subheading 3.7.** for the CDCE procedure *[2]*). The target sequence copy number can be quantified by measuring the areas under the separated peaks, which represent wild-type homoduplex (A_w), mutant homoduplex (A_m), and wild-type/mutant heteroduplexes (A_h). The equation to be used for the quantification is as follows:

$$([A_w + A_h/2] / [A_m + A_h/2]) \times 10^x \text{ copies/μL}) \times df = \# \text{ of target sequence copies/μL},$$

in which 10^x copies/μL is the concentration of the internal standard stock used for PCR (10^4/μL) and df is a dilution fold of the original sample (10-fold).
3. Add the internal standard at a desired fraction to the restriction-digested DNA sample.

3.3. Target Sequence Enrichment

1. Add each Probe 1 and Probe 2 to a probe/target sequence molar ratio of 5×10^4 to the restriction enzyme digested sample (with added internal standard), and 20X SSPE to a final concentration of 6X. Mix thoroughly and distribute into several 1.5-mL microcentrifuge tubes (≈0.5 mL/tube).
2. Place the sample tubes containing the mixture in a boiling water bath for 2 min, followed by immediate chilling in an ice bath for 10 min. Incubate the chilled sample in a thermomixer at 58°C for 2 h. This is the probe-target hybridization temperature for the target *APC* sequence (*see* **Note 6**).
3. Add 0.4 mg of MPG beads/10^8 copies of the target sequence and incubate the suspension in a rotating thermomixer (1000 rpm) at 50°C for 1 h. Gather the beads to the side of each sample tube by placing a magnet against the wall of the tube. Remove the solution and combine all the beads into one tube.
4. Resuspend the beads in the washing buffer at a concentration of 10 mg/mL and incubate in a rotating thermomixer (1000 rpm) at 50°C for 5 min. Remove the buffer and retain the beads. Repeat this washing step three times.
5. Elute the target sequence from the beads: Incubate the beads at 20 mg/mL in dH$_2$O for 2 min at 70°C. Magnetically separate the eluate from the beads and transfer the eluate into a fresh tube. Repeat this step one more time and combine the eluates.
6. Reduce the volume of the eluate to about 10 μL by speed-vacuum centrifugation.
7. Add the 10X reannealing buffer to the reduced-volume eluate to a final concentration of 1X and incubate at 55°C for 16 h. During incubation, the wild-type/mutant heteroduplex DNA fragments are formed.
8. Desalt the reannealed sample by drop dialysis: Float a 0.025-μm membrane filter on the surface of 0.1X TE buffer in a container (e.g., a plastic Petri dish) and place the container on a stir plate. Transfer the sample on the floating membrane filter by pipeting. Dialyze the sample on the membrane filter against 0.1X TE buffer by stirring the buffer with a stir bar for 2 h.
9. Reduce the volume of the desalted sample to approx 10 μL by speed-vacuum centrifugation.
10. The procedures described in this subheading allow the enrichment of a desired target relative to nontarget fragments in a pool of genomic DNA restriction fragments. Although the values may vary for different targets, a typical enrichment (fold) and yield for the target *APC* sequence are about 10^4 and 70%, respectively. Thus, this procedure can reduce the DNA sample size of 600 μg to 60 ng where over 10^8 copies of the target sequence are present. The reduced sample volume is suitable for both pre-PCR mutant enrichment (*see* **Subheading 3.5.**) and artificial clamp ligation (*see* **Subheading 3.4.**).
11. For DNA sequences with a natural clamp, proceed to **Subheading 3.5., step 1**; otherwise proceed to **Subheading 3.4., step 2**.

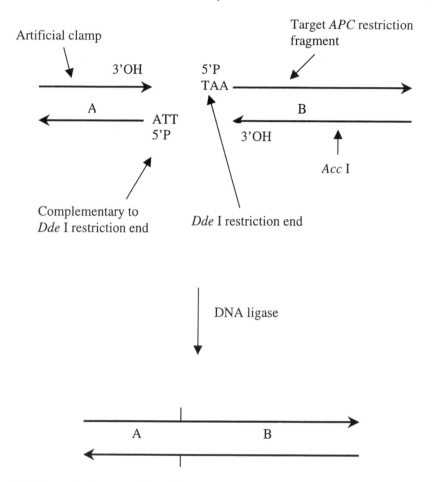

Fig. 3. Schematic diagram of the artificial clamp ligation procedure. The prepared artificial clamp contains an end that is complementary to the *Dde*I restriction end of the target *APC* sequence. The clamp and target restriction fragments are brought together as they base pair at their complementary termini and the covalent bonds can then be formed by DNA ligase.

3.4. Artificial Clamp Attachment

1. Although about 9% of the human genome can be directly studied by CDCE, the artificial clamp attachment procedure described in this subheading opens up an additional 89% of the human genome for CDCE analysis. This method is based on high-efficiency DNA ligation in which an artificial clamp is ligated to the restriction-digested ends of a target sequence (*see* **Fig. 3**). *This procedure is presently experimental. We hope the description will aid its development to a robust technique.*

2. Artificial clamp preparation (*see* **Note 7**): mix Clamp 1 and Clamp 2 in 0.1X TE buffer at a clamp concentration of 5×10^{12} mol/µL. Incubate at 95°C for 5 min and 70°C for 30 min to allow the two strands to hybridize to each other.

3. Add 10X NEBuffer #3 to a final concentration of 1X and 5 U of *Dde*I to the target sequence-enriched sample. Bring the total reaction volume to 15 μL with dH$_2$O and mix thoroughly. Incubate at 37°C for 4 h (*see* **Note 8**).

4. Add 2 μL of the prepared artificial clamp mixture, 10X ligation reaction buffer to a final concentration of 1X, and 1 U of T4 DNA ligase to the *Dde*I restriction-digested sample. Bring the total reaction volume to 30 μL with dH$_2$O and incubate at 16°C for 16 h.

5. Desalt the sample by drop dialysis (*see* **Subheading 3.3.**, **step 8**).

6. Reduce the volume of the sample to approx 10 μL by speed-vacuum centrifugation. Proceed to **Subheading 3.5.**, **step 2**.

3.5. Pre-PCR Mutant Enrichment

1. For DNA sequences with a natural clamp: to the target sequence-enriched sample, add both 10X NEBuffer #4 and 100X BSA to a final concentration of 1X. Add 5 U of *Acc*I and *Sau*3AI and mix thoroughly. Incubate at 37°C for 16 h and proceed to **step 3** (*see* **Note 9**).

2. For DNA sequences attached to an artificial clamp: add 10X NEBuffer #4 to a final concentration of 1X and 5 U of *Acc*I. Mix thoroughly and incubate at 37°C for 16 h.

3. Desalt the restriction-digested sample by drop dialysis against 0.1X TBE (*see* **Subheading 3.3.**, **step 8**) (*see* **Note 10**).

4. Reduce the volume of the sample to approx 4 μL by speed-vacuum centrifugation.

5. Prepare a 20 cm-long coated capillary of 540 μm id with a detection window 7 cm away from the anodic end (*see* **ref. 2** for details about the coating of capillaries). The detection window can be made by peeling off approx 0.5 cm of the outer surface of the capillary. Place the prepared capillary on a CDCE instrument and insert a portion of the capillary (near the cathodic end) in a water jacket (equipped with stainless steel tubing to hold the capillary) which is connected to a constant temperature circulator.

6. Set the circulator to the optimal temperature for the separation of mutant/wild-type heteroduplexes from wild-type homoduplex (*see* **Subheading 3.3.**, **step 7**) (*see* **Notes 11** and **12**).

7. Replace the 5% linear polyacrylamide matrix (about 35 μL) within the capillary (*see* **ref. 2** for details about the preparation of the matrix) (*see* **Note 13**).

8. Transfer 4 μL of the sample (**step 4**) by pipet into a piece of Teflon tube (1 cm long and 0.5 mm id). Remove the buffer reservoir away from the cathodic end of the capillary and mount the Teflon tube onto the capillary end. Bring the buffer reservoir to the end of the Teflon tube and electrokinetically inject the sample at 80 μA for 2 min. Remove the Teflon tube from the capillary end and perform electrophoresis at 80 μA (*see* **Note 13**).

9. Stop electrophoresis just before mutant heteroduplexes start to reach the anodic end of the capillary and remove the buffer reservoir from the capillary end. Place a platinum wire into 10 μL of 0.8X TBEB elution buffer in a 0.5-mL microcentrifuge tube and simultaneously dip the anodic end of the capillary in the tube. Electroelute mutant heteroduplexes at 80 μA into the elution buffer for 15–20 min (*see* **Notes 14** and **15**).

10. Dialyze the eluted sample by drop dialysis against 0.1X TBE (*see* **Subheading 3.3.**, **step 8**).

11. Reduce the volume of the sample to approx 4 μL by speed-vacuum centrifugation.

12. Remove the water jacket from the capillary and replace the matrix inside the capillary (*see* **step 7**).

13. Electroinject the sample into the capillary (*see* **step 8**) and perform electrophoresis at 80 μA and at room temperature (RTCE).

14. Stop electrophoresis just before the target sequence in double-strand form to reach the anodic end of the capillary. Electroelute the target sequence at 80 μA in 10 μL of 0.4X TBEB buffer for 5 min (*see* **Notes 14** and **15**).

15. Reduce the volume of the eluted sample to 5 μL by speed-vacuum centrifugation. Proceed to **Subheading 3.6., step 3**.

3.6. High-Fidelity PCR

1. For mutation detection of fractions down to 5×10^{-4}: measure the copy number of the target *APC* sequence in the sample (**Subheading 3.1., step 8**) as described in **Subheading 3.2.3**.
2. Add 25 μL of 2X PCR master mix, the internal standard mutant at a desired fraction (*see* **Subheading 3.2.2., step 6**), and 2 μL of *Pfu* DNA polymerase to 1 μg of the sample. Bring the total reaction volume to 50 μl with dH$_2$O and mix thoroughly. Proceed to **step 4**.
3. For mutation detection of fractions down to 10^{-6}: Add 5 μL of 2X PCR master mix and 0.4 μL of *Pfu* DNA polymerase to the mutant-enriched sample and mix thoroughly.
4. After a brief centrifugation, transfer the PCR mixture to either a 10- or 50-μL glass capillary tube by capillary action and seal both ends of the tube by heating in a gas flame.
5. Amplify the target sequence with an appropriate number of cycles to convert all the primers into product (*see* **Notes 16** and **17**). For the target *APC* sequence, each PCR cycle proceeds in the order of 10 s at 94°C, 20 s at 50°C, and 20 s at 72°C with 2 min at 94°C and 2 min at 72°C before and after the desired number of PCR cycles.
6. Cut both ends of the glass capillary tube with a glass-cutter and transfer the PCR product into a microcentrifuge tube. During the last few PCR cycles, the abundant mutant sequences form heteroduplex dsDNA with the excess wild-type fragments.
7. Incubate the PCR product at 72°C for 20 min with additional *Pfu* DNA polymerase (0.2 μL of *Pfu*/10 μL of PCR product) (*see* **Note 18**).

3.7. Post-PCR Mutant Enrichment

1. Prepare CDCE set up as described in **Subheading 3.5., step 5** with a 21 cm long coated capillary of 75 μm id.
2. Replace the 5% linear polyacrylamide matrix (about 2 μL) within the capillary.
3. Remove the buffer reservoir from the cathodic end of the capillary and place both a platinum wire and the capillary end into a microcentrifuge tube containing the PCR product (diluted 10-fold in H$_2$O). Electrokinetically inject the PCR product at 2 μA for 30 s into the capillary. Remove the sample tube from the capillary end and reinsert the capillary end into the buffer reservoir. Perform electrophoresis at 9 μA.
4. Stop the electrophoresis just before the PCR-amplified mutant heteroduplexes (*see* **Subheading 3.6., step 4**) reach the anodic end of the capillary and remove the buffer reservoir from the capillary end. Electroelute the heteroduplexes in 10 μL of 0.2X TBEB elution buffer at 9 μA for about 2–3 min using a platinum wire (*see* **Subheading 3.5., step 9**).
5. Take 5 μL of the electroeluted sample and add 5 μL of 2X PCR master mix and 0.4 μL of *Pfu* DNA polymerase. Perform the PCR (*see* **Subheading 3.6., steps 4–7**). A typical mutant enrichment is approx 20-fold.
6. Repeat **steps 2–5**. Another 5-fold mutant enrichment can be repeated, bringing the total enrichment to approx 100-fold (*see* **Note 19**).
7. Take 1 μL of the PCR product and add 5 μL of 2X PCR master mix, 3.6 μL of dH$_2$O, and 0.4 μL of *Pfu* DNA polymerase. Amplify the target sequence with 3 PCR cycles to convert all the mutant sequences into homoduplexes (*see* **Subheading 3.6., steps 4–7**).

3.8. Mutational Spectra Display

1. Prepare the CDCE apparatus with a 33 cm long coated capillary of 75 μm id and a 19 cm long water jacket, as described in **Subheading 3.5., step 5**. Set the water jacket circulator to the optimal temperature for separation of mutant from wild-type homoduplexes.

2. Replace the 5% linear polyacrylamide matrix within the capillary.
3. Electroinject the PCR product, diluted 10-fold in H_2O, and perform electrophoresis at 5 µA.
4. Measure the individual mutant fraction. The measurement is done by comparing the ratio of the area under each mutant peak to the area under the internal standard peak added to the sample (*see* **Subheading 3.2.3.**, **step 3** or **Subheading 3.6.**, **step 2**).

3.9. Individual Mutant Purification for Sequencing

1. Purify each separated mutant homoduplex by PCR and by CDCE (*see* **ref. 2**).
2. Identify each purified mutant by sequencing.

4. Notes

1. Care should be taken to maximize the transferred volume of the supernatant and to avoid the transfer of the pellet at the bottom of the tube and the top layers of the supernatant.
2. Avoid over-drying the DNA spool since a spool air-dried too long may be difficult to dissolve in 0.1X TE buffer.
3. The established protocol for target sequence enrichment requires a prior restriction digestion to liberate a target sequence desired from genomic DNA, as described in **Subheading 3.3.** A set of endonucleases for this restriction digestion should be selected to minimize the cost since digestion of genomic DNA from 10^8 cells (≈600 µg) for mutation detection at fractions down to 10^{-6} requires at least 600 U of the endonucleases. For the target *APC* sequence, *Hae*III and *Xba*I were selected to liberate the sequence-embedded 482-bp fragment (*APC* cDNA-bp 8422–8903) from genomic DNA at as low a cost as any other pair of restriction endonucleases available commercially.
4. When primers are completely depleted during PCR, the copy number of the sequence amplified is equivalent to the initial copy number of the primers. The expected concentration of the PCR-amplified mutant internal standard is 10^{11} copies/µL upon complete conversion of the primers into product.
5. The amplification efficiency of the wild-type and mutant internal standard sequences must be the same. Unequal amplification efficiency of the wild-type and mutant will introduce errors in the measurement of the target sequence copy number and mutant fractions in the samples *(16)*.
6. An optimal hybridization temperature needs to be determined for each probe-target sequence set. This temperature can be determined experimentally by measuring the target sequence recovered at different temperatures tested. For the measurement of the target sequence copy number, *see* **Subheading 3.2.3.**, **steps 1** and **2**.
7. One end of an artificial clamp prepared must be complementary to the restriction end of a target sequence (*Dde*I restriction end for the target *APC* sequence, *see* **Fig. 3**).
8. Incubation longer than 4 h and at temperatures above 37°C may reduce the ligation efficiency of an artificial clamp to desired target restriction ends. Such incubation conditions should be avoided.
9. A second set of restriction endonucleases is necessary to excise the target sequence from the somewhat longer restriction fragment containing the target sequence (*see* **Note 3**). The cost of restriction endonucleases is no longer an important factor since the sample has been enriched for the target sequence by 10^4-fold (*see* **Subheading 3.3.**). Restriction digestion of a 482-bp fragment with *Acc*I and *Sau*3AI liberates a 271-bp fragment (*APC* cDNA-bp 8434–8704, *see* **Fig. 2**) which is suitable for pre-PCR mutant enrichment by CDCE.

10. The ionic strength of the sample buffer must be below that of 1X TBE for the electrokinetic injection of an entire DNA sample. Ions are preferentially loaded before DNA onto the capillary.

11. An optimal CDCE separation temperature for mutant enrichment is the one at which mutant/wild-type heteroduplexes are well separated from the wild-type homoduplex. Such a CDCE separation temperature can be determined by CDCE test runs with fluorescein-labeled PCR product containing the wild-type homoduplex, an artificially-created mutant homoduplex, and wild-type/mutant heteroduplexes.

12. Changes in the acrylamide concentration of the matrix, the length of separation zone, and the field strength of electrophoresis can affect CDCE separation efficiency *(17)*. Modifying these CDCE operating conditions may be necessary to obtain the desired degree of CDCE separation (the same applies to **Subheadings 3.7.** and **3.8.**).

13. Take care to avoid introducing air bubbles into the capillary while performing **Subheading 3.5., steps 7** and **8**.

14. The electroelution time of the mutants separated from the wild-type by CDCE followed by RTCE is determined empirically. This electroelution time can be determined by CDCE and RTCE test runs with fluorescein-labeled PCR product (*see* **Note 11**).

15. The combination of CDCE and RTCE separation and electroelution of mutants allows about 100–200-fold mutant enrichment. CDCE is used to separate mutant/wild-type heteroduplexes from the wild-type homoduplex, whereas RTCE is used to separate the CDCE-eluted mutant heteroduplexes from residual wild-type sequences. The residual wild-type sequences, which can comigrate with the mutant heteroduplexes in the CDCE, could have been generated by incomplete restriction digestion (**Subheading 3.5., steps 1** or **2**) or during the duplex reannealing step (**Subheading 3.3., step 7**).

16. For mutation detection at fractions down to 5×10^{-4} in DNA sequences without a neighboring natural clamp, an artificial clamp can be attached to a desired sequence by a GC-primer *(14)* during this PCR.

17. The efficiency and conditions of PCR need to be determined experimentally for each target sequence. Thus, the number of PCR cycles necessary to convert all the primers into product depends on a sequence of interest and starting copy numbers. Avoid applying more PCR cycles than necessary since it may cause PCR product degradation.

18. To reduce the amount of PCR-generated byproducts, post-PCR incubation is necessary for some target sequences.

19. For mutation detection at fractions down to 5×10^{-4}, 100-fold post-PCR mutant enrichment allows CDCE visualization of mutants, separated from each other. A typical mutant with an initial fraction of 10^{-4} is enriched to a fraction of 10^{-2} using the method described in **Subheading 3.7.** For mutation detection at fractions down to 10^{-6}, a 100-fold pre-PCR mutant enrichment (*see* **Subheading 3.5.** and **Note 15**) in addition to post-PCR mutant enrichment allows visualization of the CDCE-separated mutants at initial fractions of 10^{-6} or higher.

Acknowledgments

This work was supported by grants from the National Institute of Environmental Health Sciences: Center Core Grant P30-ESO2109, Mutagenic Effects of Airborne Toxicants in Human Lungs P01-ESO7168, and Superfund Hazardous Substance Basic Research Program P42-ESO4675.

References

1. Khrapko, K., Hanekamp, J. S., Thilly, W. G., Belenkii, A., Foret, F., and Karger, B. L. (1994) Constant denaturant capillary electrophoresis (CDCE): a high resolution approach to mutational analysis. *Nucleic Acids Res.* **22,** 364–369.

2. Khrapko, K., Coller, H. A., Li-Sucholeiki, X.-C., André, P. C., and Thilly, W. G. (2001) High resolution analysis of point mutations by constant denaturant capillary electrophoresis (CDCE), in *Capillary Electrophoresis of Nucleic Acids*, Vol. 2 (Mitchelson, K. R., and Cheng, J., eds.), Humana Press, Totowa, NJ, pp. 57–72.

3. Khrapko, K., Coller, H. A., André, P. C., Li, X.-C., Hanekamp, J. S., and Thilly, W. G. (1997) Mitochondrial mutational spectra in human cells and tissues. *Proc. Natl. Acad. Sci. USA* **94,** 13,798–13,803.

4. Coller, H. A., Khrapko, K., Torres, A., Frampton, M. W., Utell, M. J., and Thilly, W. G. (1998) Mutational spectra of a 100-base pair mitochondrial DNA target sequence in bronchial epithelial cells: a comparison of smoking and nonsmoking twins. *Cancer Res.* **58,** 1268–1277.

5. Marcelino, L. A., André, P. C., Khrapko, K., Coller, H. A., Griffith, J., and Thilly, W. G. (1998) Chemically induced mutations in mitochondrial DNA of human cells: mutational spectrum of *N*-Methyl-*N'*-nitro-*N*-nitrosoguanidine. *Cancer Res.* **58,** 2,857–2,862.

6. Li-Sucholeiki, X.-C. and Thilly, W. G. (2000) A sensitive scanning technology for low frequency nuclear point mutations in human genomic DNA. *Nucleic Acids Res.* **28,** e44.

7. Li-Sucholeiki, X.-C., Khrapko, K., André, P. C., Marcelino, L. A., Karger, B. L., and Thilly, W. G. (1999) Applications of constant denaturant capillary electrophoresis/high-fidelity polymerase chain reaction to human genetic analysis. *Electrophoresis* **20,** 1224–1232.

8. Muniappan, B. P. and Thilly, W. G. (1999) Applications of constant denaturant capillary electrophoresis (CDCE) to mutation detection in humans. *Genetic Anal.* **14,** 221–227.

9. Bjørheim, J., Lystad, S., Lindblom, A., Kressner, U., Westring, S., Wahlberg, S., et al. (1998) Mutation analysis of *KRAS* exon 1 comparing three different techniques: temporal temperature gradient electrophoresis, constant denaturant capillary electrophoresis and allele specific polymerase chain reaction. *Mutation Res.* **403,** 103–112.

10. Tomita-Mitchell, A., Muniappan, B. P., Herrero-Jimenez, P., Zarbl, H., and Thilly, W. G. (1998) Single nucleotide polymorphism spectra in newborns and centenarians: identification of genes coding for rise of mortal disease. *Gene* **223,** 381–391.

11. Cline, J., Braman, J. C., and Hogrefe, H. H. (1996) PCR fidelity of *Pfu* DNA polymerase and other thermostable DNA polymerases. *Nucleic Acids Res.* **24,** 3,546–3,551.

12. André, P., Kim, A., Khrapko, K., and Thilly, W. G. (1997) Fidelity and mutational spectrum of *Pfu* DNA polymerase on a human mitochondrial DNA sequence. *Genome Res.* **7,** 843–852.

13. Khrapko, K., Coller, H., André, P., Li, X.-C., Foret, F., Belenky, A., Karger, B. L., and Thilly, W. G. (1997) Mutational spectrometry without phenotypic selection : human mitochondrial DNA. *Nucleic Acids Res.* **25,** 685–693.

14. Sheffield, V. C., Cox, D. R., Lerman, L. S., and Myers R. M. (1989) Attachment of a 40-base-pair G+C rich sequence (GC-clamp) to genomic DNA fragments by the polymerase chain reaction results in improved detection of single-base changes. *Proc. Natl. Acad. Sci. USA* **86,** 232–236.

15. Li, X.-C. and Thilly, W. G. (1996) Use of wide-bore capillaries in constant denaturant capillary electrophoresis. *Electrophoresis* **17,** 1884–1889.

16. Keohavong, P., Liu, V. F., and Thilly, W. G. (1991) Analysis of point mutations induced by ultraviolet light in human cells. *Mutation Res.* **249,** 147–159.
17. Khrapko, K., Coller, H., and Thilly, W. G. (1996) Efficiency of separation of DNA mutations by constant denaturant capillary electrophoresis is controlled by the kinetics of DNA melting equilibrium. *Electrophoresis* **17,** 1867–1874.

15

Toward Effective PCR-Based Amplification of DNA on Microfabricated Chips

Jerome P. Ferrance, Braden Giordano, and James P. Landers

1. Introduction

The polymerase chain reaction (PCR) has rapidly become the most valuable tool in the clinical diagnostic arsenal for determining specific diseases or detecting infectious agents. Primers, short pieces of DNA complementary to the DNA sequence of interest, are mixed with nucleotides, a small amount of template DNA from the sample of interest, and *Taq* DNA polymerase enzyme in the appropriate buffer. Using temperature cycling, a short piece of DNA (50–1000 bp in length), defined by the primers chosen, is rapidly amplified from the few initial template molecules added to the mixture. The amplification product is then analyzed using an electrophoretic separation. The entire process, shown in **Fig. 1**, includes the "sample preparation" which typically involves isolating the appropriate cells, from which the DNA is extracted prior to PCR amplification. To utilize this sequence most efficiently there should be a continuous flow from sample collection to diagnosis. This would eliminate both delays due to the transfer of material between each step and the need for intervention before the next step could begin. In the ideal world, this diagnosis would be immediate—the real goal is to decrease the time to as short as possible. Towards this end, the integration of the steps detailed in **Fig. 1** into a single platform is of obvious and critical importance (*see* **Note 1**).

1.1. Microchannels

As indicated by **Fig. 1**, decreasing the time required for electrophoretic interrogation of the PCR product results in a significant reduction in total analysis time. A reduction in separation time has been accomplished with the use of capillary electrophoresis (CE) which has also reduced the amount of sample and volume of reagents needed for a separation. Further decreases in time have resulted from the successful transfer of electrophoretic separations from the capillary to glass microchips where micron-scale troughs etched into the surface of the substrate effectively supplant the

From: *Methods in Molecular Biology, Vol. 163:*
Capillary Electrophoresis of Nucleic Acids, Vol. 2: Practical Applications of Capillary Electrophoresis
Edited by: K. R. Mitchelson and J. Cheng © Humana Press Inc., Totowa, NJ

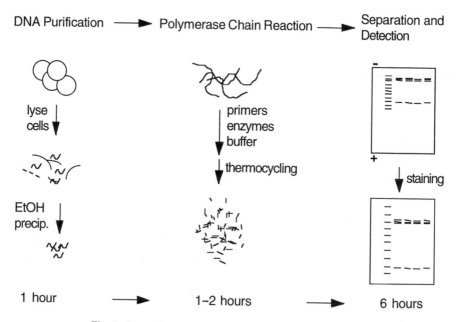

DNA Purification ⟶ Polymerase Chain Reaction ⟶ Separation and Detection

lyse cells

primers
enzymes
buffer

thermocycling

EtOH precip.

staining

1 hour ⟶ 1–2 hours ⟶ 6 hours

Fig. 1. Overall scheme for clinical and genetic analyses.

need for fused silica capillaries. Microchannels are formed when the substrate containing the etched structures, created using basic microfabrication techniques, is bonded to a second glass slide. PCR products and DNA restriction digest fragments are easily separated in microchannels coated with substances that reduce the electroosmotic flow, and filled with a sieving matrix. Separation channels 3.5 cm in length have been shown to be sufficient to identify standard fragments in the range from 70–1000 bp in 120 s (1). These devices have also been designed to allow multiple separations to be carried out in an array of channels fabricated on the same device (2).

The same advantages of reduced time, sample, and reagents brought to the separations field by miniaturization also apply to low volume PCR in capillaries. Microchip formats have also been developed for PCR where the reactions are carried out in reservoirs or microreaction chambers formed in glass, silicon, or plastic microchips. In addition, decreasing the scale of PCR allows the reaction to be carried out more efficiently, producing more product in less time with less side reactions. Both capillaries and microchip devices have reduced the time needed for PCR but cycle times concomitant with the fast separations now possible are still being developed.

Although capillaries have proven useful for small scale PCR, true integration of the PCR and fast separation steps will require a microchip device where continuous flow of the PCR products to the separation channels is achievable on a single coordinated platform (see Note 2). With this type of platform, integration not only of the last two steps shown in Fig. 1 is possible, but total integration of the complete process. Development of integrated electrophoretic microchips capable of sorting cells, DNA extrac-

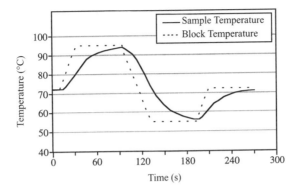

Fig. 2. A single temperature cycle for a typical PCR thermocycler showing melt, annealing, and extension temperatures. The actual temperatures experienced by the PCR solution are also included, and show the thermal lag due to heat transfer through the polypropylene tubes.

tion, PCR and electrophoresis will eliminate the need for much of the hardware associated with these methods, allowing the eventual design of portable point-of-service instruments which execute analysis on these microchips.

The ability to execute fast (second time-scale) electrophoretic separations of DNA on microchips creates a significant time discontinuity between PCR reactions, which may take a few hours, and the electrophoretic separation. Although microchip-based PCR has been reported in the literature *(3–5)*, there still exists a gap between the rapid electrophoretic separations of DNA on microchips and the ability to carry out fast and effective PCR on a microchip in a practical, cost-effective manner. This chapter details past and current work toward reducing the scale and, thus, the reaction duration of PCR methods. Efforts toward the wedding of both PCR reactions and electrophoretic separation and detection processes on a single device are also underway. The results of these initial attempts are presented along with a discussion of the barriers which will have to be circumvented or surmounted in order for an integrated technology to be realized (*see* **Note 3**).

1.2. PCR in Small Volumes

PCR in reaction volumes of 25–50 µL is routinely carried out in thermocyclers that hold small polypropylene tubes. The typical PCR process requires that the sample be cycled through three temperatures as shown in **Fig. 2**, where the denaturation (melt), annealing, and extension temperatures are reached and maintained for the requisite times. At the denaturation temperature (~95°C), the complementary strands of DNA are separated without significantly affecting the activity of the *Taq* DNA polymerase, which is extraordinarily heat-stable. Reducing the temperature to within the range 48–74°C allows the primers to anneal with the template. The optimum annealing temperature is determined empirically by the length and sequence of the primers being used. The temperature is then increased to the optimum temperature for the enzyme to extend the primers. The reaction times that are typically in the range of 5–10 s for the denaturation and annealing steps, and 1–4 min for the extension step are determined more by the apparatus than by the biochemistry of the process. The volume of the solution and the transfer of heat through the polypropylene tubes containing the solution being the factors limiting the rates of reaction. **Figure 2** also displays the actual temperature cycle experienced by the reaction solution during a normal cycle. Note how the sample temperature lags behind that of the block with the temperature transitioning less sharp than the heating block temperature.

Wittwer et al. *(6)* have shown that by increasing the surface area and heat transfer, through the use of glass capillaries as their reaction container, a better match between the internal and external temperature cycles is provided. This allows the melt and annealing times to be decreased to less than 1 s, with elongation times also reduced to 10–20 s. The volume of reaction in these capillaries is ~10 μL, and the capillaries are heated in an air thermocycler, which is essentially a closed chamber through which forced air at different temperatures circulates. This is similar to conventional thermocycling in that the capillary environment is heated eventually heating the internal solution.

Friedman and Meldrum *(7)* also took advantage of the ratio of low volume to high surface area provided by glass capillaries in designing their microscale PCR system. Glass capillaries are coated with a transparent layer of indium-tin oxide that is used for heating the capillaries to the desired cycle temperatures. Because the coating is transparent, reactions could be monitored using fluorescence to determine when sufficient cycles had been performed to amplify the DNA sequence of interest to the desired extent. The thin film coating also acts as a temperature sensor for the system. This combination provided the opportunity to individually control the temperatures and number of reaction cycles for up to 32 capillaries. Reference samples were prepared in similar capillaries without the thin-film coating; these capillaries were thermally cycled inside a commercial air thermal cycler. Samples as small as 4 μL were successfully reacted with hold times of >1 s at 93°C, >1 s at 55°C, and 20 s at 72°C.

Oda et al. *(8)* used 500-μm rectangular glass tubes sealed on one end as microchambers for a novel approach to PCR amplification of DNA. With the premise that a noncontact method could be more easily extrapolated to the microchip format, they investigated noncontact heating using a tungsten lamp as an infrared radiation (IR) source. Wittwer et al. *(6)* used a lamp to heat the air, which then heated the whole PCR environment. The approach by Oda et al. *(8)* sought to directly heat the reaction solution itself via the red part of the spectrum from the tungsten lamp, which is absorbed maximally by water. Since the reaction solution was not heated by contact with the reaction chamber itself, the total mass of material that had to reach the desired temperatures was reduced, allowing faster slew rates than air cycling. They demonstrated that IR-mediated thermocycling of 5–15 μL reaction volumes was possible, and that as reported by others *(4)*, the amplification is somewhat less efficient than conventional thermocycling but that ample product is generated for detection by CE. As shown in **Fig. 3**, the total time required for each reaction cycle could easily be reduced to 17 s, with individual steps of 2 s at 94°C, 2 s at 54°C, and 4 s at 72°C.

The translation of PCR to microchip devices was first demonstrated by Wilding et al. *(3)* who used silicon/glass microchamber structures that hold 5–10 μL of reaction

Temperature Cycling

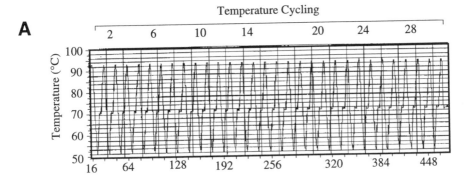

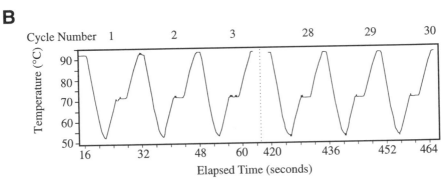

Elapsed Time (seconds)

Fig. 3. Temperature cycling with the noncontact heat source. (**A**) The reproducibility of the cycles is very good. (**B**) Over the 30 cycles, 464 s, the 17-s cycling profile is unchanged. Reprinted in part with permission from **ref. *13*,** Copyright (1998) American Chemical Society.

mixture. Although the silicon provides rapid heat transfer because of its good thermal properties, each PCR cycle is still ~3 min. Heating of the microchip is carried out by placing it in contact with a copper block attached to a Peltier temperature controller; the added mass of the heater is largely responsible for the slower cycle times. The products were analyzed on an agarose gel, and the results compared well with control reactions carried out in a normal thermocycler. Further work on these microchips investigated surface treatments of the silicon substrate to prevent protein binding and enzyme inhibition *(9,10)*; silicon dioxide-coated surfaces were found to work best in this design. Conditions for PCR in these microchambers were: 28 cycles of 15 s at 94°C, 1 min at 55°C, and 1 min at 72°C, followed by a final extension at 72°C for 10 min.

Wooley et al. *(4)* describe a different type of PCR device, referred to as a "miniature analytical thermal cycling instrument (MATCI)." This consists of a small polysilicon block heater with a chamber for holding a polypropylene insert. The volumes of the reactions are still above 10 µL, but reaction times are decreased to less than 30 s per cycle. These rapid PCR experiments consisted of 30 cycles of 96°C for 2 s, 55°C for 5 s, and 72°C for 2 s. This device was also employed in studies by

Ibrahim et al. *(11)* to measure the limits of detectable DNA. With 60 thermal cycles, initial DNA template numbers of as low as 1900 copies were successfully detected.

A different approach to the problem was taken by Kopp et al. *(5)*. Instead of having a reaction chamber, the reaction was carried out in a continuous trough etched into a glass microchip which was sealed to a second glass microchip. The microchip was placed on a heater which had different zones held at constant temperatures of 95°C, 77°C, and 60°C. The number of cycles is set by the etch pattern in the microchip and the speed of the cycle is set by the flow rate through the channel, which is controlled by the pressure applied to the inlet channel. Total reaction times of 1.5–18.7 min were used with the 20 cycles designed into their original microchip. Reaction plugs of 10 µL were used in the system to collect enough material for analysis of the products on an agarose gel. A positive feature of this system is the possibility to perform continuous samples in this device, by injecting different reaction plugs separated by small void volumes. The rapid reaction rates are possible because only the temperature of the solution has to be changed, not of the entire system. However, that a defined number of cycles is predesigned into the microchip is a major downfall as flexibility is an advantage in new reactor designs.

1.3. Submicroliter PCR

Continued improvements in new and traditional thermocyclers have decreased cycle times to less than 1 min, and reaction volumes have been reduced to the 5 µL range. However, at the same time improvements in the microchip/capillary devices have further decreased the reaction volume. For a microchip capable of integrated PCR/electrophoretic separation, the total volume of PCR product needed is minimal. In a typical microchip used for electrophoretic separations, it is possible to have volumes less than 1 nL in the injection pathway. Therefore, the total volume of the PCR reaction can be decreased significantly if it is to be directly separated in a connected channel on the same device.

Kalinina et al. *(12)* have reduced the volume of reaction by using capillaries as small as 20 µm in diameter to hold a total volume of 10 nL, allowing single copies of template DNA to be amplified and detected after 30 cycles using fluorescent energy transfer (FET). One drawback is that the small volumes and design of the system prevent further analysis of products by gel or CE to confirm the replication of the correct sized fragment. Heating in this system is carried out in an air thermal cycler, with cycles taking about 45 s.

The system of Oda et al. *(8)* has been further modified by Hühmer and Landers *(13)* to carry out reactions in 150-µm id capillaries, which hold volumes as low as 100 nL. Starting with less than 600 copies of template DNA, successful amplification of detectable product can be achieved after only 10 cycles. **Figure 4** compares the amount of PCR product after 10 cycles in this system with product generated in a traditional thermocycler. Each cycle comprises only two temperature steps for denaturation and annealing, with the extension stage taking place as the temperature ramped from the low annealing temperature (68°C) to the high denaturation temperature (94°C). The submicroliter volumes and noncontact, direct heating of the PCR solution by the IR radiation allowed cycle times to be reduced to less than 3 s in this system, as illustrated in **Fig. 5**.

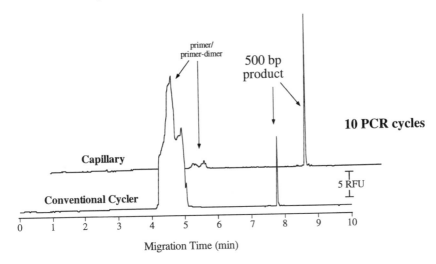

Fig. 4. A demonstration of the effectiveness of a small scale PCR amplification reaction in a capillary. In 10 PCR cycles, the concentration of primer-dimer is several orders of magnitude lower, while the signal for PCR product is significantly higher.

1.4. Integrated PCR-Electrophoretic Systems

Although smaller volume PCR has been achieved in capillaries, these approaches are not readily integrated into an electrophoretic microchip device. **Figure 6** shows the design of Wooley et al. *(4)* who demonstrated the feasibility of interfacing PCR with microchip-based electrophoresis by epoxy-bonding a polysilicon block heater to the surface of an electrophoretic microchip, and then interfacing the polypropylene insert directly with the sample channel reservoir. Once PCR is complete, the product is injected directly into the separation channel. From a practical standpoint, these chips are not easily produced because each heating block has to be individually produced and cemented to a chip. Also, the amount of PCR product generated in this system is several orders of magnitude greater than the amount needed for the separation step. Further, the integration of the steps occurring prior to the PCR is preempted by this design of a nonintegral PCR chamber.

Waters et al. *(14)* use a slightly less complicated approach, gluing glass reservoirs, which served as reaction chambers, directly onto the surface of the electrophoretic microchips above the sample channel inlets, with thermal cycling carried out by placing the entire microchip into a thermocycler. This device also encounters some of the practical problems evident with the device of Wooley et al. *(4)*. In addition, the need to place the entire microchip into a thermocycler places some additional restrictions on the use of surface coatings for the separation channel.

The current state-of-the-art for integrated PCR-electrophoretic devices has not progressed significantly beyond a point which can be described as having separate structures for the two halves of the problem interfaced in the crudest of ways. PCR amplification is actually carried out in a reservoir outside the microchip itself. An

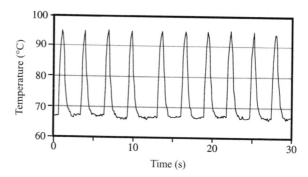

Fig. 5. Cycle times of less than 3 s were possible for two-temperature cycling in a capillary using a noncontact IR source as the heat source.

integrated PCR-electrophoretic device requires that the reaction chambers be designed and microfabricated directly into the electrophoretic microchip. The reaction chambers reported by Wilding et al. (3) are a definite step forward, but suffer from several problems. The chambers have large volumes (10 μL) and are formed of silicon wafers, which are not ideal substrates for electrophoresis because of the semiconductor nature of silicon. However, the successful construction of a device capable of sequential protein separation and labeling reactions within the same device, illustrates that the integrated concept can be reduced to practice (15). With the ability to microfabricate any number of structures into a device, the development of microchips containing the appropriately configured reaction chambers connected to separation channels is not really the limiting factor. The challenge lies in the need for accurate temperature cycling of small PCR mixture volumes in a manner that is simple, cost-effective, efficient and can be easily incorporated into instrumentation. Although such a device is dramatically more complicated than the device described for integrated protein analysis (15), the ability to accurately localize and control the temperature of nanoliter scale volumes on a microchip will empower this technology (see Note 4).

1.5. Criteria for Novel Approaches to PCR in the Microchip Format

There are a number of issues that need to be resolved in order to achieve both fully functional and fully-integrated microchips with real world applications. As discussed in **Subheading 1.4.**, the monitoring of PCR products as they are cycling has already been carried out for both μL and nL volumes. This provides the opportunity to stop any reaction at an appropriate time and to begin the analysis of the PCR products. Individual control of PCR reactions in this manner is achievable in the capillary setup developed by Friedman and Meldrum (7), where the fluorescence in each capillary can be measured independently. More importantly, the individual control of heating each capillary in this system allows different reactions performed under different conditions to be run and analyzed simultaneously. Two major drawbacks to this design however, are the expense of the individual film heaters on the capillaries and the inability to easily interface the capillaries with a microchip based separation device.

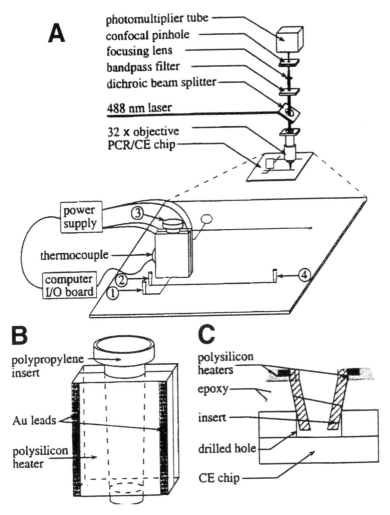

Fig. 6. The PCR-electrophoresis chip design used by Wooley et al. *(4)*. (**A**) The confocal microscope detection system and the arrangement of the PCR chamber and separation channel on the microchip. (**B**) The PCR heater design. (**C**) The chamber-chip interface. Reprinted with permission from **ref.** *4*, Copyright (1996) American Chemical Society.

In order for integrated PCR-electrophoretic microchip devices to fully realize their potential, the capability for asymmetric heating at different locations on the microchip will have to be achieved. Similar to the capillary approach, individual heaters for each region could conceivably be fashioned on the microchip. However, the approaches used for heating described thus far in the literature would not likely be acceptable because they either add thermal mass, which increases the cycle time, or they heat the entire microchip. The one exception is the noncontact heating approach where control

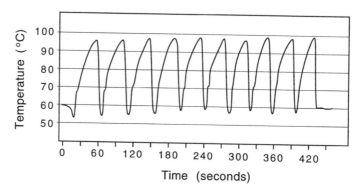

Fig. 7. Temperature cycling on a glass/PDMS microchip using a noncontact IR source. Heating rates are not as fast as those achieved in capillaries, resulting in longer cycle times.

of the radiation to distinct reaction volumes on the microchip might be achievable using fiber optic systems. The noncontact method also allows the much faster cycling times needed to match the separation times achievable on a microchip. A microchip method for a 20-long thermal cycle at 2 temperatures is shown in **Fig. 7.**

The noncontact approach to thermocycling, although ideal for the microchip, is burdened by the need for accurate thermosensing. This is particularly important with PCR where temperature is critical for amplification of the correct product. Direct measurement using a thermocouple inserted into the reaction chamber or microfabricated onto the microchip is not feasible, because the metals interfere with PCR *(8)*. However, the use of a dummy cell containing PCR buffer that is cycled along with the actual reaction chamber is possible but is cumbersome, and becomes more so as the number of reactions on a single microchip expands. Surface measurement using deposited metals is also possible, but this would again add thermal mass to the system and could increase the cycling time. There are a number of ways to monitor the temperature of a surface remotely, but the temperature of the PCR solution, not of the glass surface, is the critical parameter which must be accurately controlled. Measurement of the actual solution temperature in a noncontact manner may be achieved using a variety of methods including the use of an IR pyrometer or changes in refractive index. This latter approach has already been shown effective by Tarigan et al. *(16)* for detecting temperature changes in capillary tubes. The application of other temperature dependent properties, such as changes in light absorbance or fluorescence by including appropriate compounds in the PCR mixture can also be exploited for temperature measurement.

1.6. Capillary Analysis of Microchip PCR Products

Although it is beyond the scope of this chapter, the purification of the PCR product before separation is also an issue in these microsystems *(10)*. The high salt content of the PCR solutions may interfere with electrokinetic injection of the DNA into the separation column. Whereas pressure injection is a solution, this type of injection is much less desirable because it is harder to control. Additionally, when whole cells are used as the PCR template, some of the cellular components could interfere with the electrophoretic separations. Traditional methods used for "cleaning" PCR products involve membrane or chromatographic methods. In a chip design, it is possible to fill a connecting channel with a DNA binding matrix that could be used for purification. Another recently tested approach is to prepare a porous-membrane type of structure in the glass or silicon during microfabricated device. Khandurina et al. *(17)* have been successful in concentrating DNA before electrophoretic separation using this method.

2. Practical Applications for Microchip Devices

1. One of the well-defined applications of microchip technology will be in the diagnostic arena, with the possibility of incorporating PCR in the same device used for electrophoretic interrogation of the product providing some exciting possibilities (*see* **Note 4**).
2. These not only include PCR fragment detection, where PCR and electrophoresis are carried out sequentially in a rapid manner, but also cycle sequencing reactions which could conceivably be directly linked to microchip-based sequencing which has begun to mature *(18)*.
3. Bacterial infections as a cause of illnesses were originally detected by taking a culture and allowing it to grow for a few days to determine what bacteria were present. This has advanced to taking a culture, carrying out PCR of DNA fragments specific to the infectious agent, followed by gel electrophoresis to confirm the presence of the infectious agent. Total time for diagnosis is now typically 1 d or less once the culture reaches a clinical laboratory. The integrated PCR-electrophoresis microchip could reduce the total analysis time to several minutes, and could be performed directly at the site of patient care.
4. Viral infections would be even more important since they are only truly identifiable using PCR. The rapid diagnosis of herpes simplex virus (HSV), which causes encephalitis, would allow treatment to begin immediately, greatly affecting patient morbidity and mortality. Hofgäertner et al. *(19)* have compared the sensitivity of microchip PCR electrophoresis with detection of HSV-specific PCR products by slab-gel analysis for rapid detection of herpes simplex encephalitis. **Figure 8** shows an example of detection of an HSV-specific fragment at the femtomole level in under 90 s.
5. An extensive study with more than 200 samples of HSV infected tissues showed that the microchip approach compared well with the standard clinical assays *(19)*, which typically require >24 h for detection of the HSV. It is clear that uniting of rapid microchip-based detection with a rapid on-chip amplification of the fragment would present a new paradigm in clinical diagnostics.
6. Only a fraction of the molecular diagnostics arena involves the detection of the exogenous DNA or RNA associated with infectious agents. A larger interest in this field involves detecting mutations (gene rearrangements, base insertions, and deletions) associated with certain inherited diseases and cancer. Recently, Munro et al. *(20)* have shown that, for diagnosis of T-cell lymphoma, the information obtained from slab-gel analysis of the PCR-amplified product could, in fact, be extracted from capillary electrophoresis in a shorter time and from microchips in several seconds. Although more rapid diagnosis will inevitably improve the early detection of disease and increase patient survival rates, the opportunity to evaluate surgical tissue on-site would greatly aid surgeons in ensuring that all malignant tissue has been excised.
7. With respect to routine genotyping, the rapid screening of patients for genetic disorders beyond cancer will also be readily achievable by the medical community. This includes screening for predisposition to disease with a heritable genetic component, such as heart

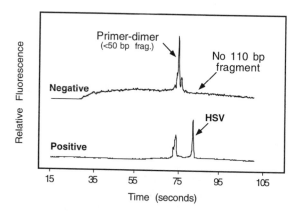

Fig. 8. Determination of positive and negative signals in less than 90 s using microchip electrophoresis of the PCR fragments from samples of HSV.

disease, obesity, and even to sensitivity to drugs or allergens—tests not carried out currently because of the laborious nature of the methodology and the associated cost.
8. Those patients with family histories, or genetically predisposed to certain conditions would benefit from such information, especially if their condition can be controlled or stabilized by diet or medication before more serious measures are required.

3. Notes

1. PCR-based DNA amplification followed by electrophoretic separation is an important method, used in a number of clinical applications. Improvements in this method, to make it faster and less costly, would expand the range of situations in which it can be applied. Significant progress on the electrophoretic step has decreased the time and expense of these separations by developing microchip-based devices. Advances in reducing the scale of PCR, by adapting this method to both capillary and microchip-based devices, has also taken place. Further advances in PCR are needed to make them compatible with the speed and volume of the electrophoresis, thus allowing a continuous flow from the amplification to the separation step. Development of a single integrated microchip device, on which both steps are carried out, will allow clinical PCR to reach its full potential.
2. The concept of an integrated PCR/electrophoretic separation chip capable of amplifying and analyzing specific DNA fragments in a very short period of time is clearly viewed as a worthwhile goal. Such microchips would have applications in a number of clinical, bioanalytical, and research areas. The concept is easy to imagine, but creation of such a device is still years away.
3. To a large extent, the science to achieve microchip-based analysis has been completed. Research has shown that PCR can be carried out in extremely small volumes and still produce the desired product. Chambers of both glass and silicon are amenable for these reactions, so that microfabrication of such devices is realistic and cost-effective (*10*). The electrophoretic separation of analytes on glass chips has also been demonstrated, and is now becoming routine procedure in the laboratory. Short cycle times for PCR have been achieved, as well as some methods for preparing the DNA for separation. Some approaches, such as monitoring the extent of PCR and temperature sensing, have been

developed and applied, but more efficient methods are still being investigated for inclusion in the next generation of devices.

4. The challenges that remain are largely in the engineering realm for the development of suitable equipment to use these microchip-based techniques. Optimized designs for each individual step have to be determined along with the best way to integrate one step into the next. Of major importance is the heating and temperature monitoring of the PCR amplification step, which is very temperature sensitive. Practical matters such as costs of the instrumentation and the microchips, size or scale of the eventual applications, time involved in pre- and postanalysis processes, number of samples or reactions to be performed simultaneously, and so on, will have to be taken into account in the development of final microchip designs. However, these issues will be resolved, allowing the development of a fully integrated microchip-based technology to begin the paradigm shift inevitable in the clinical, biomedical, and bioanalysis arenas.

References

1. Woolley, A. T. and Mathies, R. A. (1994) Ultra-high-speed DNA fragment separations using microfabricated capillary array electrophoresis chips. *Proc. Natl. Acad. Sci. USA* **91,** 11,348–11,352.
2. Freemantle, M. (1999) Downsizing chemistry. *C & E News* **77,** 27–36.
3. Wilding, P., Shoffner, M. A., and Kricka, L. J. (1994) PCR in a silicon microstructure. *Clin. Chem.* **40,** 1815–1818.
4. Woolley, A. T., Hadley, D., Landre, P., deMello, A. J., Mathies, R. A., and Northrup, M. A. (1996) Functional integration of PCR amplification and capillary electrophoresis in a microfabricated DNA analysis device. *Anal. Chem.* **68,** 4081–4086.
5. Kopp, M. T., deMello, A. J., and Manz, A. (1998) Chemical amplification: continuous-flow PCR on a chip. *Science* **280,** 1046–1048.
6. Wittwer, C. T., Reed, G. B., and Ririe, K. M. (1994) Rapid cycle DNA amplification, in *The Polymerase Chain Reaction* (Mullis, K. B., Ferré, F., and Gibbs R. A., eds.), Birkhauser, Boston, pp. 175–181.
7. Friedman, N. A. and Meldrum, D. R. (1998) Capillary tube resistive thermal cycling. *Anal. Chem.* **70,** 2997–3002.
8. Oda, R. P., Strausbauch, M. A., Huhmer, A. F. R., Borson, N., Jurrens, S. R., Craighead, J., et al. (1998) Infrared-mediated thermocycling for ultrafast polymerase chain reaction amplification of DNA. *Anal. Chem.* **70,** 4361–4368.
9. Shoffner, M. A., Cheng, J., Hvichia, G. E., Kricka, J. J., and Wilding, P. (1996) Chip PCR. I. Surface passivation of microfabricated silicon-glass chips for PCR. *Nucleic Acids Res.* **24,** 375–379.
10. Xie, W., Yang, R., Xu, J., Zhang, L., Xing, W., and Cheng, J. (2001) Microchip-based capillary electrophoresis systems, in *Capillary Electrophoresis of Nucleic Acids*, Vol. 1 (Mitchelson, K. R., and Cheng, J., eds.), Humana Press, Totowa, NJ, pp. 67–83.
11. Ibrahim, M. S., Lofts, R. S., Jahrling, P. B., Henchal, E. A., Weedn, V.W., Northrup, M. A., and Belgrader, P. (1998) Real-time microchip PCR for detecting single-base differences in viral and human DNA. *Anal. Chem.* **70,** 2013–2017.
12. Kalinina, O., Lebedeva, I., Brown, J., and Silver, J. (1997) Nanoliter scale PCR with TaqMan detection. *Nucleic Acids Res.* **25,** 1999–2004.
13. Oda, R. P, Strausbauch, M. A., Hühmer, A. F. R., Borson, N., Jurrens, S. R., Craighead, J., et al. (1998) Infrared-mediated thermocycling for ultrafast polymerase chain reaction amplification of DNA. *Anal. Chem.* **70,** 4361–4368.

14. Waters, L. C., Jacobson, S. C., Kroutchinina, N., Khandurina, J., Foote, R. S., and Ramsey, J. M. (1998) Multiple sample PCR amplification and electrophoretic analysis on a microchip. *Anal. Chem.* **70,** 5172–5176.

15. Harrison, D. J., Fluri, K., Chiem, N., Tang, T., and Fan, Z. (1996) Micromachining chemical and biochemical analysis and reaction systems on glass substrates. *Sensors and Actuators B.* **33,** 105–109.

16. Tarigan, H. J., Neill, P., Kenmore, C. K., and Bornhop, D. J. (1996) Capillary-scale refractive index detection by interferometric backscatter. *Anal. Chem.* **68,** 1762–1770.

17. Khandurina, J., Jacobson, S. C., Waters, L. C., Foote, R. S., Ramsey, J. M. (1999) Microfabricated porous membrane structure for sample concentration and electrophoretic analysis. *Anal. Chem.* **71,** 1815–1819.

18. Schmalzing, D., Adourian, A., Koutny, L., Ziaugra, L., Matsudaira, P., and Ehrlich, D. (1998) DNA sequencing on microfabricated electrophoretic devices. *Anal Chem.* **70,** 2303–2310.

19. Hofgaertner, W., Hühmer, A. F. R., Landers J. P., and Kant, J. (1999) Rapid diagnosis of HSV-induced encephalitis: Electrophoresis of PCR products on microchips. *Clin. Chem.* **45,** 2132–2140.

20. Munro, N. J., Snow, K., Kant, J., and Landers, J. P. (1999) Molecular diagnostics on microfabricated electrophoretic devices: Translating slab gel-based T- and B-cell lymphomoproliferative disorder assays from the capillary to the microchip. *Clin. Chem.* **45,** 1906–1917.

16

Low-Stringency Single Specific Primer-Amplified DNA Analysis by Capillary Electrophoresis

Michael A. Marino

1. Introduction

Polymerase chain reaction (PCR) amplification is a relatively fast, sensitive method for characterizing discrete segments of the human genome for identity testing. Most PCR-based typing assays that discriminate single nucleotide polymorphisms (SNPs) require further manipulation of the amplification product to identify the polymorphic sites. A more simplified assay, PCR products would be directly analyzed to determine whether DNA extracted from two different sources was a match or a mismatch. Examples of direct PCR assays to distinguish SNPs are allele-specific PCR *(1)*, arbitrary primed PCR (AP-PCR) *(2)*, DNA amplification fingerprinting (DAF) *(3)*, and randomly amplified polymorphic DNA (RAPD) *(4)*. AP-PCR, DAF, and RAPD involve random amplification of genomic DNA using low stringency primer annealing to generate a DNA banding pattern or profile. Sequence differences are detected without identifying the specific variations. However, the concentration of the DNA and the degree of DNA degradation can alter the profile, and the application of these techniques for human identification has not been adequately demonstrated.

Low stringency single-specific primer PCR (LSSP-PCR) *(5)* is a method that utilizes low stringency conditions for primer annealing during PCR. However, unlike the above methods, a single, discrete PCR product serves as the template for the low stringency amplification. This discrete PCR product is generated with a specific pair of primers that are annealed using high stringency conditions. The product is then amplified using one of the specific primers annealed under low stringency conditions to generate a DNA profile. LSSP-PCR has been successfully applied to the characterization of the SNP-rich human mitochondrial DNA (mtDNA) D-loop region for human identity testing *(6)*. LSSP-PCR eliminates the problem of irreproducible banding patterns that are encountered when performing the low stringency amplification directly with genomic DNA.

From: *Methods in Molecular Biology, Vol. 163:*
Capillary Electrophoresis of Nucleic Acids, Vol. 2: Practical Applications of Capillary Electrophoresis
Edited by: K. R. Mitchelson and J. Cheng © Humana Press Inc., Totowa, NJ

In the following discussion, the model system for LSSP-PCR using capillary electrophoresis (CE) is the mitochondrial DNA (mtDNA) hypervariable (HV) region *(6)*. The DNA from individuals was amplified (first step) using sequence-specific primers to produce 1021-bp fragments containing the D-loop region. Each fragment was isolated by electroelution using CE and UV detection, and subjected to a second amplification (second step) using a single primer annealed under low stringency conditions. This generated a range or profile of PCR products for each sample, which were resolved and analyzed by CE with laser-induced fluorescence (LIF) detection. Low stringency PCR methods use a single primer that is annealed at a temperature much lower than its melting temperature, resulting in the formation primer-template mismatches. Thus, the primer anneals to both DNA strands at many locations, producing a complex set of amplification products *(5)*.

2. Materials

2.1. PCR Amplification (First Step)

1. The PCR mixture: 29 µL of deionized water, 5 µL of 10X buffer, 5 µL of genomic DNA (10–20 ng), and 1 µL Ampli*Taq* polymerase (Perkin-Elmer, Foster City, CA). (*See* **Note 1**).
2. Add 5 µL of the primer pair (5 µ*M*) to the first step amplification mixture. The primers H408 (5'-CTGTTAAAAGTGCATACCGCCA-3') and L15996 (5'-CTCCACCATTA GCACCCAAAGC-3'), generate a 1021-bp region of mtDNA which includes the D-loop.
3. Thermal cycling parameters: 94°C for 5 min; 35 cycles of 94°C for 45 s, and 55°C for 60 s; 74°C for 60 s and hold at 4°C.

2.2. PCR Amplification (Second Step)

1. The second PCR mixture: 18.4 µL deionized water, 2.5 µL of 10X buffer, 20 µL of primer L15996 (12 µ*M*), 5 µL of captured 1021-bp product, and 1.6 µL Ampli*Taq* DNA polymerase.
2. Thermal cycling parameters: 94°C for 5 min; and 35 cycles of 94°C for 60 s and 30°C for 60 s using the Perkin-Elmer 9600 thermal cycler.

2.3. CE System

1. Perform all CE at constant reverse polarity voltage of 238 V/cm (approx –8 µAmps).
2. Prepare the 3.5% (w/v) polymer by adding 2.0 mL of Fragment Analysis Reagent (Applied Biosystems, Foster City, CA) to 0.4 mL of DNA Fragment Analysis Buffer (Applied Biosystems) and 1.6 mL of sterile deionized water.
3. For LIF detection, add 0.5 µL TOTO-1 (Molecular Probes, Eugene, OR) to 2.0 mL of polymer. Prepare the polymer daily and filter the solution before use with a 0.45-µm filter (Gelman Sciences, Ann Arbor, MI).
4. Capillaries consisting of fused silica with polyamide coating (Polymicro Technologies, Inc., Phoenix, AZ) with id of 100 and 50 µm are used. The total lengths will be determined by the CE system. A small section (2–4 mm) of the polyamide coating is burned off to form detection windows located approx 40 cm from the injection end of the capillary (effective length).
5. At the beginning and end of each day, condition the capillary by flushing at 2000 mbar pressure with 0.3 *N* NaOH for 30 s, deionized water for 30 s, 1 *M* HCl for 30 s, and finally deionized water for 2 min.
6. Operate the argon-ion laser (Model no. 2211-40ML, Cyonics Corporation, San Jose, CA) of the LIF detector system at 4 mW. The emission spectrum of TOTO-1 is monitored at 540 nm.

3. Methods
3.1. CE Isolation of PCR 1021-bp Fragment

1. Hydrodynamically inject the 1021-bp PCR-amplified (first step) products at 500 mbar for 0.5 min into a 100-µm id fused silica capillary.
2. Calculate the velocity (V) of the fragment by dividing the effective length of the capillary (cm) by the migration time (min) to the detector window. Then multiply V and the distance (cm) from the window to the destination vial to determine the time at which the fragment will be exiting the capillary into the destination vial (*See* **Note 2**).
3. During electrophoresis, monitor the mtDNA product at 260 nm until the 1021-bp fragment approaches the destination vial then stop the electrophoresis.
4. Replace the destination vial with a 250-µL silica conical tube containing 10 µL of 1X TBE solution, and electrophoresis the 1021-bp product into the vial. Do not use any intercalator at this time to avoid any possible interference of the intercalator with the second or low stringency PCR amplification.
5. Transfer the captured 1021-bp fragment into a sterile 0.65-mL Eppendorf tube and PCR amplified (second step) with the L15996 primer using low stringency annealing conditions.
6. Run sample blanks between different DNA samples to confirm that no contamination or carryover from the capture process remains in the system.

3.2. CE/LIF Optimization

1. The CE of LSSP-PCR products uses a 50-µm id fused silica capillary.
2. The 3.5% fragment analysis buffer now contains the intercalating dye. Load buffer into the capillary at 2000 mbar of pressure for 6 min before each sample injection.
3. Inject samples electrokinetically at –5 kV for 0.1 min.
4. Because of the increased sensitivity of the CE-LIF system using dye intercalation, the second step amplification products (LSSP-PCR products) needs to be diluted. Prepare a 1:100 stock solution of each sample. The final dilution is empirically determined by experimentation to find an appropriate dilution of the samples to achieve signal responses within the range of the detector (*see* **Note 3**).

3.3. Pattern Evaluation

1. The size range of LSSP-PCR products is determined by comparison to an external standard (low DNA mass ladder, Gibco-BRL, Gaithersburg, MD). A typical profile pattern contains products ranging in size from approx 70–800 bp.
2. Identify any significant peak present in all samples using the DNA standard to confirm its size. In the example (**Fig. 1**), the common peak was calculated to be 376 bp (marked with an * on the electropherogram). All samples exhibited this intense product peak of approx 376 bp ± 2 bp when calculated using the low DNA mass ladder.
3. This 376-bp peak which is common to all samples served as an internal marker for alignment of the electropherograms for comparison of fragment patterns (*see* **Note 4**).
4. For a control experiment that demonstrates the reliability of the procedure, two maternally related individuals should be included in any study. Since the mtDNA is maternally inherited, these samples are expected to generate the same fingerprint pattern as seen in **Fig. 2** (*see* **Note 5**).

4. Notes

1. The step one PCR product analyzed using CE-UV can be used to determine relative concentration of PCR products by comparing the peak height of each sample. The relative

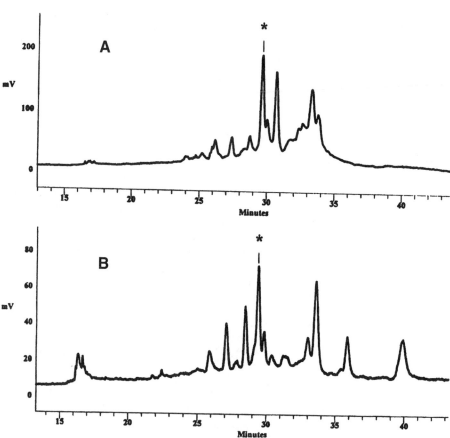

Fig. 1. LSSP-PCR product analysis. The LSSP-PCR profile of two individuals with fragments ranging from approx 70–800 bp in lengths. The electrophoresis of LSSP-PCR products
(after dilution of 1:1000) used a 50-µm id fused silica capillary and 3.5% DNA fragment analysis buffer containing the 2.5×10^{-10} M benzothiazolium-4-quinolinium dimer, TOTO-1. Electrokinetic injection of –5 kV for 6 s and LIF detection at 540 nm. The LSSP-PCR primer L15996
(5'-CTCCACCATTAGCACCCAAAGC) amplified an intense product approx 376 bp in length
(labeled *) which was found present in all samples analyzed in this study. The two individuals
(A and B) demonstrate two different fragment patterns. Most notably, individual B has two fragments with migration times greater than 35 min, whereas A has no fragments of this size.

concentrations are made uniform by adjusting the volume of captured product used for
the second step PCR amplification.
2. The migration time of the peak to be captured can be optimized for less than 15 min by
increasing the voltage.
3. Dilution of product for LIF detection can only be determined experimentally due to differences in LIF systems and intercalator used and laser condition. Generally start with a
1:1000 dilution.

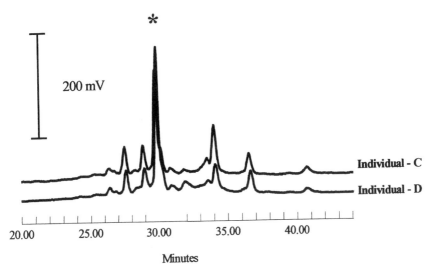

Fig. 2. LSSP-PCR profile of two maternally related individuals. The electropherograms of two maternally related individuals demonstrated the same LSSP-PCR profile. The PCR and electrophoretic conditions are the same as described for **Fig. 1.**

4. Although differences in the concentration of the 1021-bp template do not alter the pattern of each LSSP-PCR profile, the amplitude of the profile did vary. Quantification of the capture fragment and using a determined amount of template for the low-stringency amplification may eliminate this intensity difference.
5. A logical continuation of this experiment is to use other primers and determine their ability to generate different fragment patterns, as well as increase the number of samples analyzed.

Acknowledgments

I wish to acknowledge Kristal Weaver for her valued input and diligent work on the original manuscript of this study and also the Armed Forces Institute of Pathology—Center for Medical and Molecular Genetics (Rockville, MD) where the experimentation was developed. Lastly, I want to thank Transgenomic Inc. (San Jose, CA) for the continued support of this research.

References

1. Wu, D. Y., Ugozzoli, L., Pal, B. K., and Wallace R. B. (1989) Allele-specific enzymatic amplification of beta-globin genomic DNA for diagnosis of sickle cell anemia. *Proc. Natl. Acad. Sci. USA* **86,** 2757–2760.
2. Welsh, J., Honeycutt, R., McClelland, M., and Sobral, B. (1991) Parentage determination in maize hybrids using the arbitrarily primed polymerase chain reaction (AP-PCR). *Theor. Appl. Genet.* **82,** 473–476.
3. Caetano-Anõlles, G., Bassam, B. J., and Gresshoff, P. M. (1991) DNA amplification fingerprinting using very short arbitrary oligonucleotide primers. *BioTechnol.* **9,** 553–557.

4. Williams, J. G. K., Kobelick, A. R., Livak, K. J. Rafalski, J. A., and Tingey, S. V. (199) DNA polymorphisms amplified by arbitrary primers are useful as genetic markers. *Nucleic Acids Res.* **18,** 6531–6535.

5. Pena, S. D., Barreto, G., Vago, A. R., De Marco, L., Reinach, F. C., Neto, E. D., and Simpson, A. J. (1994) Sequence-specific "gene signatures" can be obtained by PCR with single specific primers at low stringency. *Proc. Natl. Acad. Sci. USA* **91,** 1946–1949.

6. Marino, M. A., Weaver, K. R., Tully, L. A., Girard, J. E., and Belgrader, P. (1996) Characterization of mitochondrial DNA using low-stringency single specific primers amplified analyzed by laser induced fluorescence—capillary electrophoresis. *Electrophoresis* **17,** 1499–1504.

17

DOP-PCR Amplification of Whole Genomic DNA and Microchip-Based Capillary Electrophoresis

Paolo Fortina, Jing Cheng, Larry J. Kricka, Larry C. Waters, Stephen C. Jacobson, Peter Wilding, and J. Michael Ramsey

1. Introduction

Universal or whole genome amplification by polymerase chain reaction (PCR) is a rapid and efficient method to generate fragments representing the target sequence, as well as to increase a limited amount of template. One of the most common PCR protocols for total genome amplification is the interspersed repetitive sequence-PCR (IRS-PCR) in which primers specific for human repeat-rich regions are used to generate PCR products between adjacent repeated sequences (1). However, although IRS-PCR across regions such as *Alu* families of human repeat has been demonstrated to be useful, the nonuniform distribution of repeat-rich region within the human genome has been a limitation. Alternative strategies have been proposed. In the primer-extension preamplification (PEP), multiple rounds of extensions with *Taq* DNA polymerase and a random mixture of 15-base oligonucleotides as primers produce multiple copies of the template present in the sample (2–5). In a more demanding protocol, called linker adaptor-PCR, *Rsa*I restricted genomic DNA fragments are ligated to *Sma*I-cut pUC plasmid. Subsequently, the inserts are amplified by PCR using the universal M13/pUC sequencing and reverse sequencing primers and then released by *Eco*RI digestion (6). The tagged random primer PCR (T-PCR) is a two-step PCR strategy which consists of a pool of all possible 3'-sequences for binding to the target DNA and a constant 5'-region for the detection of incorporated primers (7). Recently, degenerate oligonucleotide primed-polymerase chain reaction (DOP-PCR) was developed to allow random amplification of DNA from any source (8–10). DOP-PCR uses a partially degenerate sequence in a PCR protocol with two different annealing temperatures. It has been successfully applied for amplifying entire genomes such as human, mouse, and fruit fly, as well as isolated human chromosomes and cosmids (11). The technique has also been used to prepare whole chromosome paint probes (11,12) for micro-FISH assays (13–15), comparative genomic hybridization (16), to increase the amount of sample

From: *Methods in Molecular Biology, Vol. 163:*
Capillary Electrophoresis of Nucleic Acids, Vol. 2: Practical Applications of Capillary Electrophoresis
Edited by: K. R. Mitchelson and J. Cheng © Humana Press Inc., Totowa, NJ

for genotyping *(17)*, and genomic fingerprinting *(18)*. The DOP-PCR primer consists of three regions. The 5'-end carries a recognition sequence for *Xho*I (C•TCGAG), a restriction endonuclease that cuts rarely within the human genome. This sequence can be used for cloning, if desired. The sequence is then followed by a middle portion containing six nucleotides of degenerate sequence (*NNNNNN*, where *N* = A, C, G, or T in approximately equal proportions) and a 3'-end sequence containing six specific bases (ATGTGG) which primes the reaction approximately every 4 kb *(8,9)*. The principle of the technique is that at a sufficiently low annealing temperature only the six specific nucleotides included in the 3'-end of the degenerate oligonucleotide will anneal to the genomic strand allowing the primer to initiate PCR. The PCR fragments are then generated which contain the full length of the oligoprimer at one end and its complementary sequence at the other end. Subsequently, the temperature is increased to the level required for the full length of the degenerate primer to anneal. For additional details, we direct the reader to the original papers *(8,9)*.

We have adapted the DOP-PCR technique to a three-microchip format *(19)*. DOP-PCR amplified genomic DNA produced in a first silicon-glass chip is transferred to a second chip for a locus-specific, multiplex PCR of the *dystrophin* gene exons in order to detect deletions causing Duchenne/Becker muscular dystrophy (DMD/BMD). Amplicons from the multiplex-PCR are then analyzed by electrophoresis in a third microchip. The analytical performance of the microchip capillary electrophoresis (MCE) is also compared to conventional capillary electrophoresis (CE).

2. Materials

2.1. PCR Chips

2.1.1. Fabrication

1. Microfabricated, silicon-glass PCR chips (reaction vol 12 µL, surface area 210 mm^2) are manufactured using standard photolithographic procedures *(20)* (*see* **Note 1**). Each PCR chip measures 14 × 17 mm and is etched to a depth of 115 µm. To produce a "PCR friendly" surface layer, passivation is accomplished by thermal deposition of a 1000 Å thick layer of SiO$_2$ using standard deposition techniques *(21–23)*.
2. Surface-polished Pyrex glass covers (14 × 17 mm) (Bullen Ultrasonics, Inc., Eaton, OH) and the silicon chips are serially soaked in a H$_2$SO$_4$ (94%)/H$_2$O$_2$ (30%) (3/2, v/v) and then washed in deionized distilled water. Each chip is then capped with the washed Pyrex glass cover.
3. The silicon chip is placed on an aluminum plate heated to 500°C on a PC-300 insulated hot plate (Corning, Corning, NY). Temperature is monitored using a surface thermometer (Hallcrest, Glenview, IL). Glass covers placed on top of each silicon chip are anodically bonded together by applying 1000 V throughout the aluminum plate and glass cover with a current of less than 1 mA.

2.1.2. Instrumentation

1. An APH-1000M D.C. power pack (Kepco, Inc., Flushing, NY) is used to apply the required voltage.

2.1.3. Reagents

1. High molecular weight genomic DNA is extracted from human nucleated white blood cells using standard, previously described protocols *(24,25)*, commercially avail-

able kits, or with an automatic DNA extractor. DNA concentration is adjusted to 20 ng/μL (*see* **Note 2**).

2. DOP-PCR primer, (10 m*M* stock): 5'-CCGACTCGAGNNNNNNATGTGG-3' (22-mer, where *N* = A, C, G or T in approximately equal proportions) (*see* **Note 3**).

3. Duchenne/Becker Muscular Dystrophy (DMD/BMD) PCR primers *(26)*, (10 m*M* stock) are as follows:

Pm: Forward: 5'-GAAGATCTAGACAGTGGATACATAACAAATGCATG-3'
           Reverse: 5'-TTCTCCGAAGGTAATTGCCTCCCAGATCTGAGTCC-3'

Exon 3: Forward: 5'-TCATCCATCATCTTCGGCAGATTAA-3'
           Reverse: 5'-CAGGCGGTAGAGTATGCCAAATGAAAATCA-3'

Exon 6: Forward: 5'-CCACATGTAGGTCAAAAATGTAATGAA-3'
           Reverse: 5'-GTCTCAGTAATCTTCTTACCTATGACTATGG-3'

Exon 13: Forward: 5'-AATAGGAGTACCTGAGATGTAGCAGAAAT-3'
           Reverse: 5'-CTGACCTTAAGTTGTTCTTCCAAAGCAG-3'

Exon 43: Forward: 5'-GAACATGTCAAAGTCACTGGACTCCATGG-3'
           Reverse: 5'-ATATATGTGTTACCTACCCTTGTCGGTCC-3'

Exon 47: Forward: 5'-CGTTGTTGCATTTGTCTGTTTCAGTTAC-3'
           Reverse: 5'-GTCTAACCTTTATCCACTGGAGATTTG-3'

Exon 50: Forward: 5'-CACCAAATGGATTAAGATGTTCATGAAT-3'
           Reverse: 5'-TCTCTCTCACCCAGTCATCACTTCATAG-3'

Exon 52: Forward: 5'-AATGCAGGATTTGGAACAGAGGCGTCC-3'
           Reverse: 5'-TTCGATCCGTAATGATTGTTCTAGCCTC-3'

Exon 60: Forward: 5'-AGGAGAAATTGCGCCTCTGAAAGAGAACG-3'
           Reverse: 5'-CTGCAGAAGCTTCCATCTGGTGTTCAGG-3'

4. Ampli*Taq* DNA polymerase: 5 U/mL (Perkin-Elmer, Norwalk, CT).

5. 10X PCR buffer: (100 m*M* Tris-HCl, pH 9.0, 500 m*M* KCl, 15 m*M* MgCl$_2$, 1% Triton X-100 (v/v), 0.1% gelatin (w/v)) (Perkin-Elmer).

6. *Taq*Start antibody (Clontech Laboratories Inc., Palo Alto, CA).

7. TaqStart antibody dilution buffer (Clontech Laboratories Inc.).

8. 10 m*M* stocks of dATP, dCTP, dGTP, and dTTP (Amersham, Arlington Heights, IL).

9. Acrylamide (Sigma, St. Louis, MO).

10. Ammonium persulfate (Sigma).

11. *N,N,N',N'*-Tetramethylethylenediamine (TEMED) (Sigma).

12. TO-PRO (Molecular Probes, Eugene, OR).

13. 10X TBE buffer (0.89 *M* Tris-base; 0.89 *M* boric acid; 0.02 *M* EDTA, pH 8.3) (Life Technologies, Grand Island, NY).

2.2. PCR Chip Thermocycler

1. Fabrication: A custom fabricated device (Faulkner Instruments, Pitman, NJ) is used for simultaneous thermal cycling of four PCR chips. The device incorporates a 9500/071/040 Peltier heater/cooler (ITI Ferrotec, Chelmsford, MA) centrally located under an oxygen-free 40 × 40 mm copper block containing a YSI 44016 10 kΩ thermistor (Yellow Springs Instruments, Yellow Springs, OH). A constant airflow of ~40 L/min is kept under the thermal cycling device to dissipate heat.

2. Instrumentation: Airflow is monitored using an F-400 flowmeter (Gilmont Instruments, Inc., Barrington, IL). A LDC-3900 modular laser diode controller (ILX Lightwave, Boceman, MT) connects the heater/cooler to the thermistor through an RS232 interface. The laser diode controller is connected to a 486 PC through a GPIB interface and is controlled using a virtual instrument built on LabVIEW for Windows (National

Instruments, Austin, TX), which automates thermal cycling giving cycle times of approx 3 min.

2.3. Agarose Slab-Gel Electrophoresis

1. Power supply (Hoefer Scientific Instruments, Inc., San Francisco, CA).
2. Agarose gel apparatus (CBS Scientific Co, Del Mar, CA).
3. Agarose for gel electrophoresis such as Ultrapure Agarose (Gibco-BRL, Gaithersburg, MD).
4. Stock 10X ethidium bromide gel loading buffer (Sigma).
5. Stock electrophoresis running buffer: 10X TBE.

2.4. Entangled Solution Capillary Electrophoresis (ESCE)

1. System: A P/ACE system 5010 with a LIF detector (Beckman, Fullerton, CA) in the reversed polarity mode (negative potential at the injection end of the capillary column) is used for entangled solution CE. The excitation and the emission wavelengths are 488 nm and 520 nm, respectively. The external temperature of the capillary column is set at 20°C. Postrun analyses of the data are performed using the Gold Chromatography Data System (Version 8.0) *(27)*.
2. Hydroxypropyl methyl cellulose (HPMC) (Sigma, Product: H 7509). Viscosity of a 2% (w/v) aqueous solution of this polymer should be 3500–5600 cP at 20°C.
3. Surface modified fused silica capillary column DB-1, 27 cm × 100 μm (J & W Scientific, Folsom, CA).
4. Glycerol (Sigma).
5. YO-PRO-1, fluorescent dye (Molecular Probes, Eugene, OR).
6. 10X TBE buffer.

2.5. Microchip Capillary Electrophoresis

1. Microchip Fabrication: CE microchips are fabricated using standard photolithographic, wet-chemical etching and bonding techniques as previously described *(28)*. Specifically, a photomask is produced by first sputtering a 50-nm chrome film onto a glass slide and then spinning a positive photoresist (Shipley 1811) onto the chrome film. The microchip channel design (**Fig. 1A**) is produced by exposing the photoresist to a laser using a CAD/CAM laser machining system (Argon-ion, 457 nm). The photoresist is developed and the exposed chrome film is etched in $CeSO_4/HNO_3$. The design on the photomask, a simple cross with sample, buffer, sample waste, and separation channels (lengths = 0.62, 0.5, 0.65, and 3.46 cm, respectively), is transferred onto a glass microscope slide using a positive photoresist and UV exposure, and the channels are etched with a dilute, stirred HF/NH_4 solution. Thermal bonding of a cover plate to the substrate forms a closed channel network, and affixing cylindrical glass reservoirs where the channels extend beyond the cover plate with epoxy (*see* **Note 4**) provides access. Channels are ~50 μm wide and ~10 μm deep. Electroosmotic flow is minimized by covalently coating the channels with linear polyacrylamide *(29)*.
2. Instrumentation: Control of the high voltage at the fluid reservoirs is performed using a single high-voltage power supply (10A12-P4, Ultravolt) with voltage divisions in 5% increments. High-voltage relays (K81C245, Kilovac) are used to toggle between the sample loading and injection/analysis modes. The voltage control and relay switching are computer-controlled using programs written in-house in LabVIEW (National Instruments). Platinum electrodes provide electrical contacts from the power supply to the solutions in the reservoirs. Electrophoretically migrating DNA fragments are monitored using a single-point detection system (**Fig. 1B**) *(30)*.
3. In **Fig. 1B**, an argon-ion laser operating at 514.5 nm and 5 mW (Omnichrome, Chino, CA) is focused to a spot in the separation channel using a 100-mm focal length lens. The

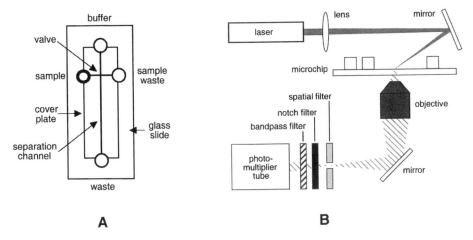

Fig. 1. (**A**) Microchip design used for electrophoretic analysis of PCR-amplified DNA. (**B**) Fluorescence-based single-point detection system used to monitor microchip capillary electrophoresis. *See* text for description.

fluorescence signal is collected by a 20X objective (0.42 NA; Newport Corp., Irvine, CA), followed by spatial filtering with a 1.0-mm pinhole (Melles Griot, Irvine, CA), and spectral filtering with a 514-nm holographic notch filter (10-nm bandwidth; Kaiser Optical Systems, Inc., Ann Arbor, MI), and a 540-nm band-pass filter (30-nm bandwidth, 540DF30; Omega Optical, Brattleboro, VT). The signal is measured with a photomultiplier tube (77348; Oriel Instruments, Inc., Stratford, CT), and is then amplified (Keithley 428) and read into the computer using a multifunction I/O card (NB-MIO-16XL-42; National Instruments).

3. Methods
3.1. DOP-PCR Amplification in a PCR Chip

1. PCR chips are filled through the entry port with 12 µL of the PCR reaction mixture. The mixture contains 9.6 ng of high molecular weight human genomic DNA (*see* Note 2), 2.5 m*M* of each dNTPs, 10 m*M* Tris-HCl (pH 9.0), 50 m*M* KCl, 1.5 m*M* MgCl$_2$, 0.1% Triton X-100 (v/v), 0.01% gelatin (w/v), 1.2 U of *Taq* DNA polymerase (Perkin-Elmer), 132 ng *Taq*Start antibody and 0.48 µL of dilution buffer (Clontech Laboratories Inc.), and 80 µ*M* of the degenerate primer (*see* Note 5).
2. A negative control PCR chip is filled with 12 µL of the PCR reaction mixture but no DNA sample.
3. DOP-PCR amplification for 1 cycle is run as follows: denature at 94°C for 8 min. Then, for 8 cycles: denature at 94°C for 1 min, anneal at 30°C for 1 min, extend at 72°C for 3 min. For the last 28 cycles: denature at 94°C for 1 min, anneal at 59°C for 1 min, and extend at 72°C for 3 min.

3.2. Agarose Gel Analysis of Amplified DOP-PCR Products

1. Gel Preparation: A 1.5% (w/v) agarose gel containing 1X TBE and standard ethidium bromide gel loading buffer should be used to examine the size range of the DOP-PCR chip amplified human genomic DNA.

2. Sample Loading and Gel Electrophoresis: 4–6 µL of the DOP-PCR products are loaded. Appropriate DNA size marker such as 125–150 ng of *Hae*III digested φX174 DNA is recommended. The gel is run at 110 V for 30–50 min. Amplicons, visualized under UV light, range in size between 200 and 1000 bp.

3.3. Locus-Specific (DMD/BMD) Multiplex-PCR in a PCR Chip

1. PCR chips are filled via the entry port with 12 µL of the PCR reaction mixture, which contains 0.24 mL of DOP-PCR amplified whole human genome DNA, 2 µM of each dNTP, 1.2 U of *Taq* DNA polymerase, 132-ng *Taq*Start antibody and 0.48 µL of dilution buffer and 1.08 µL of the mixture of 9 pairs of primers for DMD/BMD.
2. A negative control PCR chip is filled with 12 µL of the PCR reaction mixture as described above, but with no DNA sample.
3. Locus-specific PCR amplification for 1 cycle is run as follows: denature at 94°C for 5 min. Then, for 30 cycles: denature at 94°C for 10 s, anneal at 63°C for 1 min, extend at 65°C for 1 min, and finally an extension at 65°C for 10 min.

3.4. Entangled Solution Capillary Electrophoresis

1. One µL of the multiplexed locus-specific PCR products is removed from the PCR chip and diluted with 8 µL of deionized distilled water. This sample is loaded into the capillary using a 7-s injection at 5 kV. Separation of the PCR products for 11 min is performed at a field strength of 296 V/cm in 1X TBE, containing 5.0% (v/v) glycerol and 0.5% (w/v) HPMC.
2. HPMC is dissolved in the buffer using the method previously recommended *(31)*. The buffer is filtered using a 4.0-µm filter and then degassed for 15 min by sonication. TO-PRO-1, a DNA-intercalating, fluorescent dye is added to the above buffer to obtain a final concentration of 1 µM.
3. Samples are analyzed on a surface modified fused silica capillary DB-1 (J & W Scientific). The capillary column (27 cm × 100 µm) should be conditioned with 5 vol of distilled water followed by 5 vol of separation buffer, and then subjected to 30-min voltage equilibration until a stable base line is attained. The column should be washed after each run with fresh separation buffer for 42 s.

3.5. Microchip Electrophoresis

1. The microchip channels and all reservoirs are filled, by vacuum, with 3% linear polyacrylamide (LPA) in 0.5X TBE. The acrylamide is polymerized, either on or off chip, by the addition of ammonium persulfate (0.025%) and TEMED (0.15%) at room temperature (*see* **Note 6**). After the removal of polymer from the sample reservoir (polymer can be left in the other three reservoirs or replaced with 0.5X TBE), it is rinsed with 0.5X TBE and the sample to be analyzed is added. Typically, a marker DNA such as the *Hae*III digest of pBR322, at about 5 µg/mL in 0.5X TBE containing 1-µM TO-PRO, is used to align and test the integrity of the chip. PCR-amplified samples are diluted in 0.5X TBE and adjusted to contain ~1 µM TO-PRO.
2. A pinched sample loading scheme *(28)* is used to introduce the sample onto the separation channel (**Fig. 2**). In the "sample loading" mode (**Fig. 2B**), the DNA sample is electrophoretically migrated from the sample reservoir to the sample waste reservoir, filling the cross intersection. In the "injection" and "analysis" modes (**Fig. 2C, D**), the contents of the cross intersection are introduced into the separation channel and electrophoretically separated. In these modes, the potentials applied to the sample and sample waste reservoirs prevent bleeding of excess sample or sample waste

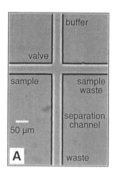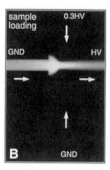

Fig. 2. CCD images of the pinched injection scheme used in MCE. (**A**) White light image of the cross-intersection and fluorescence images of the same area illustrating (**B**) sample loading, (**C**) sample injection, and (**D**) sample analysis. The potentials applied to the reservoirs are listed. Arrows depict direction of flow.

into the separation channel. DNA fragment separation is monitored using an intercalating dye, TO-PRO, and fluorescence detection, 2.5 cm from the cross-intersection (*see* **Note 7**).

4. Notes

1. Silicon-based chips for DOP-PCR and DMD/BMD multiplex PCR are from Alberta Microelectronic Center, Edmonton, Alberta, Canada.
2. Template concentrations may range from as low as 25 pg/µL to 500 ng/µL.
3. Primers can be purchased from a commercial source or synthesized by a core laboratory within a hospital/academic center.
4. Similar microchip designs have also been used to PCR amplify and electrophoretically analyze DNA on a single microchip *(32,33)*.
5. Presence of pseudogenes closely related to gene of interest, as well as the presence of other members of super gene families, might limit the efficiency of genome-wide DOP-PCR schemes.
6. Polymers generated on or off chip are equally effective for DNA separation. LPA solutions up to 6%, polymerized as described, are readily replaceable in these microchips.
7. Typically, 20 or more electrophoretic runs can be made on a microchip prepared in this manner without significant loss of resolution. Further use of the same microchip can be made for electrophoresis by replacing the sieving matrix.

Acknowledgments

This work was supported in part by NIH grant P60-HL38632 (P.F., P.W., and L.J.K.), by Oak Ridge National Laboratory Directed Research and Development (J.M.R., S.C.J., and L.C.W.) and by a Sponsored Research Agreement from Caliper Technologies Corporation to P.W. and L.J.K. and the University of Pennsylvania, with full endorsement by the Conflicts of Interest Committee of the University of Pennsylvania. P.W. and L.J.K. hold minority stock in Caliper Technologies Corporation. Lockheed Martin Energy Research Corporation manages ORNL for the U.S. Department of Energy under contract DE-AC05-96OR22464.

References

1. Nelson, D. L., Ledbetter, S. A., Corbo, L, Victoria, M. F., Ramirez-Solis, R., Webster, T. D., et al. (1989) *Alu* polymerase chain reaction: a method for rapid isolation of human-specific sequences from complex DNA sources. *Proc. Natl. Acad. Sci. USA* **86,** 6686–6690.

2. Zhang, L., Cui, X., Schmitt, K., Hubert, R., Navidi. W., and Arnheim, N. (1992) Whole genome amplification from a single cell: implications for genetic analysis. *Proc. Natl. Acad. Sci. USA* **89,** 5847–5851.

3. Snabes, M. C., Chong, S. S., Subramanian, S. B., Kristjansson, K., DiSepio D., and Hughes, M. R. (1994) Preimplantation single-cell analysis of multiple genetic loci by whole-genome amplification. *Proc. Natl. Acad. Sci. USA* **89,** 6181–6185.

4. Barrett, M. T., Reid, B. J., and Joslyn, G. (1995) Genotypic analysis of multiple loci in somatic cells by whole genome amplification. *Nucleic Acids Res.* **23,** 3488–3492.

5. Barrett, M. T., Galipeau, P. C., Sanchez, C. A., Emond, M. J., and Reid, B. J. (1996) Determination of the frequency of loss of heterozygosity in esophageal adenocarcinoma by cell sorting, whole genome amplification and microsatellite polymorphisms. *Oncogene* **12,** 1873–1878.

6. Ludecke, H. J., Senger, G., Claussen, U., and Horsthemke, B. (1989) Cloning defined regions of the human genome by microdissection of banded chromosomes and enzymatic amplification. *Nature* **338,** 348–350.

7. Grothues, D., Cantor, C. R., and Smith, C. L. (1993) PCR amplification of megabase DNA with tagged random primers (T-PCR). *Nucleic Acids Res.* **21,** 1321–1322.

8. Telenius, H., Carter, N. P., Bebb, C. E., Nordenskjold, M., Ponder, B. A. J., and Tunnacliffe, A. (1992) Degenerate oligonucleotide-primed PCR: general amplification of target DNA by a single degenerate primer. *Genomics* **13,** 718–725.

9. Telenius, H., Pelmear, A. H., Tunnacliffe, A., Carter, N. P., Behmel, A, Ferguson-Smith, M. A., et al. (1992) Cytogenetic analysis by chromosome painting using DOP-PCR amplified flow-sorted chromosomes. *Genes, Chromosomes & Cancer* **4,** 257–263.

10. Speicher, M. R., du Manoir, S., Schrock, E., Holtgreve-Grez, H., Schoell, B., Lengauer, C., et al. (1993) Molecular cytogenetic analysis of formalin-fixed, paraffin-embedded solid tumor by comparative genomic hybridization after universal DNA-amplification. *Hum. Mol. Genet.* **2,** 1907–1914.

11. Scalzi, J. M. and Hozier, J. C. (1998) Comparative genome mapping: mouse and rat homologies revealed by fluorescence in situ hybridization. *Genomics* **47,** 44–51.

12. Burkin, D. J., O'Brien, P. C., Broad, T. E., Hill, D. F., Jones, C. A., Wienberg, J., and Ferguson-Smith, M. A. (1997) Isolation of chromosome-specific paints from high-resolution flow karyotypes of the sheep (*Ovis aries*). *Chromosome Res.* **5,** 102–108.

13. Engelen, J. J., Loots, W. J., Albrechts, J. C., Plomp, A. S., van der Meer, S. B., Vles, J. S., et al. (1998) Characterization of a *de novo* unbalanced translocation t(14q18q) using microdissection and fluorescence *in situ* hybridization. *Am. J. Med. Genet.* **75,** 409–413.

14. Xiao, Y., Darroudi, F., Kuipers, A. G., de Jong, J. H., de Boer, P., and Natarajan, A .T. (1996) Generation of mouse chromosome painting probes by DOP-PCR amplification of microdissected meiotic chromosomes. *Cytogenet. Cell. Genet.* **75,** 63–66.

15. Goureau, A., Yerle, M., Schmitz, A., Riquet, J., Milan, D., Pinton, P., et al. (1996) Human and porcine correspondence of chromosome segments using bidirectional chromosome painting. *Genomics* **36,** 252–262

16. Kuukasjarvi, T., Tanner, M., Pennanen, S., Karhu, R., Visakorpi, T., and Isola, J. (1997) Optimizing DOP-PCR for universal amplification of small DNA samples in comparative genomic hybridization. *Genes, Chromosomes & Cancer* **18,** 94–101.

17. Cheung, V. G., and Nelson, S. F. (1996) Whole genome amplification using a degenerate oligonucleotide primer allows hundreds of genotypes to be performed on less than one nanogram of genomic DNA. *Proc. Natl. Acad. Sci. USA* **93**, 14676–14679.

18. Sayada, C., Picard, B., Elion, J., and Krishnamoorthy, R. (1994) Genomic fingerprinting of *Yersinia enterocolitica* species by degenerate oligonucleotide-primed polymerase chain reaction. *Electrophoresis* **15**, 562–565.

19. Cheng, J., Waters, L. C., Fortina, P., Hvichia, G., Jacobson, S. C., Ramsey, J. M., et al. (1998) Degenerate oligonucleotide primed-PCR and capillary electrophoretic analysis of human DNA on microchip-based devices. *Anal. Biochem.* **257**, 101–106

20. McGillis, D. A. (1983) In: *VLSI Technology*. (Sze, S. M., ed.), McGraw Hill, New York, NY, pp. 267–280.

21. Katz, L. E. (1983) In: *VLSI Technology*. (Sze, S. M., ed.), McGraw Hill, New York, NY, pp. 131–167.

22. Cheng, J., Shoffner, M. A., Hvichia, G. E., Kricka, L. J., and Wilding, P. (1996) Chip PCR, II. Investigation of different PCR amplification systems in micro-fabricated silicon-glass chips. *Nucleic Acids Res.* **24**, 380–385.

23. Shoffner, M. A., Cheng, J., Hvichia, G. E., Kricka, L. J., and Wilding, P. (1996) Chip PCR, I. Surface passivation of microfabricated silicon-glass chips for PCR. *Nucleic Acids Res.* **24**, 375–379.

24. Poncz, M., Solowiejzcyk, D., Harpel, B., Moroy, Y., Schwartz, E., and Surrey, S. (1982) Construction of human gene libraries from small amounts of peripheral blood: Analysis of b-like globin genes. *Hemoglobin* **6**, 27–33.

25. Sambrook, J., Fritsch, E. F., and Maniatis, T. (1989) *Molecular Cloning. A Laboratory Manual*. Cold Spring Harbor Laboratory, Cold Spring Harbor, NY.

26. Beggs, A. H., Koenig, M., Boyce, F. M., and Kunkel, L. M. (1990) Detection of 98% of DMA/BMD gene deletions by polymerase chain reaction. *Hum. Genet.* **86**, 45–48.

27. Fortina, P., Cheng, J., Shoffner, M. A., Surrey, S., Hitchcock, W. H., Kricka, L. J., and Wilding, P. (1997) Diagnosis of Duchenne/Becker muscular dystrophy and quantitative identification of carrier status by use of entangled solution capillary electrophoresis. *Clin. Chem.* **43**, 745–751.

28. Jacobson, S. C., Hergenröder, R., Koutny, L. B., Warmack, R. J., and Ramsey, J. M. (1994) Effects of injection schemes and column geometry on the performance of microchip electrophoresis devices. *Anal. Chem.* **66**, 1107–1113.

29. Hjertén, S. (1985) High performance electrophoresis: elimination of electroendosmosis and solute adsorption. *J. Chromatogr.* **347**, 191–198.

30. Jacobson, S. C. and Ramsey, J. M. (1996) Integrated microdevice for DNA restriction fragment analysis. *Anal. Chem.* **68**, 720–723.

31. Ulfelder, K. J., Schwartz, H. E., Hall, J. M., and Sunzeri, F. J. (1992) Restriction fragment length polymorphism analysis of ERBB2 oncogene by capillary electrophoresis. *Anal. Biochem.* **200**, 260–267.

32. Waters, L. C., Jacobson, S. C., Kroutchinina, N., Khandurina, J., Foote, R. S., and Ramsey, J. M. (1998a) Microchip device for cell lysis, multiplex PCR amplification, and electrophoretic sizing. *Anal. Chem.* **70**, 158–162.

33. Waters, L. C., Jacobson, S. C., Kroutchinina, N., Khandurina, J., Foote, R. S., and Ramsey, J. M. (1998b) Multiple sample PCR amplification and electrophoretic analysis on a microchip. *Anal. Chem.* **70**, 5172–5176.

18

Analysis of Triplet-Repeat DNA by Capillary Electrophoresis

Yuriko Kiba and Yoshinobu Baba

1. Introduction

Recently, much attention has been focused on triplet-repeat expansions on the human genome, because they are reported to cause a number of neurodegenerative diseases such as the familial mental retardation, myotonic dystrophy, autosomal dominant diseases, or Huntington disease, which are so called triplet-repeat diseases (1,2). A hallmark of most of these diseases is the phenomenon of "anticipation," which are not easily explained by Mendel's Laws of genetic inheritance. This phenomenon includes a parental sex bias and a "decrease in the age at onset of the disease" or severity of the disease in consecutive generations due to the tendency of the unstable triplet repeat to lengthen when passed from one generation to the next. The expansion of the triplet-repeat element is associated with the defect, in which the extent of the expansion roughly correlates with the severity of the disease symptoms. The triplet-repeat element expansions are not stable, and it is unclear precisely how the expansions are directly associated with triplet-repeat diseases. Although several expansions encode enlarged polyglutamine tracts within their encoded protein, which might be expected to alter either the charge density and pI, or the folded structure of the protein, the mechanism for the etiology and progression of most of these diseases is not yet understood.

In this chapter, we have attempted to establish an analytical technique for the investigation of the triplet-repeat expansion on human genome and DNA diagnosis of triplet-repeat diseases. Initially, denatured random sequence DNA markers are examined to understand the standard electrophoretic behavior of ssDNA fragments. We could successfully separate all of the ssDNA fragments by oligonucleotide length by capillary electrophoresis (CE) within a very short time and with very high repeatability. When four sets of single-stranded triplet-repeat oligonucleotide with a sequence (CNG)n are examined as models of triplet-repeat DNA elements and are tested under the same conditions, remarkably the mobilities of each of the triplet repeat oligonucleotides is much larger than random single-strand oligonucleotides of the same length or

From: *Methods in Molecular Biology, Vol. 163:*
Capillary Electrophoresis of Nucleic Acids, Vol. 2: Practical Applications of Capillary Electrophoresis
Edited by: K. R. Mitchelson and J. Cheng © Humana Press Inc., Totowa, NJ

shorter length *(3–6)*. This unusually high mobility of triplet-repeat oligonucleotides is considered to be an effect of a characteristic higher order structure formed by G::C base-pairing, which decreases the contour length of the triplet-repeat oligonucleotide compared to a random single-strand oligonucleotide.

2. Materials

1. The DNA molecular marker (20-bp DNA ladder) is from Funakoshi (Tokyo, Japan).
2. An intercalating dye YO-PRO-1 is purchased from Molecular Probes (Eugene, Oregon, USA).
3. Methylcellulose (4000 cP) obtained from Sigma (St. Louis, MO, USA) is used as a separation matrix in a running buffer of: 50 mM Tris-base, boric acid, pH 8.5, 0.7% methyl cellulose, 0.1 μmol/L YO-PRO-1.
4. All other chemicals are of analytical-reagent grade from Wako (Osaka, Japan).
5. Capillary: DB-17-coated capillary of 100 μm id, 0.1 mm coating phase thickness, 360 μm od, 27 cm total length, 19.1 cm effective length (J&W Scientific) for the Beckman P/ACE 2100 system.

3. Methods

3.1. Oligonucleotides and ssDNA Size Markers

1. The 20-bp DNA molecular marker ladder (200 mg/mL) contains 50 fragments from 20–1000 bp in increment each of 20 bp, and was diluted with Mili-Q water to 1–30 mg/mL. The DNAs are stored at –20°C until use.
2. The 20-bp ladder marker is heated at 94°C for 2 s to ensure full denaturation before electrokinetic injection into the capillary.
3. Ten types of ssDNA oligonucleotides are used as examples of triplet-repeat DNA elements. Four different triplet sequences with ten (CNG) repeats (CAG, CTG, CGG, CCG):
   (CCG)$_{10}$ (5.5 od, 0.19 mM) diluted to 0.6 nmol/mL with distilled water.
   (CGG)$_{10}$ (14 od, 0.49 mM) diluted to 1.6 nmol/mL.
   (CAG)$_{10}$ (18 od, 0.64 mM) diluted to 2.1 nmol/mL.
4. (CTG) repeats: 20 and 30 repeats of CTG are used in this study.
   (CTG)$_{10}$ (13 od, 0.46 mM) diluted to 1.5 nmol/mL, and when detection is by LIF diluted to 1.0 nmol/mL for PDA.
   (CTG)$_{20}$ (36.5 od, 0.65 mM) is diluted to 1.1 nmol/mL,
   (CTG)$_{30}$ (48 od, 0.65 mM) is diluted to 1.3 nmol/mL.
5. As the samples of ssDNA with random sequences, ssDNA with 20-, 40-, 60-, and 80-mer are synthesized using the sequence of *p53*, exon 7 as the template. The entire template sequence is as follows:
   5'-GTTGGCTCTGACTGTACCACCATCCACTACAACTACATGT
   GTAACAGTTCCTGCATGGGCGGCATGAACCGGAGGCCCAT-3'
   These fragments were also diluted with distilled water:
   20-mer fragment (10.5 od, 0.61 mM) diluted to 1.2 nmol/mL.
   40-mer fragment (9.5 od, 0.27 mM) diluted to 0.5 nmol/mL.
   60-mer fragment (7.0 od, 0.13 mM) to 0.4 nmol/mL.
   80-mer fragment (7.5 od, 0.11 mM) to 0.2 nmol/mL.

3.2. Capillary Electrophoresis

1. CE is performed on a Hewlett-Packard HP^{3D}CE system. DNA samples are detected by UV absorption spectroscopy at 260 nm with a photo diode array detector. The separation

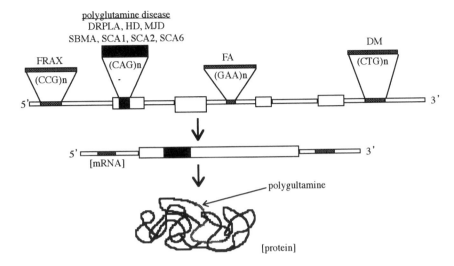

Fig. 1. Identified expanded triplets of triplet-repeat disease and their locations on the human genome. FRAX: fragile X syndrome; DRPLA: dentatorubral pallidoluysian atrophy; HD: Huntington disease; MJD: Machado-Joseph disease; SBMA: spinobulbar muscular atrophy; SCA: spinocerebellar degeneration; FA: Friedreich ataxia; and DM: myotonic dystrophy.

capillary on the HP^{3D}CE system is a DB-17-coated capillary (100 μm id, 0.1 μm coating phase thickness) with a total length 32.5 cm and an effective length 24 cm. The DNA samples are electrophoretically fractionated by size at 30°C in a running buffer 50 mM Tris-borate, pH 8.5, containing 0.7% methyl cellulose.

2. Capillary electrophoresis is also performed on a Beckman P/ACE 2100 system equipped with a laser-induced fluorescence detector (LIF). The separation capillary on the P/ACE 2100 is a DB-17-coated capillary (100 μm id, 0.1 μm coating phase thickness) with a total length of 24 cm, and with an effective length of 19.1 cm. The fluorescence of DNA intercalated with YO-PRO-1 dye, excited at 488 nm and detected at 560 nm.

3. DNA samples are introduced onto the P/ACE 2100 by electrokinetic injection into capillary at 5 kV for 10 s. Each separation experiment is carried out repeatedly for 10 times to evaluate the reproducibility of the separation of each test oligonucleotide. The DNA samples are electrophoretically fractionated by size at 30°C in a running buffer 50 mM Tris-borate, pH 8.5, containing 0.7% methyl cellulose, 0.1 μM YO-PRO-1. The electric fields applied are 100 V/cm, 200 V/cm, and 300 V/cm.

4. The inside of the capillary is washed with 50% (v/v) methanol/water solution for 3 min followed by a wash with running buffer for 3 min between each run.

3.3. Results

3.3.1. Sequence of Triplet Repeat Expansion Elements

1. The triplet-repeat disease genes identified so far can be separated into four groups according to the loci of the expansion element in the gene (*see* **Fig. 1**).
   The expansions are seen in:
   a. The 5'-untranslated region (CCG repeat), FRAX: fragile X syndrome.

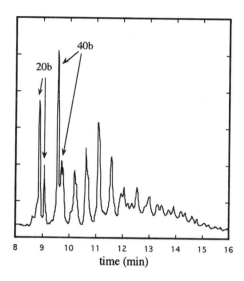

Fig. 2. CE separation of denatured 20-bp DNA ladder. Capillary: 100 μm id, 360 μm od, 27 cm total length, 19.1 cm effective length. Running buffer: 50 m*M* Tris-base, pH 7.0–7.4, 50 m*M* boric acid, 0.7% methyl cellulose, and 0.1 μmol/L YO-PRO-1. Run conditions: electric field: 100 V/cm; injection, 5 kV for 10 s; fluorescence detection: excitation 488 nm, emission 560 nm.

     b. Translated regions (CAG repeat), DRPLA: dentatorubral pallidoluysian atrophy, HD: Huntington disease, MD: Machado-Josef disease, SBMA: spinobulbar muscular atrophy, SCA: spinocerebellar degeneration.

     c. Intron regions (GAA repeats), FA: Friedreich ataxia.

     d. 3'-untranslated region (CTG repeat), DM: myotonic dystrophy.

2. As shown in **Fig. 1**, so far only 3 kinds of the CNG sequence repeat (CAG, CTG, CCG) and the GAA sequence repeats have been identified as sequences associated with triplet-repeat expansion diseases. The physiological function of such genes causing the particular disease is not yet completely understood.

3. The new method has been developed for DNA diagnosis of triplet-repeat diseases, termed "repeat expansion detection" (RED) *(4,5)*. The fragment produced by the RED reaction is a single-stranded triplet-repeat oligonucleotide. It is necessary to accurately determine the size of the fragment for the precise DNA-based diagnosis of the "triplet-repeat diseases."

4. In this study we examine whether CE can be used as an effective tool for the analysis of RED reaction (*see* **Note 1**). We examine the electrophoretic behavior of some synthetic single-stranded triplet-repeat DNA fragments of the sequence structure (CNG)n as a model of the RED products and of triplet repeat DNA, in comparison with the behavior of a random sequence ssDNA size marker.

3.3.2. Random Sequence Single-Strand Elements

1. We first examine a 20-bp DNA ladder as a control marker denatured at 94°C for 2 s to ensure single strandedness. Because the GC-contents of these fragments are each around 50%, they are considered to demonstrate a "standard" electrophoretic behavior. **Figure 2**

shows the electropherogram of the denatured 20-bp DNA ladder. Each peak is separated into two peaks representing the two complementary strands of each size of fragment, and suggesting that the dsDNA markers are completely denatured into ssDNA. The DNA fragments, with a range of 20–240 bp, are each well separated within 15 min.

2. To understand the electrophoretic behavior of DNA, we need accurate mobility data. We examine the repeatability of migration time and mobility of denatured single-stranded DNA marker by CE, using 0.7% methyl cellulose as a separation matrix for DNA separation at 100 V/cm electric field, which is supposed to be the most suitable condition for the tested DNA. The separations are performed 10 times for each sample. The mobility and its relative standard deviation (RSD) is calculated for using each migration time of the 10 runs. This allows the repeatability of ssDNA separation by CE to be determined. The results show that the maximum variation in mobility for all the fragment sizes of the denatured 20-bp DNA ladder is less than 2%, implying a very accurate reproducibility of each fragment separation.

3.3.3. Mobility of Triplet-Repeat Sequence Elements

1. We investigated the electrophoretic behavior of triplet oligonucleotides of 10 repeats of CAG, CTG, CCG, and CGG as described in **Fig. 1**, as models of the triplet-repeat disease DNA expansion elements. These sequences are each known to be the disease causing triplet-repeat expansion elements of particular genes, except for CGG.

2. **Figure 3** shows the electropherogram of triplet-repeat DNA fragment $(CCG)_{10}$, superimposed into the electropherogram of denatured 20-bp DNA ladder. It is obvious from **Fig. 3** that the migration time of triplet-repeat DNA fragment with 30-mer is much shorter than the time expected from the migration of the DNA ladder markers, with the mobility expected of a ~10–15-mer. A high electrophoretic mobility is observed for *all* of the triplet-repeat oligonucleotides tested, regardless of their sequence. This is one of the outstanding characters for triplet-repeat DNA.

3. **Figure 3** also clearly shows that an *additional* peak appears, as well as the main peak of triplet-repeat DNA. We have also evaluated the repeatability of the mobility of the triplet-repeat DNA fragments. The electrophoretic mobility and its RSD, over 10 independent and contiguous runs, is calculated to have a variation of 1% or less, indicating that each separation run can be performed with high accuracy, and that the high mobility of all triplet-repeat oligonucleotides in CE is highly reproducible.

4. The Ogston model *(6,7)* was applied for further consideration to understand such an unusual mobility of triplet repeat DNA fragments. The Ogston model is one of the most commonly used mathematical theories for the analysis of the electrophoretic behavior of small DNA fragment up to 200 bp. The mobility, μ, of a DNA molecule in a polymer solution such as the methyl cellulose polymer used here, is related to the DNA size in base-pair N, and the polymer concentration, T, as expressed as follows:

$$\ln(\mu) = \ln(\mu_0) - C*T*N \qquad (1)$$

in which, μ_0 is the mobility of DNA in a solution without polymer matrix, and C is a constant.

5. **Figure 4** shows the Ogston plot of triplet DNA fragments, $(CNG)_{10}$, and denaturated 20-bp DNA marker. The mobilities for tested triplet repeat oligonucleotides are much larger than that of random ssDNA markers of equivalent length. This suggests that the contour length of the triplet-repeat oligonucleotide becomes smaller than the contour length of the random sequence marker, because of some characteristic higher-order structures formed by G:C base pairs within triplet-repeat DNA.

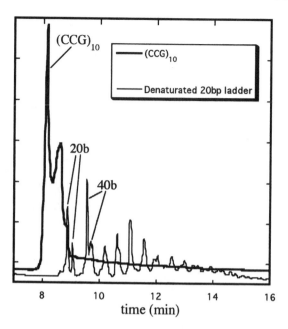

Fig. 3. CE separation of a denatured 20-bp DNA ladder and of $(CTG)_{10}$. Capillary: 100 μm id, 360 μm od, 27 cm total length, 19.1 cm effective length. Running buffer: 50 mM Tris-base, boric acid, pH 8.5, 0.7 % methyl cellulose, and 0.1 μmol/L YO-PRO-1. Run conditions: electric field: 100 V/cm; injection, 5 kV for 10 s; fluorescence detection: excitation 488 nm, and emission 560 nm.

6. The higher-order structures, which result in a contour DNA size smaller than expected for a truly single-strand molecule, have a great effect on their electrophoretic mobility and behavior. Since the oligonucleotides have a repeated sequence structure of CNG, the molecule is likely to fold back on itself to form a higher-order structure stabilized by C:G base-pairing *(8)*.

7. In addition, the different triplet-repeat sequences have differences in their mobility, probably because of the different thermal stability of the higher-order structure of each sequence due to the steric hindrance to folding imposed by the mismatched nucleotide N in each repeated triplet.

3.3.3.1. Mobility of Triplet-Oligonucleotides of Different Length

1. To get more detailed information for the electrophoretic behavior for triplet-repeat DNA, we use synthetic ssDNA as a control marker and compared the mobility of the sequence (CTG)n, which is also known to occur in a triplet-repeat disease gene.

2. Three lengths of the oligonucleotide (CTG)n of 30-, 60-, 90-mer are analyzed and their respective mobilities are compared with the mobilities of the synthetic a ssDNA marker fragment in **Fig. 5.**

3. The migration time for these triplet-repeat oligonucleotides is shorter than expected, although mobility increases with the increasing length of the oligonucleotides of (CTG)n from the 30-mer > 60-mer > 90-mer (*see* **Note 2**). Thus, one of the characteristics of such

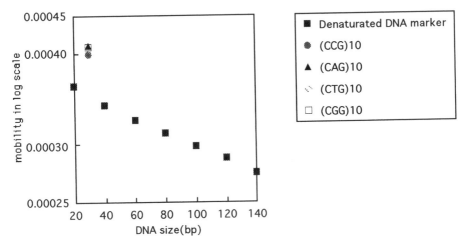

Fig. 4. Ogston plot of triplet DNA fragment (CNG)₁₀ and of a denatured 20-bp DNA marker.

triplet-repeat DNA fragments is that such fragments migrate faster than expected, from comparisons with truly single-strand random DNA sequence of the same length, which with random sequence cannot form stable, compact fold-back structures.

4. The electrophoretic behavior of 10X, 20X, and 30X repeats of the (CTG)n repeat element are analyzed simultaneously. As anticipated, the mobility of the (CTG)$_{20}$ and (CTG)$_{30}$ oligonucleotides are also much larger than that of expected by the Ogston model (*7*).

3.3.4. Model of Folded Structure of Triplet-Repeat Oligonucleotides

1. Recent biomedical research has shown that triplet-repeat DNA with the sequence structure (CNG), forms specific higher-order structures, which effect on the genomic expression and is related to a particular kinds of diseases. Some hairpin-fold models have been proposed involving ssDNA regions formed with 10X triplet-repeats elements.
2. Single-strand oligonucleotides of 10 repeats of the CTG sequence is thought to form two kinds of structures in the solution:
   a. The first is a stable fold-back structure, with the longest possible duplex stem made by base-pairing between G:C to form duplex and with a mismatch pair of T::T intervening between the duplex pairs, which because of its compact folded structure, the flexibility of DNA is reduced although the persistence length increases, which reduces the contour length of the DNA.
   b. The second form is a quasi-stable structure to account for the minor component with different electrophoretic behavior from the stable structure. This form could be the result of a shift of the duplex stem to involve one less G:C base-pair and result also in a short protruding sticky-end, possibly permitting weak alternating dimerization between oligonucleotides (*8*).
3. Similar sets of stable structures can also be considered to form with other sequence triplet-repeat oligonucleotides tested. Thus, the difference sequences of each triplet-repeat allows the several different higher-order structures to form, which differ in the compaction of their hairpin structure depending on the steric hindrance imposed by the mismatched bases in the duplex region and the size of the single-strand fold back loop.

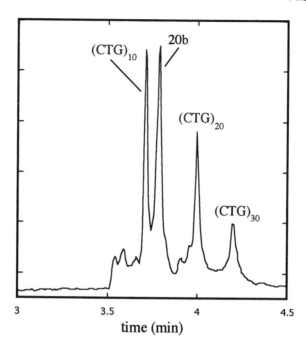

Fig. 5. CE separation of synthetic 20-nt single-stranded oligonucleotide and of $(CTG)_{10}$, $(CTG)_{20}$, and $(CTG)_{30}$. Capillary: 100 μm id, 19.1 cm effective length. Running buffer: 50 mM Tris-base, boric acid, pH 8.5, 0.7% methyl cellulose, and 0.1 μmol/L YO-PRO-1. Run conditions: electric field: 100 V/cm; injection, 5 kV for 10 s and with fluorescence detection: excitation 488 nm, emission 560 nm.

4. Notes

1. Interestingly, we can discover evidence of the formation of the higher-order structure by ssDNA using some experimental data obtained by CE. It is obvious that capillary electrophoresis can be an effective tool for the analysis of higher order structure of DNA in addition to physical techniques such as nuclear magnetic resonance (NMR), and thermal melting analysis.
2. Based on these experiments, we believe that CE could be an effective tool for the DNA diagnosis of triplet-repeat diseases in the near future *(5)*. Successful amplification and isolation of expanded triplet-repeat DNA elements of target genes would provide the basis for the accurate measurement of the length of the expanded triplet-repeat DNA with the understanding and expectation of their precise higher-order structure which will result in a measurable increase in electrophoretic mobility. An estimation of the molecular mass of the expansion region element using an independent technique such as matrix-assisted laser desorption ionization-time of flight (MALDI-TOF) spectroscopy *(9)* would also be advantageous for full characterization of the expansion elements.

References

1. Baba, Y. (1996) Analysis of disease-causing genes and DNA-based drugs by capillary electrophoresis. *J. Chromatogr. B* **687,** 271–302.

2. Paulson, H. L. and Fischbeck, K. H. (1996) Triplet repeat diseases. *Ann. Rev. Neurosci.* **19,** 79–107.
3. Kiba, Y. and Baba, Y. (1999) Capillary electrophoretic behavior and conformational analysis of triplet-repeat DNA. *Bunseki Kagaku* **48,** 193–203.
4. Kiba, Y. and Baba, Y. (1999) Unusual capillary electrophoretic behavior of triplet repeat DNA. *J. Biochem. Biophys. Methods* **41,** 143–151.
5. Kiba, Y. and Baba, Y. (submitted) A large DNA fragment migrates much faster than a small DNA fragment in polymer solution: Capillary electrophoresis as a new tool to elucidate the higher-order structure of human genes.
6. Kiba, Y., Ninomiya, M., and Baba, Y. (1999) DNA separation by capillary electrophoresis. *Chromatography* **20,** 27–35.
7. Viovy, J.-L. and Heller, C. (1996) Principles of size-based separations in polymer solutions, in *Capillary Electrophoresis in Analytical Biotechnology* (Righetti, P. G., ed.), CRC press, Boca Raton, Chap. 11., pp. 477–508.
8. Kasuga, T., Cheng, J., and Mitchelson, K. R. (2001) Magnetic bead-isolated single-strand DNA for SSCP analysis, in *Capillary Electrophoresis of Nucleic Acids*, Vol. 2 (Mitchelson, K. R. and Cheng, J., eds.), Humana Press, Totowa, NJ, pp. 135–147.
9. Marzilli, L. A., Koertje, C., and Vouros, P. (2001) Capillary electrophoresis — mass spectrometric analysis of DNA adducts, in *Capillary Electrophoresis of Nucleic Acids*, Vol. 1 (Mitchelson, K. R., and Cheng, J., eds.), Humana Press, Totowa, NJ, pp. 395–406.

19

Chemical Mismatch Cleavage Analysis by Capillary Electrophoresis with Laser-Induced Fluorescence Detection

Jicun Ren

1. Introduction

The chemical mismatch cleavage (CMC) analysis was first described by Cotton et al. (1) and has successfully been used for detection and identification of mutations in several genes implicated in causing human genetic disorders (2–7). Compared with some current methods, such as denaturing gradient gel electrophoresis (8) and single-stranded conformation polymorphism (SSCP) which detects polymorphism in short DNA fragments, CMC has a higher diagnostic sensitivity and can analyze larger DNA fragment lengths (9).

The CMC method involves the formation of heteroduplex dsDNA by annealing ssDNA from the wild-type and the mutant allele. The two alleles can either be derived from a heterozygous DNA sample or by combining two samples, one with wild-type DNA and one containing the mutant allele. During denaturation and reannealing, four duplexes are formed. Two of these will be the wild-type and the variant homoduplexes, and the other two are heteroduplexes. The nucleotide base at the mismatch site of the heteroduplex can react either with hydroxylamine or osmium tetroxide, which modify unpaired cytosine (C) or thymine (T) residues, respectively. DNA is then cleaved at the modified base(s) by piperidine, and the cleavage products are size separated by denaturing gel electrophoresis and detected using autoradiography (1,2), or by fluorescence (5–7,10). The method can efficiently detect a single point mutation as well as a small insertion or deletion.

Capillary electrophoresis (CE) has emerged as an attractive alternative to gel electrophoresis techniques for the analysis of DNA fragments with the advantages of speed, automation, and small sample requirements, combined with high separation efficiency (11–16). Low viscosity sieving media like short-chain linear polyacrylamide (SLPA) have the additional advantages of efficient filling of small dimension (<75 µm) capil-

From: *Methods in Molecular Biology, Vol. 163:*
Capillary Electrophoresis of Nucleic Acids, Vol. 2: Practical Applications of Capillary Electrophoresis
Edited by: K. R. Mitchelson and J. Cheng © Humana Press Inc., Totowa, NJ

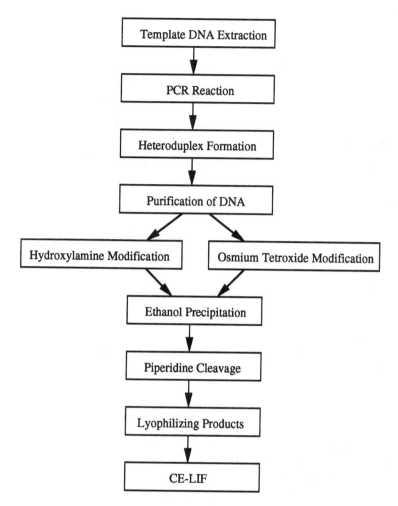

Fig. 1. A schematic protocol for CMC analysis using CE-LIF.

laries with sieving medium *(17–19)*. Small dimension capillaries ensure efficient heat dissipation and thereby fast analysis without loss of resolution. When coupled to laser-induced fluorescence detector (LIF), CE analysis is characterized by unsurpassed sensitivity. CE-LIF has been used for analysis of the CMC products and has successfully detected the T833C and G919A mutations of the *cystathionine β-synthase* gene (*CBS*) *(20)*. The basic protocol of CMC analysis based on CE is demonstrated in **Fig. 1**. Compared with the conventional CMC analysis based on slab-gel electrophoresis, the CE format affords relatively high resolution of ssDNA (down to 1 nt), precise size assessment of CMC products, sensitive detection with small sample requirement and fast analysis. Thus, CMC combined with CE-LIF is suitable for the screening of known mutations that give expected CMC products. It can also be used to detect unknown mutations, the location of which is indicated by the fragment size.

2. Materials
2.1. DNA Extraction
1. Whole blood samples.
2. The QIAquik Blood Kit (HT Biotechnology Ltd., UK), (*see* **Note 1**).
3. Eppendorf centrifuge with cooling.

2.2. PCR Reaction
1. Reaction tubes (thin-walled, Gene Amp).
2. Fluorescein-labeled primers. For example: the following primers are used for amplification of the exon 8 of *CBS* gene containing T883C and G919A mutations:
   5'-fluorescein-ACTGGCCTTGAGCCCTGAA-3' (forward), and
   5'-fluorescein-AGGCCGGGCTCTGGACTC-3' (reverse).
3. 10X PCR buffer (HT Biotechnology Ltd., UK): 100 mM Tris-HCl, pH 9.0, 500 mM KCl, 15 mM MgCl$_2$, 0.1% (w/v) gelatin and 1% Triton X-100, stored at –20°C.
4. 1.25 mM dNTP (dATP, dGTP, dCTP and dTTP), stored at –20°C.
5. Super *Taq* DNA Polymerase (HT Biotechnology Ltd., UK).
6. dd-Water: double-distilled and purified on a Milli-Q Plus water purification system (Millipore, Bedford, MA).
7. PCR thermocycler (Perkin-Elmer).

2.3. Formation and Purification of Heteroduplex DNA
1. Buffer A: 3 mM Tris-HCl, pH 7.7, 1.2 M NaCl, and 3.5 mM MgCl$_2$.
2. QIAquick PCR Purification Kit (QIAGEN Co. Germany).
3. TE buffer (1X): 10 mM Tris-HCl, 1 mM EDTA, pH 8.0.

2.4. Chemical Cleavage Reactions (see Note 2)
1. 4 M Hydroxylamine (Sigma Chemical Co.) adjusted with diethylamine to pH 6.0. The stock hydroxylamine solution should be aliquoted into 1-mL Eppendorf tubes and stored at –70°C; it is stable for up to 3 mo under these conditions.
2. 4% Osmium tetroxide solution (Sigma Chemical Co.). This stock solution should be stored at 4°C, and is stable for at least 3 wk.
3. *N*-(2-hydroxyethyl)piperazine-*N'*(2-ethane sulfonic acid) (HEPES).
4. 0.3% Osmium tetroxide solution containing 2% pyridine, 0.5 mM EDTA and 5 mM HEPES, pH 8.0 (use fresh immediately after dissolution).
5. tRNA (Sigma Chemical Co.): 315 mg/L sol, stored at –20°C.
6. 1 M Piperidine (make up fresh).
7. Stop buffer: 0.3 M sodium acetate, 0.1 mM EDTA, pH 5.2.
8. 100% Ethanol (keep at –20°C).
9. 70% Ethanol (keep at –20°C).

2.5. Measurement of Size of Chemical Cleavage Products
1. Fluorescein-labeled DNA marker standard (50–500 bp).
2. Formamide (AR grade).

2.6. CE Procedure
1. The <75-μm id polyacrylamide-coated capillary has an effective length of 30–40 cm (*19,21*).
2. CE instrument with LIF detector and an argon-ion laser with 488-nm emission.

3. Short-chain linear polyacrylamide *(21)*.
4. TBE buffer (1X): 89 m*M* Tris-base, 89 m*M* H_3BO_3, 2 m*M* EDTA, pH 8.3.
5. 7 *M* Urea in 1X TBE buffer, pH 8.3.

3. Methods

3.1. DNA Extraction

1. Collect whole blood samples (at least 200 μL).
2. Extract the DNA from whole blood using the "QIAquik Blood Kit" according to the instructions from the manufacturer (QIAGEN) (*see* **Note 1**).
3. Store the purified DNA at 4°C.

3.2. PCR Reaction

1. 100 μL PCR mixture contains: 10 μL of 10X PCR buffer, 10 μL of 1.25 m*M* dNTP, 2 μL of 10 μ*M* each of the forward and reverse primers, 1 μL *Taq* DNA polymerase (1 U), 5 μL DNA (at least 100 ng), and 72 μL of dd-water.
2. The PCR reaction requires an initial denaturation at 94°C, 4 min, followed by 36 cycles of: denaturation: 94 °C for 30 s, annealing: 58°C for 40 s, extension: 72°C for 20 s, and a final extension: 72°C for 7 min.

3.3 Formation of Heteroduplex DNA

1. Mix 100 μL of PCR product with 60 μL buffer A.
2. Heat the mixture at 96°C for 6 min.
3. Followed by DNA strand annealing at 42°C for 1.5 h.

3.4. Purification of Heteroduplex DNA

1. Transfer the heteroduplex DNA solution into a 1-mL Eppendorf tube, then add 2.5X volume of 100% ethanol (cold) and then mix gently.
2. Place the mixture into refrigerator and stand it for 1 h at –20°C.
3. Recover PCR product by centrifugation at 10,000*g* for 10 min at 4°C (*see* **Note 3**).
4. Wash the DNA pellet once with 500 μL of 70% ethanol by centrifugation at 10,000*g* for 10 min at 4°C.
5. Further purify the heteroduplex DNA using a QIAquick PCR Purification Kit according to the instructions of the manufacturer. The final volume of the purified DNA sample is 50 μL.

3.5. Hydroxylamine Modification Reaction

1. Mix 5 μL of DNA with 20 μL of 4 *M* hydroxylamine, (*see* **Note 4**).
2. Incubate the mixture at 37°C for 60 min.
3. Add 200 μL of stop buffer to the mixture and transfer it to ice to stop the modification reaction.
4. Precipitate the DNA with ethanol:
   a. Add 2.5X (the aqueous volume) of cold 100% ethanol and incubate for 1 h at –20°C.
   b. Recover the DNA by centrifugation at 8000*g* at 4°C for 10 min.
   c. Rinse the DNA pellet twice with 500 μL of cold 70% ethanol.

3.6. Osmium Tetroxide Modification Reaction

1. Mix 5 μL of DNA with 18 μL of 0.3% osmium tetroxide solution and 2 μL of 315 mg/L tRNA, (*see* **Note 4**).
2. Incubate the mixture at 25°C for 20 min.

3. Add 200 μL of stop buffer to the mixture and transfer it to ice to stop the modification reaction.
4. Precipitate the DNA with ethanol.
   a. Add 2.5X (the aqueous volume) of cold 100% ethanol and incubate for 1 h at –20°C.
   b. Recover the DNA by centrifugation at 8000g at 4°C for 10 min.
   c. Rinse the DNA pellet twice with 500 μL of cold 70% ethanol.

3.7. Piperidine Cleavage Reactions

1. Dissolve the DNA pellets from the hydroxylamine or the osmium tetroxide modification reactions into 50 μL of 1 M piperidine (*see* **Note 5**).
2. Incubate the mixture at 90°C for 30 min.
3. Transfer the mixture to ice and then lyophilize the samples under vacuum, (*see* **Note 6**).

3.8. CE Procedure

1. Prepare a 6% SLPA solution in 1X TBE buffer containing 7 M urea as a denaturing sieving medium (*see* **Note 7**).
2. Use 1X TBE as the electrophoresis separation buffer.
3. Rinse the new polyacrylamide-coated capillary with the separation buffer for 5 min.
4. A thin capillary (<75 μm id) should be used when LIF detection is compatible with the concentration of cleavage products.
5. Fill capillary with the 6% SLPA solution.
6. Introduce the DNA samples by electrokinetic injection, at –2 kV, for 6 s.
7. Run the CE at reverse polarity at constant electric field at –20 kV under denaturing buffer conditions at 25°C, typically for 20–25 min (*see* **Note 8**).
8. Replace the sieving medium in capillary between each run.

3.9. Detection of Chemical Cleavage Products

1. Add 12 μL 80% formamide to the lyophilized product and then dissolve it completely.
2. Heat the product at 95°C for 5 min and then cool it in ice water for 10 min.
3. Run the CE procedure at –20 kV under denaturing buffer conditions at 25°C, typically for 20–25 min (*see* **Note 8**).
4. Identify the genotype of the sample according to its electropherogram (*see* **Note 9**).
5. **Figures 2** and **3** show the typical electropherograms of the cleavage products derived from T833C and G919A genotypes of *CBS* gene, respectively (*see* **Note 10**).

3.10. Measurement of Size of Chemical Cleavage Products

1. Mix the DNA marker and the purified PCR product and dissolve into 12 μL of 80% formamide at 95°C for 5 min and then cool it in ice water for 10 min.
2. Run the CE procedure at –20 kV under denaturing buffer conditions at 25°C, typically for 20–25 min (*see* **Note 8**).
3. Regress the data and obtain the equation of the calibration curve (fragment size vs relative migration time) for the marker using PCR product as internal standard.
4. Calculate the size of the cleavage products on the basis of their relative migration times and the equation of the calibration curve (*see* **Note 11**).

4. Notes

1. The high-quality template DNA needs to be used in PCR reaction. We suggest that DNA is extracted from whole blood using the QIAquik Blood Kit, or classic methods such as phenol/chloroform extraction and ethanol precipitation.

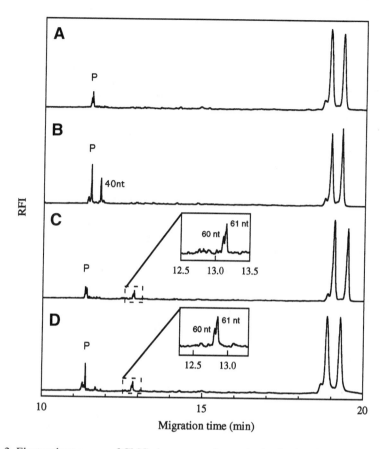

Fig. 2. Electropherograms of CMC cleavage products obtained using hydroxylamine as the chemical probe. PCR products are treated for 60 min at 37°C with 3.8 *M* hydroxylamine. **(A)** Wild-type sample; **(B)** T833C mutant heterozygous sample; **(C)** G919A mutant heterozygous sample; and **(D)** mixture (at ratio 1:1) of wild-type DNA sample and G919A mutant homozygous sample, (*see* **Note 12**). The electrophoresis was carried out under denaturing conditions at 25°C and the applied voltage was –20 kV. The letter P denotes peaks derived from residual primers and degradation products; RFI, relative fluorescent intensity. Reproduced from *Clinical Chemistry*, 1996; **44**, 2108–2114.

2. **SAFETY WARNING:** *Both osmium tetroxide and piperidine are toxic chemicals. Both skin and eye contact must be avoided completely. The handling and chemical cleavage reactions should be performed under a fume hood. The supernatants of ethanol precipitation after osmium tetroxide modification should be carefully collected for safe disposal.*
3. Purification of the heteroduplex DNA should be performed in two steps. The first step involves the removal of the large amount of salts present in heteroduplex solution by simple ethanol precipitation, as the heteroduplex solution contains a large amount of salts which greatly reduce the recovery of DNA using QIAquik PCR Purification Kit directly from heteroduplex formation buffers. The second step involves using the QIAquik PCR

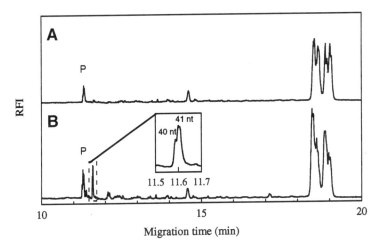

Fig. 3. Electropherograms of CMC cleavage products obtained by osmium tetroxide as chemical probe. The PCR products were treated for 20 min at 25°C with 0.32% OsO₄. Traces are: (**A**) a wild-type DNA sample; and (**B**) a T833C mutant heterozygous sample. The electrophoresis was carried out under denaturing conditions at 25°C and the applied voltage was –20 kV. The letter P denotes peaks derived from residual primers and degradation products; RFI, relative fluorescent intensity. Reproduced from *Clinical Chemistry*, 1996; **44**, 2108–2114.

Purification Kit for the removal of the fluorescent materials and fluorescein-labeled primers from the desalted DNA. The labeled materials interfere with the detection of the cleavage products.

4. The conditions used during the chemical cleavage reactions significantly affect the yield of the chemical cleavage reaction products and thus the detection of the cleavage products. We recommend using conditions for the chemical modification reactions that are in the range:
   Hydroxylamine modification: 3.0–3.8 *M* hydroxylamine at 35–37°C, for 45–75 min. Osmium tetroxide modification: 0.2–0.4% osmium tetroxide at 25–30°C, for 15–25 min.

5. The chemical modifications of unpaired nucleotides and the piperidine cleavage reactions can conveniently be performed in PCR apparatus, which offers precise temperature control.

6. In the conventional CMC protocol, after piperidine cleavage the cleaved products are collected by ethanol precipitation. In order to avoid loss of DNA during ethanol precipitation step, we suggest that the products from piperidine cleavage should be lyophilized rather than precipitated, which is suitable for the treatment of a large number of samples.

7. SLPA has low viscosity and is a good sieving medium used in DNA separation. Commercial CE instruments use automated pumps and capillaries can be easily filled and SLPA medium replacement can be easily carried out. High resolution of ssDNA (down to 1 nt) can be obtained using 6% SLPA as the sieving medium. Other commonly used media such as poly(ethylene oxide), cellulose derivatives, and long chain polyacrylamides have high intrinsic viscosity and significantly poorer DNA resolution compared to SLPA sieving medium. The hydroxypropyl methyl cellulose and poly(ethylene oxide) media are not generally suitable for commercial CE instruments as high concentrations are required for high-resolution performance which precludes easy filling of the capillary.

8. The electrophoresis current should be about 10–20 μA to avoid Joule heating.
9. The yields of the cleaved products depend significantly on the mismatch nucleotides occurring in the heteroduplexes, and in the adjacent sequence environments of the mismatch sites. Some studies including ours show that the G/T mismatch from a G-A, or a C-T mutation is inactive or only weakly reactive to osmium tetroxide modification.
10. The uncleaved heteroduplex product is used as an internal standard. The use of the relative migration time instead of the migration time can improve the reproducibility and precision of the size measurement of cleavage products.
11. The sequence-specific migration prevents the identification of the exact localization of the mutation site by CMC. The mutation location should be subsequently determined by DNA sequencing.
12. The CMC analysis can be used for the detection of homozygous mutant samples. Here, heteroduplex formation is obtained by mixing (1:1) the patient's DNA sample with a known wild-type DNA sample to create heteroduplex molecules containing mismatch sites.

References

1. Cotton, R. G. H., Rodrigues, N. R., and Campbell, R. D. (1988) Reactivity of cytosine and thymine in single-base pair mismatches with hydroxylamine and osmium tetroxide and its application to the study of mutations. *Proc. Natl. Acad. Sci. USA* **85,** 4397–4401.
2. Rodrigues, N. R., Rowan, A., Smith, M. E. F., Kerr, I. B., Bodmer, W. F., Gannon, J. V., and Lane, D. P. (1990) P53 mutation in colorectal cancer. *Proc. Natl. Acad. Sci. USA* **87,** 7555–7559.
3. Forrest, S. M., Dahl, H. H., Howells, D. W., Dianzani, I., and Cotton, R. G. H. (1991) Mutation detection in phenylketonuria by using chemical cleavage of mismatch: Importance of using probes from both normal and patient samples. *Am. J. Hum. Genet.* **49,** 175–183.
4. Roberts, R. G., Bobrow, M., and Bentley, D. R. (1992) Point mutation in the dystrophin gene. *Proc. Natl. Acad. Sci. USA* **89,** 2331–2335.
5. Verpy, E., Biasotto, M., Meo, T., and Tosi, M. (1994) Efficient detection of point mutation on color-coded strands of target DNA. *Proc. Natl. Acad. Sci. USA* **91,** 1873–1877.
6. Rowley, G., Saad, S., Giannelli, F., and Green, P. M. (1995) Ultrarapid mutation detection by multiplex, solid-phase chemical cleavage. *Genomics* **30,** 574–582.
7. Verpy, E., Biasotto, M., Brai, M., Misiano, G., Meo, T., and Tosi, M. (1996) Exhaustive mutation scanning by fluorescence-assisted mismatch analysis discloses new genotype-phenotype corrections in angioedema. *Am. J. Hum. Genet.* **59,** 308–319.
8. Khrapko, K., Hanekamp, J. S., Thilly, W. G., Belenkii, A., Foret, F., and Karger, B. L. (1994) Constant denaturant capillary electrophoresis (CDCE): a high resolution approach to mutational analysis. *Nucleic Acids Res.* **22,** 364–369.
9. Ellis, T. P., Humphrey, K. E., Smith, M. J., and Cotton, R. G. H. (1998) Chemical cleavage of mismatch: a new look at an established method. *Hum. Mutat.* **11,** 345–353.
10. Haris, I. I., Green, P. M., Bentley, D. R., and Giannelli, F. (1994) Mutation detection by fluorescent chemical cleavage: application to hemophilia B. *PCR Meth. Appl.* **3,** 268–271.
11. Grossman, P. D. and Soane, D. S. (1991) Capillary electrophoresis of DNA in entangled polymer solution. *J. Chromatogr.* **559,** 257–266.
12. Gelfi, C., Orsi, A., Leoncini, F., and Righetti, P. G. (1995) Fluidified polyacrylamide as molecular sieves in capillary zone electrophoresis of DNA fragments. *J. Chromatogr. A* **689,** 97–105.
13. Chang, H. T. and Yeung, E. S. (1995) Poly(ethyleneoxide) for high-resolution and high-separation of DNA by capillary electrophoresis. *J. Chromatogr. B* **669,** 113–123.

14. Hebenbrock, K., Williams, P. M., and Karger, B. L. (1995) Single strand conformational polymorphism using capillary electrophoresis with two-dye laser-induced fluorescence detection. *Electrophoresis* **16**, 1429–1436.

15. Ren, J., Deng, X., Cao, Y., and Yao, K. (1996) Analysis of DNA fragments and poly-merase chain reaction products from Tx gene by capillary electrophoresis with a laser-induced fluorescence detector using non-gel sieving media. *Anal. Biochem.* **233**, 246–249.

16. Ulvik, A., Ren, J., Refsum, H., and Ueland, P. M. (1997) Simultaneous determination of methylenetetrahydrofolate reductase C677T and factor V G1691G genotypes by mutagenically separated PCR and multiple-injection capillary electrophoresis. *Clin. Chem.* **44**, 264–269.

17. Ren, J., Ulvik, A., Ueland, P. M., and Refsum, H. (1997) Analysis of single-strand confor-mation polymorphism by capillary electrophoresis with laser-induced fluorescence detec-tion using short-chain polyacrylamide as sieving medium. *Anal. Biochem.* **245**, 79–84.

18. Ren, J., Ulvik, A., Refsum. H., and Ueland, P. M. (1999) Application of short chain polydimethylacrylamide as a sieving medium for the electrophoretic separation of DNA fragments and mutation analysis in uncoated capillary. *Anal. Biochem.* **276**, 188–194.

19. Ren, J. and Ueland, P. M. (1999) Temperature and pH effects on single strand conforma-tion polymorphism analysis by capillary electrophoresis. *Hum. Mutat.* **13**, 458–463.

20. Ren, J., Ulvik, A., Refsum, H., and Ueland, P. M. (1998) Chemical mismatch cleavage combined with capillary electrophoresis: detection of mutations in exon 8 of the cys-tathionine β-synthase gene. *Clin. Chem.* **44**, 2108–2114.

21. Ren, J. (2001) SSCP analysis by capillary electrophoresis with laser-induced fluorescence detector, in *Capillary Electrophoresis of Nucleic Acids*, Vol. 2 (Mitchelson, K. R., and Cheng, J., eds.), Humana Press, Totowa, NJ, pp. 127–134.

IV

PRACTICAL APPLICATIONS FOR ANALYSIS OF RNA AND GENE EXPRESSION

20

Quantitation of mRNA by Competitive PCR Using Capillary Electrophoresis

Stephen J. Williams and P. Mickey Williams

1. Introduction

The high sensitivity of the reverse transcriptase-polymerase chain reaction (RT-PCR) allows the amplification of low abundance mRNA transcripts, but the biomedical research and diagnostics community have increasing interest in the ability to accurately quantify the amounts of mRNA in cell samples. Although theoretically it should be possible to calculate the starting copy number from a consideration of the efficiency of the reverse transcription step of the reaction and the number of PCR cycles performed, it is widely recognized that this approach is flawed and can lead to highly inaccurate results *(1)*.

The accumulation of PCR product during the amplification phase of the reaction can be described by the equation,

$$[P]_n = [X] (1 + E)^n \tag{1}$$

in which, n is the number of PCR cycles, $[P]_n$ is the PCR product concentration after n cycles, [X] is the starting copy number after the reverse transcription step (starting concentration of cDNA) and E is the efficiency of the amplification. To accurately calculate [X], E must be known and be constant from reaction-to-reaction. This is not always the case, since E is sensitive to variations in the reaction conditions and to the presence of inhibitors in the reaction. Small changes in E can yield large changes in the final product concentration, due to the exponential nature of the PCR amplification process. In addition, if enough cycles are performed the PCR reaction enters the "plateau" phase, whereby $[P]_n$ approaches a maximum value and **Eq. 1** is no longer valid. The exponential amplification aspect of the PCR reaction can shut down in this manner, if the supply of primers and/or the nucleotides are depleted, or if the DNA polymerase ceases to function effectively (for example, due to denaturation). In the plateau phase, if renaturation of the duplex product becomes dominant over primer annealing,

From: *Methods in Molecular Biology, Vol. 163:*
Capillary Electrophoresis of Nucleic Acids, Vol. 2: Practical Applications of Capillary Electrophoresis
Edited by: K. R. Mitchelson and J. Cheng © Humana Press Inc., Totowa, NJ

this also effectively competes out the initiation of amplification and prevents further efficient product accumulation.

The reverse transcription step of RT-PCR is also associated with an efficiency factor:

$$[X] = [R] \, E' \tag{2}$$

in which, [R] is the concentration of the expressed message, and E′ is the amplification efficiency for transcription. E′ is also subject to considerable variation and is particularly sensitive to salt concentration. Fortunately, many of the practical problems associated with RT-PCR can be accounted for by the incorporation of an internal standard into the reaction mix, to provide a reference to correct for reaction conditions and for variation of E and E′. Quantitative competitive RT-PCR (QC RT-PCR) is a technique, when performed under carefully optimized conditions, that leads to accurate and reproducible quantitation of mRNA.

1.1. Quantitative Competitive RT-PCR

Traditionally, slab gels have been used to separate and quantitate QC RT-PCR reaction products. More recently, capillary electrophoresis (CE) incorporating laser-induced fluorescence (LIF) has emerged as an attractive alternative. CE-LIF offers many advantageous features: ease of automation, fast analysis times, and high sensitivity. The use of capillaries filled with noncrosslinked, replaceable gels ensures that run-to-run migration times of the analytes are reproducible, because each run is performed in fresh gel with identical formulation.

QC RT-PCR requires the development of an "internal standard" (also termed, "competitor," "control") that acts as a internal control or reference for the efficiency of the reaction. Ideally, the standard should be RNA rather than DNA sequence to account for the variability in the reverse transcription step of the reaction. For accurate results, the internal standard RNA must have a number of key features:

1. The primer-annealing sequence must be identical to that of the target sequence.
2. The standard must be largely similar in sequence to that of the target RNA, with a small deletion or insertion to allow electrophoretic size discrimination.
3. Above all, the standard must have an amplification efficiency equal to that of the target. The methods used for generating the internal standard are not reviewed in this chapter. Instead the reader is referred to the detailed protocols supplied by O'Connell et al. *(1)*.

QC RT-PCR analysis is generally performed by undertaking a series of several replicate RT-PCR reactions, with each reaction containing equal amounts of total RNA, while the amount of the internal standard added into the series of replicates is varied. The reactions are made replicate by performing the process under identical conditions, and stopping all of the reactions at the same number of PCR cycles. The amplified products are analyzed using CE or slab-gel electrophoresis. The internal standard template competes with the target template for the reaction components such as, the primer, the *Taq* DNA polymerase, the nucleotides, and so on. When the initial concentration of the target exceeds that of the standard, the predominant PCR product is derived from the target sequence. In the reverse case, when the concentration of the standard exceeds that of the target, the standard PCR product predominates. At the point where the starting concentrations of target and standard templates are identical, equal amounts of target and internal standard PCR product are generated. At a given cycle number, plotting the log of the ratio of the concentration of internal standard and target PCR products, (log [(standard)/(target)]), vs the log of the initial RNA internal standard concentration added to the reactions, (log [internal standard]), yields a straight line graph. The gradient of the line is equal to 1 when the reaction efficiencies of the target and standard are equal. The initial starting concentration of mRNA can then be determined by extrapolating to the x axis from the y value of zero (log 1 = 0). **Figure 1** illustrates this approach.

Ideally, precise quantitation involves validating the system with a range of known input copy numbers for both target and standard sequences. Given the correct choice of an internal standard, an attractive feature of QC RT-PCR is that the unknown mRNA may be quantified during either the log-linear phase of the PCR reaction, or during the plateau phase of the reaction. This is advantageous because it is unnecessary to verify that the electrophoretic analysis is performed with a product generated at a particular number of cycles that lies in the exponential region of the amplification. Instead, one can assay the product at some high cycle number (for example, 40 cycles) without regard for the reaction kinetics at this point. Indeed, allowing the reactions to proceed to this later plateau phase ensures there is enough product to prevent any detection sensitivity problems.

1.2. CE Analysis of QC RT-PCR Products

Most QC RT-PCR assays result in amplification of products in the 100–500-bp range. Given that the target and internal standard sequences should match one another closely in terms of sequence, and therefore length, CE methodologies adopted for QC RT-PCR analysis should have the capability of separating DNA fragments up to 500 bp, with approx 10–20-bp resolution. The choice of the capillary sieving polymer is an important decision that considers speed of analysis, ease of use, and desired separation performance.

A distinction between noncrosslinked gel sieving matrices employed in CE can be made on the basis of the ease of replacement of the gel in the capillary. Gels, such as 0.75% (w/v) hydroxyethyl cellulose (HEC), can generally be pumped in and out of the capillary at pressures < 20 PSI. Whereas higher viscosity gels, for example the high molecular weight linear polyacrylamides, require either polymerization *in situ* or much higher pressure (>100 PSI) to replace them in the capillary. In addition, many of the low viscosity gels do not require polymerization by the user, which can be a significant advantage in terms of ease and speed of matrix preparation and the speed of capillary regeneration between runs. In these cases, it is necessary only to dissolve a known amount of the polymer in the appropriate buffer solution. Generally speaking, it is possible to use the low viscosity gels for separation of double-stranded PCR products in the size range and with the resolution appropriate for QC RT-PCR analysis.

1.2.1. Capillary Coating

The modification of the internal surface of a fused silica capillary to suppress electroosmotic flow is critical for high efficiency separation of DNA *(2)*. The most common and most successful treatment has been covalent attachment of polyacrylamide to

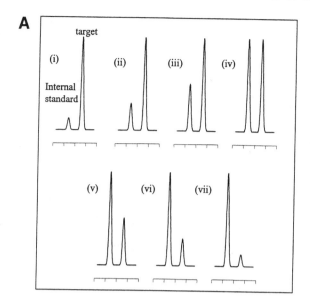

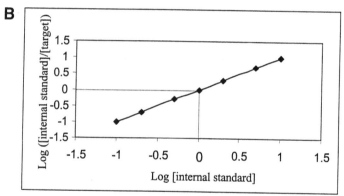

Fig. 1. The QC RT-PCR method. (A) Model electropherograms obtained when the ratio of [internal standard]:[target] is: (i) 1:10; (ii) 1:5; (iii) 1:2; (iv) 1:1; (v) 2:1; (vi) 5:1, and (vii) 10:1. (B) Graphical analysis to determine the concentration of starting target RNA for the model system depicted in (A).

the surface. A detailed protocol for this type of surface derivitization is included in the methods section. It should be noted that some replaceable polymer sieving matrices are considered to be "self-coating." Certain modified polyacrylamides, for example, can absorb to the capillary wall, and provide effective suppression of the electroosmotic flow. The absorbed polymer can be either the sieving polymer itself or a low concentration of an additional polymer that is added to the separation matrix. In either case, ample flushing of the bare fused silica capillary with the matrix prior to the separation usually ensures a good wall coating.

1.2.2. Signal Detection

The sensitivity of UV detection suffers in CE because of the short optical pathlength (~75 µm) provided by the internal diameter of the capillary. Without the use of large reaction volumes and desalting procedures to allow large amounts of product to be loaded on capillary UV detection will most likely provide insufficient sensitivity for QC RT-PCR product analysis. LIF is the preferable detection technique. Fluorescent labels can be incorporated into PCR products through the use of fluorescent primers, or alternatively post-PCR labeling is possible by adding a fluorescent intercalating dye to the CE sieving matrix.

1.3. Future Development of QC RT-PCR

The methodologies presented in this chapter allow the reader to perform accurate and reproducible quantitation of mRNA using QC RT-PCR, with analysis using CE-LIF (*see* **Note 1**). The benefits of using CE-LIF as an end point detection technique lie in the speed, the sensitivity and the ease of automating the assay procedure. For example, commercial 96-capillary array instruments are available for high-throughput applications in clinical diagnostics. In the future, rapid multiplexed PCR product sizing may be accomplished on miniaturized, disposable electrophoresis chips, which are already under development. We also expect to see fully integrated microfabricated devices incorporating sample preparation, PCR amplification and electrophoretic sizing and detection *(3)*. For QC RT-PCR such devices are attractive since they minimize user intervention, allow automation of all steps in the process, and speed the "sample-to-answer" times, potentially to within several minutes.

2. Materials

2.1. RNA Extraction

1. RNA extraction is most easily accomplished using one of the readily available commercial kits. For example, RNeasy kits (QIAGEN. Inc., Valencia, CA) provide sufficiently pure RNA for RT-PCR analysis.
2. Most kits are based on the original protocols of Chirgwin et al. *(4)*, Chomczynski and Sacchi *(5)* or Boom et al. *(6)*, which may be employed if desired.

2.2. RT-PCR Reagents

1. As with RNA extraction methods, many sources are available for the necessary RT and PCR enzymes. We prefer to perform both RT and PCR reactions in a single tube, rather than as separate reactions. The advantage of a single tube reaction is that errors and contamination, associated with sample transfer, are eliminated.
2. Several single tube reaction strategies are available: use a single enzyme which possesses both a thermally stable RT and DNA polymerase activity, for example, *rtTh* (Applied Biosystems, Foster City, CA), or use a mixture of two enzymes such as *MuLV* RT and AmpliTaq Gold (Applied Biosystems) which provide thermally stable RT and DNA polymerase activity, respectively. Both approaches have been successfully employed for QC RT-PCR.
3. Additional reagents necessary for combined RT-PCR reactions, such as nucleotides, buffers, and so on can also be purchased in kit form, for example, GeneAmp *rtTh* RT-PCR kit (Applied Biosystems).

2.3. Capillary Electrophoresis

1. The CE analysis of QC RT-PCR reactions should be performed on a system equipped with LIF detection (argon-ion laser excitation is assumed in the following protocol description).
2. Fused silica capillaries (Polymicro Technologies Inc., Phoenix, AZ) are typically 40 cm long with a 75 μm id.
3. Linear polyacrylamide derivatization of the capillary surface requires hydrochloric acid, sodium hydroxide, acetic acid, acetonitrile, 3-methacryloxypropyltrimethoxysilane, acrylamide, ammonium persulfate (APS) and TEMED (Sigma St Louis, MO).
4. Hydroxypropyl methyl cellulose (HPMC) can be obtained from Sigma.
5. The DNA intercalating dye, thiazole orange (TO), can be purchased from Aldrich (Aldrich Chemical Company, Milwaukee, WI).
6. Tris/Borate/EDTA (1X TBE: 89 mM Tris-base, pH 8.5, 89 mM boric acid, 2 mM EDTA) can be prepared from the individual components, or be purchased as a prepared 10X stock solution (Research Genetics, Huntsville, AL).

3. Methods

The following protocols guide the reader through setup, execution, and analysis of a successful QC RT-PCR experiment. The generation of the RNA internal control is not discussed in detail. Instead, the reader is referred to O'Connell et al. *(1)*.

3.1. RNA Extraction

1. RNA is prepared from the tissue or cells of interest using commercial mRNA isolation kits, following the suggested protocol of the manufacturers (*see* **Note 2**).

3.2. Reverse Transcriptase-Polymerase Chain Reaction

The steps outlined below are specific to a *rtTh*, single tube, single enzyme reaction kit.

1. Using a commercially available RT-PCR kit, prepare a master mix of reaction components on ice, as described in the manufacturer's supplied protocol.
2. Split the master mix into the desired number of aliquots. The exact number of RT-PCR reactions performed will depend on the precision of quantitation required, and on any prior knowledge of the initial RNA concentration.
3. In the absence of information about the latter, it is advisable to span a wide range of concentrations at which the internal standard is added to the reaction mixtures. A suitable range might cover 5 orders of magnitude (i.e., 10^2–10^7 copies), in which case 5X RT-PCR reactions would need to be performed.
4. Spike each RT-PCR reaction with the internal standard at the relevant concentration followed by a constant amount of total RNA sample (~50 or 100 ng).
5. RT is performed at 60°C for 30 min to generate cDNA products from the mRNA templates.
6. The RT reaction is immediately followed by the PCR process, in which sufficient product is generated for detection. The cycle temperatures are generally, denaturation at 95°C for 15 s, annealing at 55–65°C for 30 s, extension at 72°C for 30 s (*see* **Note 3**).
7. The number of PCR cycles performed varies, but generally 40 cycles are sufficient even for the detection of initially single copy mRNA species (*see* **Note 4**).

3.3. CE Analysis of RT-PCR Reactions

1. Derivitization of the capillary surface to suppress electroosmotic flow can be accomplished as follows (**steps 1–4**). Flush the capillary with 1 M HCl for 1 h. Rinse the capillary with 20 capillary volumes of water and then flush with 1 M NaOH for a further hour.

2. Prepare a solution of 0.4% (v/v) 3-methacryloxypropyltrimethoxysilane (Bind Silane), 0.2% acetic acid in acetonitrile, and flush the capillary for 3 h. This results in a coating of bifunctional reagent on the inner capillary surface.

3. Prepare a 3% (w/v) acrylamide solution in water and degas the solution under moderate vacuum (~15 mm/Hg) for 1 h. Prepare a fresh 10% (w/v) solution of APS in water.

4. Add 10 µL of the APS and 5 µL of TEMED to 1 mL of the degassed acrylamide solution and then immediately flush the capillary with 20 capillary volumes of this solution. Cap the capillary ends to prevent evaporation and leave the solution overnight to polymerize onto the capillary surface. Following the polymerization, rinse the capillary with water and store it in a dry place until it is required.

5. Dissolve HPMC (0.5% w/v) in a 1X TBE (89 mM Tris-base, 89 mM boric acid, 2 mM EDTA, pH 8.5) buffer with overnight stirring at room temperature (*see* **Note 5**). Filter the sieving matrix through a 1.2-µm membrane filter. While gently stirring the polymer solution, add the TO intercalator dye to a final concentration of 1 µM (*see* **Note 6**).

6. Degas the polymer solution under 30 mm Hg vacuum prior to use. The capillary is filled with the polymer under pressure for each run.

7. Samples can be injected into the capillary using either hydrodynamic injection or electrokinetic injection (*see* **Note 7**). The capillary should be refilled with fresh polymer for each injection. Typical injection conditions would be 5 s at 0.5 psi for a hydrodynamic injection, and 10 s at 5 kV for an electrokinetic injection.

8. Electrophoresis of the samples should be performed at approx 300 V/cm, which for a 40-cm capillary corresponds to an applied voltage of 12 kV.

3.4. Data Analysis

1. Tabulate the target and the internal standard peak areas for each RT-PCR reaction electropherogram. The identities of peaks representing the target and internal standard can be confirmed by CE analysis of individual reactions containing only one of the products. In addition, the comparison of the migration times of these peaks with the migration times of a DNA size ladder, run under identical separation conditions, provides further confirmatory evidence (*see* **Note 8**).

2. If the size difference between the internal standard and target product is small (as it should be to ensure matched amplification efficiencies), then it is unnecessary to correct for the length dependence of dye intercalation. By a similar argument, migration time normalization of the peak areas is also expected to have only a minor effect on the accuracy of quantitation (*see* **Note 9**).

3. Plot a graph of the log of the ratio of the internal standard peak area/target peak area (*y* axis) vs the log of the initial internal standard concentration (*x* axis). A linear plot with a slope close to 1 should be obtained (*see* **Note 10**).

4. The initial starting concentration of the mRNA can be determined by extrapolating to the "*x*" axis from a "*y*" value of zero (log 1 = 0) (*see* **Fig. 1**). If the curve is nonlinear, the amplification efficiency of the target and control do not have a constant relationship, and thus accurate quantitation is not possible (*see* **Note 11**).

4. Notes

1. The commercialization of "real-time" or "kinetic" PCR reactions allows the user to quantitate mRNA levels using RT-PCR, but without the need for electrophoretic separation of the reaction products. The technique reduces the time and effort, but requires dedicated instrumentation to perform the assay. Currently, two commercial systems are available, from Applied Biosystems and from Roche Molecular Biochemicals (Indianapolis, IN).

Briefly, the cycle-by-cycle accumulation of PCR product is monitored in real-time by following the degradation of a dual-labeled *Taq*Man® fluorescent probe. The 5' nucleolytic activity of *Taq* DNA polymerase cleaves the probe, allowing the fluorescent labels to become spatially separated and thus increasing fluorescent signal. Input mRNA levels can be calculated at a threshold fluorescence level based on a correlation between the mRNA concentration and PCR cycle number. Increase in product concentration can also be monitored using the incorporation of fluorescent DNA intercalators.

2. The method of choice should allow the RNA to be fully purified, free from contaminating ribonuclease activity, proteins and genomic DNA. Obviously, ribonuclease activity will degrade the RNA. Contaminating proteins may inhibit the efficiency of enzymatic amplifications and genomic DNA may serve as a target for amplification, confounding the quantitative results.

3. When using *MuLV* or *AMV* reverse transcriptase in a "2 enzymes—1 tube reaction," the temperatures for the reverse transcription step are usually set at 48°C. The optimum primer annealing temperature is determined by the specific reaction makeup and the sequences of the primers, and should be optimized for each new primer set and target mRNA.

4. Since quantitation is possible and accurate even in the plateau phase of the PCR reaction, it is advisable to perform QC RT-PCR analysis in this region of the amplification reaction to minimize sensitivity problems during product detection.

5. A variety of different polymer sieving matrices provide good separation of dsDNA in the size range 50–500 bp. Linear polyacrylamide and derivatives of polyacrylamide are perhaps the most widely used. Premixed solutions, optimized for a particular DNA size range, can be purchased commercially. In-house derivitization of these solutions is not recommended because of the inherent variable nature of this type of polymerization process. Cellulose derivatives, such as HPMC or HEC, are also highly effective for separation of DNA of this size, and need only be dissolved in the appropriate buffer system prior to use. The removal of fine, undissolved particles of the matrix celluloses by passage through a 1.2-μm membrane filter under low pressure is necessary for uniform electrophoresis runs.

6. The detection limit for TO is around 4 amol of dsDNA per fragment band. TO is efficiently excited by the 488-nm line of the argon-ion laser. Fluorescence can be collected at 520 nm with a suitable band-pass filler.

7. Electrokinetic injection imparts a bias on the sampling, since the analytes are injected into the capillary based on their electrophoretic mobility, given by the relationship:

$$Q = \mu_{ep} r^2 E C t$$

in which, Q is the quantity of the analyte injected, μ_{ep} is the analyte electrophoretic mobility, r is the radius of the capillary, E is the applied field strength, C is the concentration of the analyte, and t is the time of the injection. In addition, the quantity of analyte injected is influenced by the overall salt concentration of the sample because of the sample-stacking phenomenon *(7)*. Whereas the peak area of an individual analyte will be affected by the above factors, the relative peak area (target/internal standard) will not change. In this respect, the internal standard acts as a control for the variability of the injection. Note that this ratio is also self-correcting for other separation parameters, such as the run-to-run variability of intercalator dye incorporation. In the event that detection sensitivity is lacking, the sample can be desalted and injected electrokinetically to improve the signal strength. The easiest method of desalting is using membrane dialysis *(8)*. Removing salts from a reaction product can increase the amount of product that can be electrokinetically loaded and hence the signal by up to two orders of magnitude.

8. If the target is expressed in low levels, <10 copies/cell, the likelihood of non-specific amplification products is increased, and the resulting electropherogram can be difficult to interpret due to the multiple product peaks. Many potential steps can be taken to minimize the nonspecific amplification, such as selecting an alternative primer set, hot-start PCR, nested primer reactions, increasing the annealing temperature, and so on. In a complex electropherogram, the identity of a peak may be ascertained if pure target and standard RNA can be analyzed individually. Pure target RNA can be produced by conventional molecular biology techniques.

9. To normalize for the length dependence of dye intercalation, multiply the area of the internal standard peak by the ratio of the number of target base pairs/the number of standard base pairs.

10. The gradient of the plot of log ([standard]/[target]) vs log [initial standard] gives information on the relationship of the amplification efficiencies of target versus standard. A slope equal to 1 is generally accepted as evidence that both target and standard have matched efficiencies. However, Hayward et al. *(9)* have noted that in certain cases the value of the gradient in these plots can be misleading. A more rigorous approach to demonstrate equal amplification efficiencies involves performing multiple RT-PCR reactions containing standard and target at known concentration. Reactions are stopped at various cycle numbers (for example every 3 cycles from cycle 25–50) and the products analyzed by CE.

11. A possible source of error during QC RT-PCR analyses is the formation of heteroduplex molecules during the PCR stage of the reaction. A heteroduplex is a duplex formed between either strand of the target and the complementary strand of the standard product. Errors in quantitation can occur if the heteroduplex comigrates with either the target peak, or with the internal standard peak, resulting in an over-estimation or under-estimation of a particular product. Errors due to the formation of a heteroduplex are more likely in situations where the ratio of internal standard to target is very high, or very low (i.e., a large excess of either product). CE peaks with shoulders, or with non-Gaussian shapes can be indicative of the formation of heteroduplexes.

References

1. O'Connell, J., Goode, T., and Shanahan, T. (1998) Quantitative measurement of mRNA expression by competitive RT-PCR, in *Methods in Molecular Biology, Vol. 92: PCR in Bioanalysis* (Meltzer, S. J., ed.) Humana Press, Inc., Totowa, NJ, pp. 183–193.
2. Chiari, M. and Cretich, M. (2001) Capillary coatings: choices for capillary electrophoresis of DNA, in *Capillary Electrophoresis of Nucleic Acids*, Vol. 1 (Mitchelson, K. R. and Cheng, J., eds.), Humana Press, Totowa, NJ, pp. 125–138.
3. Xie, W., Yang, R., Xu, J., Zhang, L., Xing, W., and Cheng, J. (2001) Microchip-based capillary electrophoresis systems, in *Capillary Electrophoresis of Nucleic Acids*, Vol. 1 (Mitchelson, K. R. and Cheng, J., eds.), Humana Press, Totowa, NJ, pp. 67–83.
4. Chirwgwin, J. J., Przbyla, A. E., MacDonald, R. J., and Ruter, W. J. (1979) Isolation of biologically active ribonucleic acid from sources enriched in ribonuclease. *Biochemistry* **18,** 5294–5299.
5. Chomczynski, P. and Sacchi, N. (1987) Single step method of RNA isolation by acid guanidinium thiocyantae-phenol-chloroform extraction. *Anal. Biochem.* **162,** 156–159.
6. Boom, R., Sol, C. J., Salimans, M. M., Jansen, C. L., Wertheim-van Dillen, P. M., and van der Noordaal, J. (1990) Rapid and simple method for purification of nucleic acids. *J. Clin. Microbiol.* **28,** 495–503.

7. Olechno, J. D. and Nolan, J. A. (1996) Injection methods in capillary electrophoresis, in *Capillary Electrophoresis in Analytical Biotechnology* (Righetti, P. G., ed.), CRC Press, New York, pp. 61–99.

8. Ruiz-Martinex, M. C., Salas-Solano, O., Carrilho, E., Kotler, L., and Karger, B. I. (1998) A sample purification method for rugged and high-performance DNA sequencing by capillary electrophoresis using replaceable polymer solutions. A: development of the cleanup protocol. *Anal. Chem.* **70,** 1516–1527.

9. Hayward, A. L., Oefner, P. J., Sabatini, S., Kainer, D. B., Hinojos, C. A., and Doris, P. A. (1998) Modeling and analysis of competitive RT-PCR. *Nucleic Acids Res.* **26,** 2511–2518.

21

Quantitative RT-PCR
from Fixed Paraffin-Embedded Tissues
by Capillary Electrophoresis

Giorgio Stanta, Serena Bonin, and Maurizio Lugli

1. Introduction

It is possible to analyze RNA extracted from paraffin embedded-tissues that have been fixed in formalin *(1–3)*. In fact, all biopsy or surgical tissues are routinely fixed and paraffin-embedded and conserved for many years for future examination, and present a very valuable medical and research resource for epidemiological studies, and for the development of public health policies. The specific analysis of RNA remaining embedded in carefully preserved tissues facilitates the utilization of human pathological tissue samples for such diagnostic purposes.

Some of the methods for the quantitative extraction of RNA from microdissection of fixed tissues *(2,4)* propose to utilize single 6–8-μm histological sections dissected from the paraffin blocks. The RNA is chemically isolated with standard procedures and is then first transcribed into cDNA and then subsequently amplified with the polymerase chain reaction (PCR) as double-strand DNA. These methods are derived from techniques used to quantify RNA from fresh tissue samples *(5,6)*. However, the quantitation of RNA from paraffin-embedded tissues present specific problems that cannot be resolved with a competitive analysis, because of RNA degradation may have occurred in the preserved samples. Since degradation is presumably random in such tissues, the relative amounts of a target RNA and several internal RNA "standards" may be able to be used to infer the relative ratios of each of the RNA species. These relative amounts of RNA may then be related to amounts known to occur in fresh tissues. We describe a relative, quantitative analysis of specific RNA sequences from human tissues that have been fixed and then paraffin-embedded, with the use of capillary electrophoresis (CE) technology.

For each gene specific mRNA, there is linear relationship which must be established between the initial amount total RNA and the amount of specific gene product, after reverse transcriptase-PCR (RT-PCR). This involves the determination of the

From: *Methods in Molecular Biology, Vol. 163:*
Capillary Electrophoresis of Nucleic Acids, Vol. 2: Practical Applications of Capillary Electrophoresis
Edited by: K. R. Mitchelson and J. Cheng © Humana Press Inc., Totowa, NJ

range of the linear relationship for the log of the initial concentration of total RNA and the log of the concentration of specific product *(7,8)*. Briefly, the method involves the reverse transcription of RNA with *Avian Myeloblastosis Virus Reverse Transcriptase* *(AMV-RT-ase)* using gene specific antisense primers, in which a variable number of cycles of PCR amplification are performed. The number of cycles must be in the range of linearity between the log of the amount of amplification product and the number of cycles. The level of degradation of each different RNA sample can be different. For this reason, it is necessary in the application of quantitative RT-PCR in paraffin-embedded tissues to standardize the quantity of specific gene product by comparison with the expression at mRNA level of another "housekeeping gene" like *β-actin* or *glyceraldehyde 3-phosphate dehydrogenase* *(GAPDH)*. The amplification products are quantified by CE with laser-induced fluorescence (LIF) detector. The methods here reported are already described in other publications of the authors *(2,4,7,8)*.

2. Materials

1. Digestion solution (100 mL) for RNA: consists of 40 mL of 4 M guanidine thiocyanate, 3 mL of 1 M Tris-HCl pH 7.6, 2.4 mL of 30% sodium N-Lauryl sarcosine, and 54.6 mL of diethyl pirocarbonate (DEPC)-treated H_2O (*see* **Note 1**).
2. β-Mercapto-ethanol: 0.28 μL must be added per 100 μL of the digestion solution just before use.
3. Proteinase K: 20 mg/mL: solubilize the protein in 50% DEPC-treated water and 50% glycerol, and store at –20°C until use.
4. Phenol equilibrated in water.
5. Glycogen: a solution of 1.0 mg/mL in water. Store in aliquots at –20°C until use.
6. Tris-EDTA buffer, 1X: 10 mM Tris-HCl, pH 8.0, 1 mM EDTA. Make a 10X stock solution and dilute with DEPC water for use. Store the stock at room temperature.
7. 50 mM Tris-HCl, pH 8.0 (stock solution 1M). Store the stock at room temperature.
8. Ethanol for DNA precipitation (100 mL): 95% ethanol (96.7 mL) with the addition of 3 M sodium acetate pH 7 (3.4 mL), store at –20°C until use.
9. AMV-RT-ase: obtain from Biochemicals suppliers, e.g., Promega.
10. AMV buffer 1X: 50 mM Tris-HCl, pH 8.3, 50 mM KCl, 10 mM MgCl$_2$, 10 mM DTT, 0.5 mM spermidine. This buffer is stored as 5X stock solution in small aliquots at –20°C until use.
11. The dNTPs are aliquoted together in a stock solution 10 mM each and stored at –20°C.
12. Oligonucleotides: both sense and antisense PCR primers are made as stock solutions at 30 pmol/μL and stored in small aliquots at –20°C.
13. *Taq* DNA polymerase can be obtained from Pharmacia BioTech.
14. PCR buffer 1X: 10 mM Tris-HCl, pH 8.3, 55 mM KCl. The buffer is stored as a 10X stock solution at –20°C.
15. eCAP® LIF FLUOR dsDNA 1000 Kit: (part number 477407, Beckman Coulter, CA). This kit contains: eCAP® Coated Capillary 47 cm × 100 μm id, eCAP® Gel Buffer, and EnhanCE® Intercalator.

3. Methods
3.1. Cutting Paraffin Sections

1. Sections 6–8 μm thick are cut with a microtome. It is not necessary to change the blade for every sample to ensure cleanliness because the paraffin in the tissue cleans the microtome blade at every cut.

2. We routinely clean the blade after each sample with xylene to eliminate paraffin residues. Place the cut sections into sterile 1.5-mL Eppendorf tubes (4).

3.2. Microdissection

1. A very easy and inexpensive method can be used to separate in a specimen tissue from different surrounding tissues (e.g., tumor tissue from surrounding normal tissues).
2. The histological sections must be cut in sequence, the first section is stained immediately with hematoxilin and eosin to allow for tissue recognition and for the specimen location to be determined. The immediately subsequent sections are separated under a stereomicroscope with a sterile needle and the desired tissue component collected.

3.3. Deparaffinization

1. The first step for the RNA extraction from paraffin-embedded tissues is the elimination of the paraffin fixative. Paraffin is solubilized using an organic solvent such xylene. The tissue is then washed several times with ethanol to completely eliminate residual xylene, as even small amounts will block the activity of the enzymes used in the subsequent steps (4).
2. Place a maximum of five sections (each of 6–8 μm) in a 1.5-mL Eppendorf tube, and solubilize the paraffin by washing twice with 1 mL of xylene for 5 min at room temperature. Pellet the tissue section by centrifugation at 7000g after each wash, and carefully remove the solvent by micropipet.
3. Wash the tissues with 1 mL of absolute ethanol for 10 min, and once with 95% ethanol to remove residual xylene. Pellet the tissue section by centrifugation at 7000g after each wash and carefully remove the ethanol by micropipet.
4. Air-dry the tissue in a thermoblock at 37°C with the tube lids open for about 20 min.

3.4. Protein Digestion

1. The proteins must be removed from the preserved tissue by digestion with proteolytic enzymes (2,4).
2. Add 1 vol (100–300 μL) of digestion solution containing β-mercapto-ethanol to each dried section. This is usually from 100–300 μL, but depends on the quantity of tissue.
3. Add proteinase K to a final concentration of 6 mg/mL (*see* **Note 2**).
4. Incubate the tissue overnight at 45°C with gently swirling in a thermomixer.

3.5. RNA Extraction

1. Add 1 vol of a 70:30 (v/v) mixture of phenol-water:chloroform to each tube.
2. Mix the emulsion thoroughly by vortexing for several times, then place the tube in ice for 15 min.
3. Then centrifuge at top speed (7000g) in a microcentrifuge for 20 min at 4°C.
4. Collect the upper aqueous phase into a new tube. Take care to avoid the proteinaceous interface between the two phases.

3.6. RNA Precipitation

1. The small amounts of RNA can be concentrated by precipitation with isopropanol. Glycogen is added as an inert "carrier" to aid precipitation of the RNA.
2. Add 5 μL of glycogen (1 mg/mL) and 1 vol of isopropanol to the aqueous phase and keep the solution at –20°C for 48 h. A long incubation at –20°C is necessary to obtain an efficient and quantitative precipitation of the fragmented RNA species.
3. Pellet the RNA by centrifugation in a microcentrifuge at 7000g for 20 min at 4°C.
4. Wash the pellet in 100 μL of 75% ethanol stored at –20°C. The RNA pellet can be safely stored at –80°C for very long time with 75% ethanol until use.

5. Prior to decanting the ethanol, pellet the RNA by centrifugation at 7000g in a microcentrifuge for 5 min, and air-dry the pellet.
6. Resuspend the RNA pellet in 25 µL of DEPC-treated water and store at –80°C until use. About 1–2 µg of total RNA is isolated from one tissue section of 1 × 1.5 cm and 6 µm thick.

3.7. mRNA Quantification

1. Messenger RNA can be quantitatively extracted and estimated from paraffin-embedded tissues with RT-PCR and CE analysis.
2. Two oligonucleotides are used, one in mRNA sense and one in antisense orientation. The latter is used for the reverse transcription. The sense and the antisense oligonucleotides must locate in two successive exons of the gene, to allow for amplification of a "short" product. The short product is distinguished from potential "long" amplification products from contaminant genomic DNA, which contains introns separating the exon sites.
3. The segments of mRNA chosen for studies are usually very short, between 75 and 100 bases, because the degradation of the RNA in tissue samples would preclude accurate quantification if longer products were chosen (*see* **Note 3**).
4. The amplification conditions must be determined very carefully because a relative quantitation is only possible when the log of total RNA has a linear relationship with the log of quantity of the specific gene amplification product. These conditions depend on the amount of RNA and the number of amplification cycles.
5. Moreover, the RNA degradation must be normalized using as comparison a well-expressed "housekeeping gene" mRNA *(7,8)*.

3.7.1. Reverse Transcription

1. In order to use similar quantities of target RNA for each sample estimation, total RNA is measured by UV absorption in a spectrophotometer (*see* **Note 4**).
2. Then the different samples are diluted to the same concentration of total RNA.
3. In a final vol 10 µL, the RNAs are incubated at 42°C for 60 min with 2.5 U of AMV-Reverse Transcriptase (Promega) in 1X AMV buffer, 1 mM dNTPs and 15 pmol (0.5 µL) of downstream antisense primer.

3.7.2. cDNA Amplification

1. Add 40 µL of a master mix containing 1X PCR buffer, lacking Mg ion (because already present in the AMV buffer), 15 pmol of upstream primer and 1.2 U of Ampli*Taq* Polymerase.
2. Denature the DNA for 3 min at 95°C in a Thermocycler. Many suitable machines are available, but cyclers with fast thermal ramp rates are preferable.
3. Carry out thermal cycling for: 5 cycles of 95°C/1 min, 55°C/1 min, 72°C/1 min followed by between 30–55 cycles of 95°C/30 s, 55°C/30 s, 72°C/30 s.
4. In the case when oligonucleotides are rich in C+G or poor in C+G, the cycling conditions should be changed by evaluating the specific melting temperature.

3.7.3. Capillary Electrophoresis

1. CE analysis on P/ACE 5010 (Beckman Coulter, Fullerton, CA) is used with LIF detection. Separations are carried out in the reversal polarity mode (anode at the detector side).
2. The LIF detection uses an argon-ion source (excitation: 488 nm, emission: 530 nm).
3. The analysis of PCR products is conducted using the eCAP dsDNA 1000 Kit (Beckman Coulter), comprising, a coated capillary (100 µm id and 47 cm), filled with Tris-borate-EDTA buffer containing a replaceable linear polyacrylamide (LPA) sieving matrix.

4. The unlabeled PCR products are visualized directly by the addition of 0.4 µg/mL of the fluorescent intercalator EnhanCE (Beckman Coulter) to the buffer system.
5. The capillary is set at 20°C during the run. Samples are injected hydrodynamically for 90 s at 0.5 psi, and separation is carried out at 15 kV for 30 min.
6. Since the PCR fragments amplified from specific cDNA templates are of different lengths, several different PCR amplification samples can be co-injected and are analyzed simultaneously in the same capillary.
7. Data is collected and analyzed using Gold Software version 8.1 (Beckman Coulter). This software allows us to standardize the data for each gene product directly, relative to the estimate of expressed *β-actin (9)*. We use this to estimate the level of degradation of RNA in the tissue samples as general index of RNA degradation.
8. The logarithms of the estimated quantity of amplified cDNA (peak area) and the logarithms of the concentrations of total RNA submitted to RT-PCR or the number of amplification cycles are compared by a linear regression method. A correlation coefficient is calculated from the regression that can be used to normalize the estimate of specific gene expression.

3.8. Search for Linearity Conditions

1. A linear relationship is found to exist between the log of the quantity of total RNA and the log of the specific amplified gene product being examined *(7,8)*.
2. The conditions for linear amplification of RNA extracted from formalin-fixed and paraffin-embedded tissue must be determined empirically, taking into account that most of the RNA is highly degraded.
3. Since genes are expressed to different extents in tissues, the specific gene products must be amplified using different numbers of cycles for detection. For paraffin-embedded tissues, RT-PCR amplification usually remains linear for between 30 and 60 cycles of amplification for most (low-medium) abundance mRNA sequences.
4. The second step of the determination of the linear amplification phase of the RT-PCR reaction is to test the chosen number of cycles with increasing quantities of cDNA that is synthesized by starting with between 8 and 2000 ng of total RNA for 100 µL of amplification solution. The logarithms of the peak area of the different amplified cDNA product behave in a linear relationship for most of the mRNAs, in the range between the log of 125–1000 ng of total RNA.

3.9. Standardization of the Extent of RNA Degradation

1. Because of the possible different levels of RNA degradation, comparison among samples cannot be made directly. The results must be standardized on the basis of quantitation of a well-expressed, reference mRNA in every tissue, such as that of *β-actin*, or *GAPDH*.
2. In order to standardize the data, we calculate the mean value of the peak area of the reference mRNA for all the cases studied. The mean value is then divided by the results of the peak area for the reference sequence of each sample and the ratio multiplied by the peak area value of the specific mRNA to compare *(7,8)*.
3. For quantitation purposes, we use common master-mixes for all the steps of the case study analysis. The procedure of standardization of RNA degradation can be performed also for each sample by the coamplification of the reference and the target RNAs *(9)*, with a direct standardization.

4. Notes

1. Note that the guanidine thiocyanate is toxic, manipulate it under a chemical hood.
2. The digestion is performed in the presence of 1 *M* guanidine thiocyanate to protect RNA. Although proteinase K remains active throughout the digestion process and does not self-

digest appreciably, high concentrations of proteinase K are needed for efficient digestion of the tissue proteins *(10)*.

3. RNA extracted from paraffin-embedded tissues using this method is highly degraded, with fragments ranging from 100 to 200 bases. The level of degradation is variable from sample to sample, depending on the conditions used for tissue fixation and paraffin-embedding. For an efficient RT-PCR analysis is advisable to only amplify fragments of 100 bases or less.

4. The spectrophotometer quantitation at 260 nm ([RNA] µg/µL = Absorbance $_{260}$ × 40 × 10^{-3} × Dilution factor) is not accurate because of the high dilution of the RNA samples obtained from paraffin-embedded tissues. Although only an approximate quantitation of the total extracted RNA is determined by UV absorption, this does not prevent accurate RT-PCR quantification of particular genes, because all of the results are standardized by reference to defined endogenous RNA products.

References

1. Jackson, D. P., Quirke, P., Lewis, F., Boylston, A. W., Sloan, J. M., Robertson, D., and Taylor, G. R. (1989) Detection of measles virus RNA in paraffin-embedded tissue. *Lancet* **I**, 1391.
2. Stanta, G. and Schneider, C. (1991) RNA extracted from paraffin-embedded human tissues is amenable to analysis by PCR amplification. *BioTechniques* **11**, 304–308.
3. von Weizsacker, F., Labeit, S., Koch, H. K., Oehlert, W., Gerok, W., and Blum, H. E. (1991) A simple and rapid method for the detection of RNA in formalin-fixed, paraffin-embedded tissues by PCR amplification. *Bioch. Biophys. Res Comm.* **174**, 176–180.
4. Stanta, G., Bonin, S., and Perin, R. (1998) RNA extraction from formalin-fixed and paraffin-embedded tissues, in *Methods in Molecular Biology: RNA Isolation and Characterization Protocols* (Rapley, R. and Manning D. L., eds.), Humana Press, Totowa, NJ, pp. 23–26.
5. Gilliland, G., Perrin, S., and Franklin-Bunn, H. (1990) Competitive PCR for quantitation of mRNA, in *PCR protocols: A Guide to Methods and Applications* (Innis, M. A., Gelfand, D. H., Sninsky, J. J., and White, T. J., eds), Academic Press, San Diego, CA, pp. 60–69.
6. Rappolee, D. A., Wang, A., Mark, D., and Werb, Z. (1989) Novel method for studying mRNA phenotypes in single or small numbers of cells. *J. Cell. Biochem.* **39**, 1–11.
7. Stanta, G., Bonin, S., and Utrera, R. (1998) RNA quantitative analysis from fixed and paraffin-embedded tissues, in *Methods in Molecular Biology: RNA Isolation and Characterization Protocols* (Rapley, R., and Manning D. L., eds.), Humana Press, Totowa, NJ, pp. 113–119.
8. Stanta, G. and Bonin, S. (1998) RNA quantitative analysis from fixed and paraffin-embedded tissues: membrane hybridization and capillary electrophoresis. *BioTechniques* **24**, 271–276.
9. Lu, W., Han, D. S., Yuan, J., and Andrieu, J. M. (1994) Multi-target PCR analysis by capillary electrophoresis and laser-induced fluorescence. *Nature* **368**, 269–271.
10. Fisher, J. A. (1988) Activity of proteinase K and RNase in guanidinium thiocyanate. *FASEB J.* **2**, A1126.

22

Differential Display Analysis by Capillary Electrophoresis

Xilin Zhao and Kimberly S. George

1. Introduction

Differential display polymerase chain reaction (DD-PCR), originally described by Liang and Pardee in 1992 *(1)*, has been utilized to study gene expression events in a myriad of biological processes. These processes include cell differentiation, hormonal regulation of gene expression, apoptosis, and carcinogenesis *(2–11)*. The technique involves extraction of RNA from two or more tissue samples, then reverse transcription of the total RNA into single-strand cDNA using a 3' anchored oligo-dT primer. The single-strand cDNA is converted by PCR amplification into double-strand cDNA using the same oligo-dT primer and a random arbitrary primer, and represents a subset of the total mRNA. Standard protocols usually describe the incorporation of a radiolabeled nucleotide into the PCR reaction product, which allows the PCR products to be visualized by autoradiography following separation by polyacrylamide gel electrophoresis. The resulting differential banding patterns generated from two or more samples are compared to identify differentially expressed cDNA fragments.

Since DD was introduced, numerous technical modifications and improvements have been reported including nonisotopic and fluorescent methodologies *(12–20)*. One fluorescent adaptation of the standard DD-PCR (FDD-PCR) reaction has been developed that takes advantage of the specificity, selectivity, and differential fluorescence of three dye labeled oligo-dT primers *(21)*. This fluorescent protocol offers a number of advantages including: higher throughput, the ability to simultaneously screen and compare banding patterns from three different primer combinations, and decreased time and expense compared to traditional isotopic methods. FDD-PCR has been successfully applied to the analysis of gene expression in a wide range of organisms and tissues. This range includes bacteria, established cell lines, primary human cells and tumor samples, and the

From: *Methods in Molecular Biology, Vol. 163:*
Capillary Electrophoresis of Nucleic Acids, Vol. 2: Practical Applications of Capillary Electrophoresis
Edited by: K. R. Mitchelson and J. Cheng © Humana Press Inc., Totowa, NJ

tissue obtained in a variety of disease states, such as viral infection, breast cancer, and psychiatric diseases.

Methods for DD using either isotopic or fluorescent labels have been developed for the analysis of DD-PCR products using slab-gel electrophoresis, which allows for the simultaneous comparison of numerous samples. Adaptation of the FDD-PCR method to a capillary electrophoresis (CE) format however, offers several further advantages for analysis such as automated sample loading and more uniform electrophoresis. Although DD-PCR bands are readily recoverable from slab gels, this is not easily achieved from the capillary separated formats. At present, the capillary format offers the potential for the fraction collection of bands of interest. Experiments have demonstrated that FDD-PCR can be easily adapted to the CE format of the Perkin-Elmer ABI 310 Genetic Analyzer. The analysis of FDD-PCR products can be performed with equal precision and reproducibility on both the ABI 377 DNA sequencer and the ABI 310 Genetic Analyzer *(22)*. Both the DNA sequencer and capillary methodologies offer a higher throughput means for sample screening than the conventional radioactive DD-PCR. A prototype CE system that allows for high precision fraction collection of microliter volumes of fluorescent PCR products using a sheath flow collection device has been described by Müller et al. *(23)*. Suitable modifications to CE systems that would allow for fraction collection of FDD-PCR products would eliminate the need for gel-based fragment isolation and would greatly increase the analytical throughput of FDD-PCR *(24)*. Coordination of slab gel and CE methods offers the best of both worlds, high throughput and precision in detection of differentially displayed fragments by CE, and the ease to isolate particular differentially displayed fractions of interest from gels. In this chapter, the protocols for the use of the Perkin-Elmer ABI 310 Genetic Analyzer (CE system) for the detection and identification of FDD-PCR generated EST patterns are described.

2. Materials

2.1. RNA Isolation and Purification

All chemicals and reagents are molecular biology grade.

1. TRIzol LS (Life Technologies, Gaithersburg, MD).
2. Chloroform.
3. Isopropyl alcohol.
4. 95% (v/v) Ethanol.
5. 70% (v/v) Ethanol solution.
6. DEPC-treated water (Quality Biologicals, Gaithersburg, MD).
7. Agarose.
8. Formamide.
9. 37% (v/v) Formaldehyde stock.
10. TBE buffer: 89 mM Tris-base, pH 8.3, 89 mM boric acid, 10 mM EDTA.
11. RNA Loading buffer: 50% glycerol, 1 mM EDTA, 0.4% bromophenol blue.
12. Ethidium bromide.
13. Amplification grade DNase I (Life Technologies).

Table 1
Sequences of the Arbitrary Primers

Primer	Sequence
OP-01	5'-TACAACGAGG-3'
OP-02	5'-TGGATTGGTC-3'
OP-03	5'-CTTTCTACCC-3'
OP-04	5'-TTTTGGCTCC-3'
OP-05	5'-GGAACCAATC-3'
OP-06	5'-AAACTCCGTC-3'
OP-07	5'-TCGATACAGG-3'
OP-08	5'-TGGTAAAGGG-3'
OP-09	5'-TCGGTCATAG-3'
OP-10	5'-GGTACTAAGG-3'
OP-11	5'-TACCTAAGCG-3'
OP-12	5'-CTGCTTGATG-3'
OP-13	5'-GTTTTCGCAG-3'
OP-14	5'-GATCAAGTCC-3'
OP-15	5'-GATCCAGTAC-3'
OP-16	5'-GATCACGTAC-3'
OP-17	5'-GATCTGACAC-3'
OP-18	5'-GATCTCAGAC-3'
OP-19	5'-GATCATAGCC-3'
OP-20	5'-GATCAATCGC-3'
OP-21	5'-GATCTAACCG3'
OP-22	5'-GATCGCATTG-3'
OP-23	5'-GATCTGACTG-3'
OP-24	5'-GATCATGGTC-3'
OP-25	5'-GATCATAGCG-3'
OP-26	5'-GATCTAAGGC-3'

2.2. Fluorescent Differential Display

2.2.1. Reverse Transcription

1. Superscript II RNase H-reverse transcriptase (200 U/µL) (Life Technologies) (*see* **Note 1**).
2. 5X First Strand Buffer: 250 mM Tris-HCl, pH 8.3, 375 mM KCl, 15 mM MgCl$_2$.
3. 25 mM MgCl$_2$.
4. 0.1 M Dithiothreitol.
5. dNTP mix: 10 mM each of dATP, dCTP, dGTP, and dTTP.
6. DEPC-treated water (Quality Biologicals).
7. HPLC purified degenerate sequence anchor primers (50 µM) (V = A, C, or G degeneracy) (*see* **Note 2**):
   5'-6-carboxyfluorescein (6-FAM) labeled oligo-dT$_{12}$VA;
   5'-4,7,2',4',5',7'-hexachloro-6-carboxyfluorescein (HEX) labeled oligo-dT$_{12}$VC;
   5'-4,7',2',7'-tetrachloro-6 carobxyfluorescein (TET) labeled oligo-dT$_{12}$VG primers.

2.2.2. PCR Amplification

1. 10X PCR buffer II: 100 mM Tris-HCl, pH 8.3, 500 mM KCl.
2. 25 mM MgCl$_2$.
3. dNTP mix: 100 μM each of dATP, dCTP, dGTP, and dTTP.
4. 10 μM of Arbitrary decamer primers, *see* **Table 1** (Operon, Alameda, CA).
5. Ampli*Taq* DNA Polymerase (5 U/μL) (Perkin-Elmer, Norwalk, CT). Store at –20°C but not in a frost-free freezer.
6. Centrisep Columns (Princeton Separations, Adelphi, NJ).

2.3. Capillary Electrophoresis

1. Formamide.
2. Blue Dextran.
3. GeneScan TAMRA-500 molecular mass standard (PE Biosystems, Foster City, CA).
4. POP-4 Polymer (PE Biosystems).

3. Methods

3.1. RNA Isolation and Purification

1. Sample Homogenization: Homogenize solid tissue samples using a glass homogenizer in 0.75 mL of TRIzol LS per 50–100 mg of tissue.
2. Cells: cells grown in monolayer cultures can be homogenized directly by adding 0.3–0.4 mL of TRIzol LS per 10 cm^2. Cells grown in suspension should be collected by centrifugation and then the cell pellet lysed by repetitive pipeting in 0.75 mL of TRIzol per 5 – 10 × 10^6 animal, plant, or yeast cells, or per 1 × 10^7 bacterial cells. Incubate the homogenized samples for 5 min at 15–30°C.
3. Add 0.2 mL of chloroform per 0.75 mL of TRIzol LS reagent. Cap the tubes securely and shake vigorously for 15 s. Incubate the samples for 2–15 min at 15–30°C, then centrifuge samples at no more than 12,000g for 15 min at 2–8°C to pellet cell debris.
4. Transfer the upper aqueous phase of the lysate to a clean tube. Then precipitate the RNA by adding 0.5 mL of isopropyl alcohol per 0.75 mL of TRIzol LS reagent used in the initial homogenization. Incubate the samples at 15–30°C for 10 min.
5. Centrifuge the samples at 2–8°C for 10 min at a speed no greater than 12,000g. Gently remove the supernatant. Wash the RNA pellet once in 1 mL of 75% ethanol per 0.75 mL of TRIzol LS reagent used in the initial homogenization. Mix the sample by vortexing and centrifuge at a speed no greater than 7500g for 5 min at 2–8°C.
6. Wash the pellet again in 1 mL of 95% ethanol per 0.75 mL of TRIzol LS reagent used in the initial homgenization. Mix the samples by vortex mixing and then repellet by centrifugation at a maximum of 7500g for 5 min, at a temperature of 2–8°C.
7. Briefly, air-dry the RNA pellet, then dissolve the RNA in 50 μL of RNase-free water.
8. Quantitate the amount of recovered RNA by determining the absorbance at A$_{260/280}$.
9. Incubate 10–100 μg of total RNA with 10 U/μL of RNase-free DNase in 10 mM Tris-HCl, pH 8.3, 50 mM KCl, 1.5 mM MgCl$_2$ for 15 min at 37°C.
10. Repeat the TRIzol LS extraction of the DNase I treated RNA pellet recovered in **step 9** as outlined in **steps 1–7**.
11. Evaluate the integrity of the RNA on a 1.0% agarose gel containing 7% formaldehyde by loading 2 μg of DNase treated RNA per lane (*see* **Note 3**).
12. The purified RNA samples may be stored at –70°C until use (*see* **Note 3**).

3.2. Fluorescent Differential Display

3.2.1. Reverse Transcription

1. Set up three separate reverse-transcription reactions for each RNA sample in three separate PCR tubes, each tube containing one of the three different oligo-dT$_{12}$ anchored primers (*see* **Note 4**). Each 20-µL reverse transcription reaction contains:

5X First Strand buffer	4 µL
0.1 *M* DTT	2 µL
dNTP mixture (10 m*M*)	1 µL
DNase-treated total RNA (0.1 µg/µL)	3 µL
Anchored oligo- dT$_{12}$ primer (50 µ*M*)	1 µL
Superscript II RNase H-reverse transcriptase	1.5 µL
DEPC-treated H$_2$O	7.5 µL

2. Add 3 µL of the mRNA, 1 µL of the anchored oligo-dT$_{12}$ primer and 7.5 µL of DEPC-treated water to a PCR tube. Heat the tube at 65°C for 5 min to denature the RNA. Cool the mixture on ice then add 8.5 µL of the master mix containing the remaining components.
3. Incubate the complete reaction at 42°C for 60 min.
4. Inactivate the Superscript II RNase H-reverse transcriptase enzyme by heating at 95°C for 5 min.
5. Keep reactions on ice, or at 4°C prior to proceeding to the PCR reaction. Reactions can be stored at –20°C for up to 1 mo (*see* **Note 5**).

3.2.2. PCR Amplification

1. Set up three separate reactions for each anchored oligo-dT and arbitrary primer combination. Each 20-µL PCR reaction contains:

10X PCR II buffer	2 µL
25 m*M* MgCl$_2$	1.36 µL
dNTP mixture (100 µ*M*)	0.67 µL
Anchored oligo- dT$_{12}$ primer (50 µ*M*)	1 µL
Arbitrary primer (10 µ*M*)	1 µL
Ampli*Taq* DNA Polymerase (5 U/µL)	0.5 µL
RT-mix from reverse transcription	2 µL
DEPC-treated water	11.47 µL

2. PCR amplification reactions are performed in a Perkin-Elmer Model 9600 thermocycler. The reaction is for 40 cycles of: 94°C for 30 s, 40°C for 2 min, and 72°C for 30 s, followed by a final extension step at 72°C for 5 min.
3. Following PCR amplification, mix 5 µL of each of the three differentially labeled fluorescent PCR products for a final volume of 15 µL. Free the cDNA reaction products of unincorporated primers using Centrisep columns (Princeton Separations, Adelphi, NY) (*see* **Note 6**).
4. Prepare the Centrisep columns by rehydrating each column with 800 µL of DEPC-treated water. Allow columns to stand at room temperature for 2 h prior to use. Remove the top and bottom caps from the columns, place the column in a 2-mL collection tube and centrifuge at 750*g* for 2 min.
5. Add 15 µL of the PCR product to the center of each column, place the column in a fresh 1.5-mL tube, and centrifuge at 750*g* for 2 min. Dry the specimens in a speed vacuum until completely dry.

6. Resuspend the purified cDNA products in 6 μL of DEPC-treated water. Store reactions at –20°C (*see* **Note 5**).

3.3. Capillary Electrophoresis of FDD-PCR Samples

1. Prepare samples for electrophoresis by mixing 0.5 μL of the mixed PCR products with 8 μL of a 10:1 formamide-TAMRA-500 molecular mass standard mix (*see* **Note 7**).
2. Denature the cDNA samples for 2 min at 90°C.
3. Place the samples on ice for 5 min.
4. Load the samples into a sample rack and place on the Perkin-Elmer ABI model 310 Genetic Analyzer.
5. The electrophoresis parameters are dependent on the length of the capillary and the gel matrix employed (*see* **Note 8**). For a 47-cm capillary (36 cm to detector) and POP-4 polymer use the following run parameters (*see* **Note 9**):

   Injection voltage: 15,000 V
   Injection time: 5 s
   Electrophoresis voltage: 15,000 V
   Electrophoresis time: 30 min
   Electrophoresis temperature: 60°C
   Virtual filter C

6. It is not necessary to run the electrophoresis for greater than 30 min as the majority of FDD-PCR products are less than 800 bp.
7. Although conventional slab-gel electrophoresis of the DD-PCR products allows for direct sample comparison of samples in the different gel lanes, the FDD-PCR products run on the ABI 310 must be compared to one another using GeneScan 2.1 software. Experiments have demonstrated that both methods provide equal resolution and comparable results *(22)*. Analysis of FDD-PCR products using the ABI 310 offers an alternative to laboratories that want to perform differential display that do not have access to an ABI 377. An example of the data generated using the ABI 310 and the GeneScan 2.1 software is presented in **Fig. 1**.
8. The ABI 310 Genetic Analyzer lacks fraction collection capabilities, therefore, differentially expressed fragments cannot be isolated. Preliminary experiments using two different prototype CE instruments capable of fraction collection have shown that differential bands can be identified, isolated, and reamplified, however, standard protocols for this application have not been established.

4. Notes

1. Comparison of many different reverse transcriptase enzymes in our laboratory has revealed that Superscript II RNase H-reverse transcriptase demonstrates a higher reproducibility and maximizes the yield of transcripts. This difference is observed as an increase in the number of both differential and nondifferential fragments in the downstream analysis.
2. The fluorescently labeled oligo-dT primers must be HPLC purified to avoid background fluorescence on both the ABI 310 and ABI 377 fluorescence analysis systems.
3. Any RNA extraction procedure can be utilized provided it produces high quality full-length mRNA. In our laboratory, alternative RNA extraction protocols have been utilized with equal success. RNA quality can be assessed using a number of methods, however, a denaturing RNA gel is the best method *(25)*. High-quality RNA is characterized by sharp, intact 28S and 18S ribosomal bands and a minimum of smearing should be evident above,

Mr (bp)

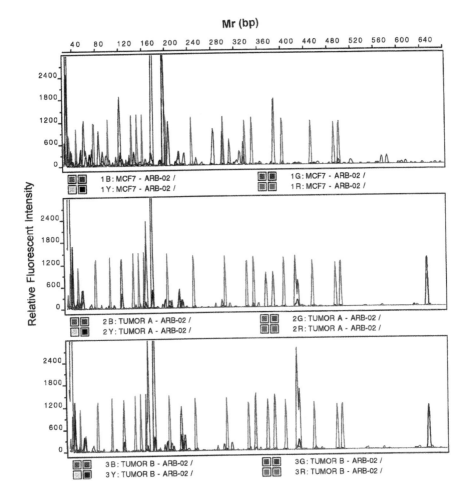

Fig. 1. GeneScan electropherograms of FDD-PCR samples run using the ABI 310 Analyzer. The FDD-PCR patterns were generated with arbitrary primer OP-02 and cDNA from MCF7 cells (*top*), a breast cancer cell line, mouse mammary tumor samples A (*middle*) and B (*bottom*), respectively. The size of fragments displayed across the horizontal axis at the top of the electropherogram is in base pairs, and the relative fluorescence signal (peak) intensity is displayed along the vertical axis. A representative, differential display fragment of approx 360 bp is illustrated, and can be seen in the mouse mammary tumor B sample (*bottom*).

between, and below the ribosomal bands. Store RNA in aliquots at –70°C for up to 1 yr. Aliquot the RNA in single use quantities to avoid repeated freezing and thawing.

4. One of the main criticisms of DD is the lack of reproducibility of the reactions. A number of factors can contribute to reaction variability including, the quality of the reagents, variation in the pipeting volumes, contamination or impurity of RNA samples, and so on (*18*). To improve the reproducibility of the DD-PCR reaction we recommend the following measures:

a. Use uniform reagents and avoid mixing reagents from the different lot numbers.
b. To avoid pipeting and volume errors, utilize large volume master mixes as much as possible, set up all samples for a particular primer combination simultaneously.
c. Perform all reactions in duplicate, include positive and negative controls with all reactions. We recommend the use of two independent negative control reactions. (1) a "reagent control" with water in the place of cDNA to assess to purity of the reagents, and (2) a "no RT control" with RNA in place of the cDNA to evaluate the level of DNA contamination in each RNA sample.

5. Lyophilized FDD-PCR products can be stored at –70°C for up to 1 mo. FDD-PCR products that have been resuspended in DNase-free water are stable for 1 wk at 20°C.

6. It is essential to remove any residual unincorporated fluorescent primer from the cDNA products. Unincorporated fluorescent primer would contribute to the background fluorescence during the DD analysis.

7. The amount of sample loaded onto the capillary is a function of the volume of the denatured sample in the mixture, the injection voltage, and the injection time. Each of the primers has a different fluorescent moiety with a very narrow absorption spectrum. Overloading the column with sample results in a strong fluorescent signal that can interfere with the discrimination between individual fluorescence signals due to spectral overlap of the dyes. Although the parameters described for use are optimal for good signal to noise ratios, sample volumes as much as 1 μL can be analyzed.

8. Most FDD-PCR products range in size of 100–800 bp.

9. Alternative sieving polymers were tested for CE separations, however the POP-4 polymer provided the best separation for this particular application. Importantly, each polymer requires slightly different electrophoresis parameters that must be determined empirically.

References

1. Liang, P. and Pardee A. B. (1992) Differential display of eukaryotic messenger RNA by means of the polymerase chain reaction. *Science* **257,** 967–971.

2. Amson R. B., Nemani, M., Roperch J. P., Israeli, D., Bougueleret, L., et al. (1996) Isolation of 10 differentially expressed cDNAs in p53-induced apoptosis: activation of the vertebrate homologue of the Drosophila seven in absentia gene. *Proc. Natl. Acad. Sci. USA* **93,** 3953–3957.

3. Asling, B., Dushay, M. S., and Hultmark, D. (1995) Identification of early genes in the *Drosophila* immune response by PCR-based differential display: the Attacin A gene and the evolution of attacin-like proteins. *Insect Biochem. Mol. Biol.* **25,** 511–518.

4. Blok, L. J., Kumar, M. V., and Tindall, D. J. (1995) Isolation of cDNAs that are differentially expressed between androgen-dependent and androgen-independent prostate carcinoma cells using differential display PCR. *Prostate* **26,** 213–224.

5. Chapman, M. S., Qu, N., Pascoe, S., Chen, W-X., Apostol, C., Gordon, D., and Miesfeld, R. (1995) Isolation of differentially expressed sequence tags from steroid-responsive cells using mRNA differential display. *Mol. Cell Endocrinol.* **27,** R1–7.

6. Donohue, P. J., Alberts, G. F., Guo, Y., and Winkles, J. A. (1995) Identification by targeted differential display of an immediate early gene encoding a putative serine/threonine kinase. *J. Biol. Chem.* **270,** 10,351–10,357.

7. Lehar, S. M., Nacht, M., Jacks, T., Vater, C. A., Chittenden T., and Guild, B. C. (1996) Identification and cloning of EI24, a gene induced by p53 in etoposide-treated cells. *Oncogene* **12,** 1181–1187.

8. Liang, P., Averboukh, L., Keyomarsi, K., Sager, R., and Pardee, A. B. (1992) Differential display and cloning of messenger RNAs from human breast cancer versus mammary epithelial cells. *Cancer Res.* **52**, 6966–6968.

9. Maser, R. L. and Calvert, J. P. (1995) Analysis of differential gene expression in the kidney by differential cDNA screening, subtractive cloning, and mRNA differential display. *Semin. Nephrol.* **15**, 29–42.

10. Shibahara, K., Asano, M., Ishida, Y., Aoki, T., Koike, T., and Honjo, T. (1995) Isolation of a novel mouse gene MA-3 that is induced upon programmed cell death. *Gene* **166**, 297–301.

11. Zimmerman J. W. and Schultz, R. M. (1994) Analysis of gene expression in the preimplantation mouse embryo: use of mRNA differential display. *Proc. Natl. Acad. Sci. USA* **91**, 5456–5460.

12. An, G., Luo, G., Veltri, R. W., and O'hara, S. M. (1996) Sensitive, nonradioactive differential display method using chemiluminescent detection. *BioTechniques* **20**, 342–346.

13. Bauer, D., Muller, H., Reich, J., Reidel, H., Ahrenkeil, V., Warthoe, P., and Strauss, M. (1993) Identification of differentially expressed mRNA species by an improved display technique (DDRT-PCR). *Nucleic Acids Res.* **21**, 4272–4280.

14. Callard, D., Lescure, B., and Mazzolini, L. (1994) A method for the elimination of false positives generated by the mRNA differential display technique. *BioTechniques* **16**, 1096–1103.

15. Chen, Z., Swisshelm, K., and Sager, R. (1994) A cautionary note on reaction tubes for differential display and cDNA amplification in thermal cycling. *BioTechniques* **16**, 1002–1006.

16. Haag, E. and Raman, V. (1994) Effects of primer choice and source of Taq DNA polymerase on the banding patterns of differential display. *BioTechniques* **17**, 226–228.

17. Ito, T., Kito, K., Adati, N., Mitsui, Y., Hagiwara H., and Sakaki, Y. (1994) Fluorescent differential display: arbitrarily primed RT-PCR fingerprinting on an automated DNA sequencer. *FEBS Lett.* **351**, 231–236.

18. Liang, P., Averboukh L., and Pardee, A. (1993) Distribution and cloning of eukaryotic mRNAs by means of differential display: refinements and optimization. *Nucleic Acids Res.* **21**, 3269–3275.

19. Lohmann, J., Schickle, H., and Bosch, T.C. (1995) REN display, a rapid and efficient method for non-radioactive differential display and mRNA isolation. *BioTechniques* **18**, 200–202.

20. Mou, L., Miller, H., Li, J., Wang E., and Chalifour, L. (1994) Improvements to the differential display method for gene analysis. *Biochem. Biophys. Res. Commun.* **199**, 564–569.

21. Jones, S. W., Cai, D., Weislow O. S., and Esmaeli-Azad, B. (1997) Generation of multiple mRNA fingerprints using fluorescence-based differential display and an automated DNA sequencer. *BioTechniques* **22**, 536–543.

22. George, K. S., Zhao, X., Gallahan, D., Shirkey, A., Zareh, A., and Esmaeli-Azad, B. (1997) Capillary electrophoresis methodology for identification of cancer related gene expression patterns of fluorescent differential display polymerase chain reaction. *J Chromatogr. B* **695**, 93–102.

23. Müller, O., Foret, F., and Karger, B. L. (1995) Design of a high-precision fraction collector for capillary electrophoresis. *Anal. Chem.* **67**, 2974–2980.

24. Magnúsdóttir, S., Heller, C., Sergot, P., and Viovy, J.-L. (2001) Collection of capillary electrophoresis fractions on a moving membrane, in *Capillary Electrophoresis of Nucleic Acids*, Vol. 1 (Mitchelson, K. R. and Cheng, J., eds.), Humana Press, Totowa, NJ, pp. 323–331.

25. Farrell, R. E. (1993) RNA isolation strategies, in *RNA Methodologies: A Laboratory Guide for Isolation and Characterization*. Academic Press, San Diego, CA, pp. 47–91.

V

PRACTICAL APPLICATIONS FOR DNA SEQUENCE ANALYSIS

23

Capillary Array Electrophoresis Analyzer

Masao Kamahori and Hideki Kambara

1. Introduction

The Human Genome Project is a challenging and important project for the future of mankind *(1)*. It has driven the development of various analytical instruments as well as methods *(2)*. The development of an automated DNA sequencer with a fluorescence detection system has been a key factor for the success of the genome project. The first generation of DNA sequencer has used a slab gel. The DNA sequencing throughput per system was only 20 kb/d when the genome project started in 1990 *(3)*. It was proposed that it would take more than thirty years to sequence the entire human genome without a drastic improvement in DNA sequencing technology. The desired throughput of the new DNA sequencer would be 1 Mb/d *(4)*.

DNA sequencing by capillary electrophoresis (CE) seems to be very attractive because of its high-analysis speed and convenient handling properties, such as the ease of sample injection and a very high speed in base reading *(5,6)*. However, the actual base reading throughput of a conventional CE apparatus is not large because it has only one electrophoresis migration lane. A system capable of massive parallel DNA sequencing in a short period of time has to be developed for realization of a very high throughput, such as 1 Mb/d. Various DNA sequencing technologies including, sequencing by hybridization *(7,8)*, ultra-thin slab-gel electrophoresis *(9)*, and capillary array gel electrophoresis (CAE) *(10–18)* have been developed for the realization of high-throughput DNA sequencing. Among these, CAE seems the most promising technology to achieve the high-throughput DNA analysis target. Importantly, a capillary array system can also be successfully applied to other associated genomic analyses such as rapid, high-throughput DNA screening *(19)*, gene expression profiling *(20)*, and DNA band sorting *(21)*. These other techniques are also becoming very important techniques for post-genome investigations, such as for drug discovery and for DNA diagnostics *(22,23)*.

From: *Methods in Molecular Biology, Vol. 163:*
Capillary Electrophoresis of Nucleic Acids, Vol. 2: Practical Applications of Capillary Electrophoresis
Edited by: K. R. Mitchelson and J. Cheng © Humana Press Inc., Totowa, NJ

1.1. Requirements of Capillary Array Systems

Three capillary array systems are presented in this chapter including, CAE systems using a sheath-flow technique *(10,11)*, a multiple on-column capillary array system combined with side-entry laser irradiation *(12)*, and a capillary array system with a DNA fragment sorting device *(21)* (*see* **Note 1**).

1.1.1. Relationship Between Fragment Mobility and Gel Matrix

The first methods used to obtain a rapid gel electrophoresis for DNA sequencing was a slab-gel system with optimized conditions for fast electrophoresis *(24)*. The migration time of a DNA fragment depends on the concentration of the crosslinked acrylamide gel, the electric field strength, the gel temperature, and the migration path length. The migration speed of a DNA fragment is proportional to the applied electric field strength, however a very high electric field produces a large amount of Joule heat, which causes DNA band broadening and destruction of the gel. Migration lanes that occupy a small cross-sectional area avoid many of the above drawbacks. Therefore, a capillary gel or an ultra-thin slab gel is used to obtain a rapid DNA analysis. Besides increasing the electric field, the optimization of electrophoresis conditions is also important to obtain a high-speed analysis.

The dependence of migration time on the gel concentration is shown for several different sizes of DNA fragments in **Fig. 1**. The time required for a DNA migration increases with gel concentration; therefore, a low gel concentration is better for a rapid DNA analysis. The use of a low concentration gel, which has a large pore size, is limited by two factors. The first one is the difficulty in producing a slab gel with an acrylamide concentration lower than 3% T (total concentration), because polymerization does not occur uniformly across such a wide area. This problem can be overcome by the use of a capillary tube as the gel supporter. The second factor is the reduction of DNA band spacing by the use of low gel concentration, which reduces the fragment resolution.

The gel concentration dependence of DNA band spacing for various DNA fragment lengths is shown in **Fig. 2**. Although the band spacing is dependent on gel concentration at a low gel concentration, it is almost independent of the gel concentration at a high gel concentration, over 6% T acrylamide for DNA fragments over 300 bases. Since the width of the DNA band does not change much with different gel concentrations, low gel concentration is usually superior to high concentration for rapid gel electrophoresis and usually has sufficient band resolution. Both capillary gel electrophoresis and slab gels have many similar separation characteristics and employ similar gel formulations. For both a low gel concentration is generally better for a high-speed analysis. An acrylamide concentration as low as 4% T (total concentration) and 5% C (crosslinker) and including 7 *M* urea is used for DNA sequencing. An unpolymerized premix of this formula has a low viscosity and can easily be injected into a capillary to be polymerized. Gel-filled capillaries can be used repeatedly. More recently, a low concentration polymer gel has been used instead of the premix, which is easily injected into a capillary and is used without polymerization in the capillary *(25)*. The use of polymer gel makes it possible to operate a fully automatic system including the automatic replacement of the gel in the capillary after each electrophoresis run.

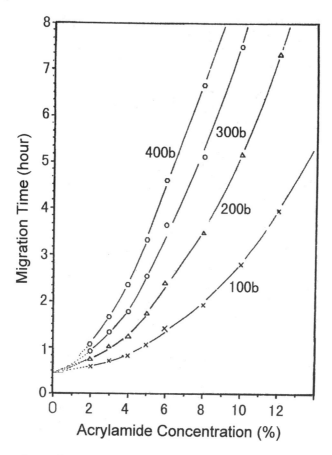

Fig. 1. Dependence of the migration time of DNA fragments on acrylamide concentration. The migration time of fragments increases approximately in proportion to the square of the acrylamide concentration, which indicates that the use of low gel concentration is superior for a rapid electrophoresis than a high concentration gel. A slab gel of 22-cm migration path length is used with an electric field strength of 40 V/cm.

1.2. Fluorescence Detection

The speed of DNA fragment migration and the base reading in a conventional slab-gel electrophoresis run is about 1 base/min. In the new rapid capillary gel electrophoresis system, the target throughput is one order of magnitude higher than the conventional DNA sequencer, 400–1000 kb/d! In order to realize this target throughput, the base reading speed must be greater than 10 bases/min (1 base/6 s). Because at least 10 data points are necessary to record one DNA fragment peak in an electropherogram, fluorescence signal measurements must be made every 0.6 s. Usually, capillaries with an inner diameter smaller than 0.1 mm are used for sequencing. Since the width of a DNA band is typically about 0.4 mm *(26)*, the volume of one band is about 2 nL in such a capillary. Assuming that the DNA

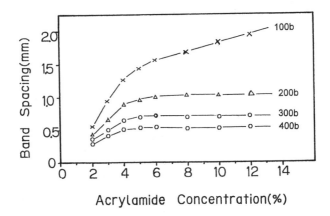

Acrylamide Concentration(%)

Fig. 2. Dependence of DNA band spacing on acrylamide concentration. For DNA fragments as long as 400 bases the band spacing during a gel electrophoresis is almost constant at 4% T acrylamide. This observation indicates that it is possible to run the electrophoresis rapidly (high electric field) without losing resolution by using a low acrylamide concentration gel.

fragment concentration in one band is 10^{-9} M, this rate of detection requires about 10^{-18} mol of DNA fragment to be color selectively detected in 0.6 s. This requires a highly sensitive detection system as well as an efficient laser irradiation system.

1.2.1. Fluorescence Emission

In a fluorescence detection system, the emitted fluorescence photon number along the irradiated region per second is "I." "I" is represented by **Eq. 1**, in which Io, ε, φ, 1, s, and M are respectively, the initial laser flux density, the molar absorption coefficient, the quantum yield, the length of the irradiated region, the surface area to be irradiated, and the molar concentration.

$$I = Io\varepsilon\phi ls M \qquad (1)$$

Only a part of the emitted fluorescence can be detected by a detector, which is mostly determined by a solid angle of the detection optics. The solid angle for the fluorescence detection "Ω" is represented by **Eq. 2** using an image magnification factor, "m" of the optical system, and an F value (= f/D), where f and D are the focal length and the diameter of the lens, respectively.

$$\Omega = m^2 / (16\, F^2(m + 1)^2) \qquad (2)$$

Assuming that the concentration of the target DNA fragment is 10^{-9} M, the number of fluorophore molecules in an irradiated region in a capillary (0.075 mm id, 0.1 mm od) is calculated to be 2.6×10^5. Argon-ion laser (488 nm) of 10 mW emits about 2.5×10^{16} photon/s. When the laser beam is focused to a 0.1 mm diameter, the flux density of the laser is 3.2×10^{18} photon/s/mm^2. As an absorption coefficient and a quantum yield of a fluorophore are about 80,000 cm$^{-1}M^{-1}$ and 0.5, respectively, the emission from the irradiated region is estimated to be 2.3×10^7 photon/s.

When the F value of a lens is 1, the number of photons collected by the detecting optics with m = 1 is about 3.6×10^5/s. The number of photoelectrons per s collected by the detector is about 32,000, by assuming that the transmission efficiency of an optical filter and the quantum yield of the photodetector (CCD) are 0.3 and 0.3, respectively. When a photomultiplier tube (PMT) is used as a detector, the quantum yield is 0.05. In a rapid analysis, the detection is carried out every 0.6 s. This means about 19,000 photoelectrons have to be detected within this brief detection period. If the color selective detection is carried out, the number of photoelectrons will be much smaller because some of the fluorophore tags cannot be excited as efficiently as the above estimations. The number of emitted photoelectrons also becomes much smaller if the laser beam scanning irradiation is used.

1.2.2. Arrangement of Detectors

The duty cycle of the detection is determined by the ratio of the inner diameter of a capillary and the length of irradiated region (= outer diameter × number of capillaries). Consider the case of 100 capillaries with an outer diameter of 0.15 mm and an inner diameter of 0.075 mm. If the duty cycle of the laser irradiation is 1/200 then the number of photoelectrons, which can be detected with a PMT coupled with a lens having a large F value of 0.6, becomes approx 44. This can be improved by raising the laser flux density until the photobleach of fluorophore occurs during the measurement period. Generally, a fluorophore having an emission at a long wavelength region has a small photobleach cross-section *(27)* and gives a large total emission before it is photobleached.

An efficient way of laser irradiation is the side-entry laser irradiation where a laser is introduced from the direction parallel to the capillary array plane. By this configuration, the laser irradiation efficiency increases by more than one order of magnitude and an intense fluorescence emission signal can be obtained. Although the configuration is good for procuring a high sensitivity, it is not easy to make this configuration in a capillary array system, because the laser beam is scattered and diffracted by capillaries and would not transmit easily through the array. There are two possible ways to overcome the difficulty: a "sheath-flow" method and a "multiple-focusing" method as explained later in this **Subheading 2.3.**

2. Materials

2.1. Capillaries

1. Capillaries are supplied from Poly-Micro Technology in Arizona (*see* **Note 2**).
2. For a sheath-flow system, the inner and the outer diameters of capillaries were 0.075 mm and 0.20 mm, respectively. They are coated with black polyimid film to decrease the fluorescence emission from the capillary edges.
3. For the on-column detection system, the capillaries have an inner diameter of 0.075 mm and an outer diameter of 0.35 mm.
4. The black polyimid coat is removed in the irradiation area to create an optical window for excitation and signal measurement.
5. Acrylamide is obtained from Gibco-BRL (NY). A crosslinked acrylamide gel (4% T, 5% C) is produced in a capillary of 50 cm, at a low temperature of 10°C, and under high pressure

to prevent the bubble formation in the polymerizing gel. The gel-filled capillaries are then cut into 25 cm to be arrayed for DNA analysis (*see* **Note 3**).

2.2. Fluorophore Detection

1. In a CAE system the analysis of all four dideoxy-terminated fragment species must be done in each individual electrophoresis lane because of the "capillary to capillary" differences in fragment migration. Consequently, four-color tagging of the fragments is used.
2. Generally the emission maximum wavelengths of fluorophores are very close to the optimum excitation wavelengths, therefore one laser can excite only one fluorophore species efficiently.
3. The energy transfer phenomenon can be used to improve the excitation efficiency for the fluorophores, which have the excitation maximum far from the laser wavelength. Energy transfer fluorophores increase the difference between the excitation maximum and the emission maximum of fluorophores. Four energy-transfer dyes or "Big Dyes" from both Amersham-Pharmacia Biotech and Perkin-Elmer-Applied BioSystems Division (PE-ABD) respectively give twice the detection sensitivity of conventional color tags. DNA sequencing samples are generated with the Sanger method DNA sequencing kits from PE-ABD (Foster City, CA), or from Amersham-Pharmacia Biotech (Buckinghamshire, England). The sequence fragment production is carried out according to the protocols recommended by the manufacturers.
4. The sample injection is carried out electrokinetically; therefore, ionic components other than the target fragments should be eliminated before the injection. The reaction products are precipitated in ethanol followed by dissolution in formamide prior to injection. Recently, Karger et al. *(28,29)* reported a method using poly(ethersulfone) ultrafiltration membranes and spin columns to remove template and salt from the sequencing products. This method improves both the amount of sample in each injection and the reproducibility of sequence runs.
5. As a high sensitivity is required for DNA band detection during CE, even the minor emission component, which has the same wavelength as those of fluorophores in an excitation light source, should be eliminated. A band pass filter having the transmission wavelength of the excitation light source is put on the laser source. This is useful in reducing the signal background due to the scattering of the minor components in the light source.
6. A cooled CCD camera (TEA/CCD-1024EM/1, Princeton Instruments, Inc., Trenton, NJ) is used with a focusing lens of F = 1.2 and f = 50 mm from Canon.
7. An image splitting prism coupled with four-color filters from Omega (Omega Optical, Brattleboro, VT) is used to make four-color dotted-line images of the irradiated positions. The transmission maximum wavelengths of the filters are 549, 567, 595 and 620 nm.
8. Fluorophore excitation is carried out with an argon-ion laser (488 nm, 20 mW, Spectra Physics, CA), or a YAG laser (532 nm, 15 mW, Coherent, CA).

3. Capillary Array Instruments

There are two types of laser irradiation for detecting fluorescence in capillary array instruments: a laser irradiation on columns of electrophoresis tubes and a laser irradiation at a sheath-flow region (*see* **Note 1**). A capillary array instrument with a laser scanning coupled with a confocal microscope technology has been reported by Mathies et al. *(10,11)*. In this system, a bundle of capillaries is placed on a moving stage of a confocal microscope. Fluorescence is detected with four photomultiplier with color

Capillary Array is placed

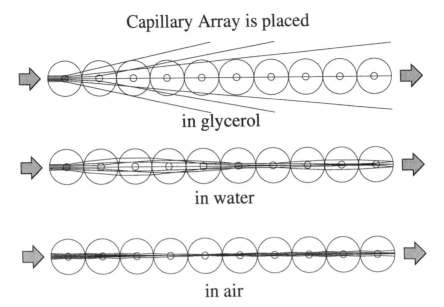

in glycerol

in water

in air

Fig. 3. The trajectory of a laser beam passing through multiple gel-filled capillaries. As the refractive index of an acrylamide gel is smaller than that of a quartz capillary, the gel tube acts as a concave lens which defocuses the laser beam. Only when the capillaries are placed in the air, and the ratio of their inner and the outer diameters is smaller than 0.2, does a laser beam pass through many gel capillaries without the beam becoming significantly defocused.

selective filters. The scan rate is 3 Hz. This type of capillary array instrument is commercialized by Molecular Dynamics (Sunnyvale, CA).

Another type of irradiation is an expanded laser beam irradiation from the forward direction of the capillary array plane, which has been proposed by Yeung et al. *(16)*. The signal detection is carried out with a CCD camera. The irradiation efficiency of the method is not high because the total amount of irradiation per capillary is low. Thus it requires a high power laser.

3.1. Side-Entry Laser Irradiation

The most efficient irradiation method is the "side-entry laser irradiation" where a laser beam is introduced from the direction perpendicular to the capillary array plane, to irradiate all the capillaries simultaneously *(12–15,17,30)*. There are two types of systems with the side-entry irradiation, "multiple on-column irradiation" and "sheath-flow irradiation."

3.2. Multiple On-Column Irradiation

The **Fig. 3** shows the trajectories of a laser irradiating capillaries placed in various conditions. When the centers of the capillaries are all positioned on one line with an accuracy of 5 m in air, a laser beam can pass through all of the migration lanes in each of the capillaries. The multiple on-column irradiation system is very simple. However, this geometry requires a precise positioning of capillaries in an on-column system. A laser irradiates the gel through the capillary tube surfaces because a laser beam is scattered and refracted by capillary tubes, which act as lenses that prevent the laser beam from passing through all the capillaries. A laser beam can be focused repeatedly by the capillaries if the ratio of the inner and outer diameter of the capillaries is less than 0.2 mm. The attenuation of the laser power by the reflection at capillary surfaces is about 5% per capillary and 20 capillaries can be irradiated without serious laser power attenuation due to the reflection. The attenuation is reduced by putting all the capillaries in water, or by having materials with the same refraction index as the capillary tubes. However, the laser diverges because gel columns inside the capillaries act as concave lenses, due to the refractive index of the separating gel being smaller than that of the quartz capillaries.

Figure 4 shows a DNA sequencing electropherogram obtained with a side-entry on-column irradiation system. The fluorescence detection is carried out with a CCD camera with an image splitting prism coupled with four-color filters.

Figure 5 shows that the focusing property of the on-column irradiation system can be improved by placing rod lenses made of quartz between every capillary (*10*). The laser focusing is greatly improved and the laser attenuation by reflection at boundaries is reduced. More than 50 capillaries can be irradiated without serious power attenuation with a side-entry on-column irradiation in the system. Optical fibers placed between the emitting points and the CCD camera can be used for the color selective fluorescence detection instead of the image splitting prism. The detecting sensitivity of the system is about 10^{-12} M of FITC-labeled DNA. This corresponds to about 1000 DNA molecules in one band.

3.3. Sheath-Flow Irradiation

Figure 6 illustrates a detailed schematic view of the sheath-flow cell and optical detection system. The DNA fragments are separated in the gel-filled capillaries and are then eluted from the ends of the capillaries into a buffer solution in an optical cell. A sheath-flow is created in the cell to carry DNA bands to the laser-irradiated region before they diffuse. The bands are laser irradiated and their signal emissions are detected.

Figure 7 shows a schematic view of the sheath-flow instrument. One obstacle to applying the side-entry irradiation through a capillary array is the presence of capillary tubes themselves at the irradiated region. If the glass capillary tubes can be removed, and the laser irradiation achieved through liquid alone then highly efficient fluorescence excitation and detection can be realized. Sheath-flow technology removes the capillary tubes from the irradiation/detection area to achieve highly efficient laser irradiation and fluorescence detection in a capillary array system (*12,13*). The flow solution is supplied from a buffer solution bottle placed 10 cm above the sheath-flow cell and the flow is created by gravity. The flow speed is about 0.08 mm/s, which is almost comparable to the migration speed of DNA fragments through the gel matrix. If the buffer flow speed is much faster than the speed of DNA migration in the gel, the

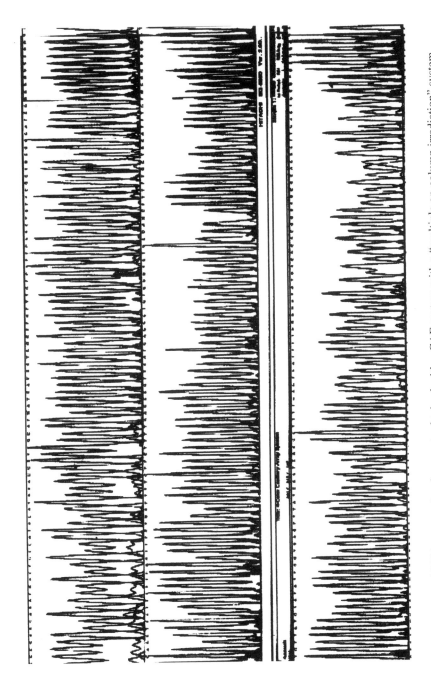

Fig. 4. Electropherogram of DNA sequencing fragments obtained with a CAE system with a 'multiple on-column irradiation" system.

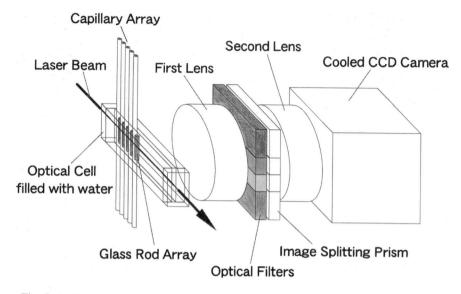

Fig. 5. A schematic view of the "multiple on-column irradiation" system. The laser beam is refocused by each of the rod lenses placed between each of the capillaries and a collimated beam is maintained across the array.

DNA concentration decreases in the sheath-flow signal detection region. In contrast a very slow sheath-flow reduces the band resolution and also makes the flow unstable.

Figure 6 also shows the principle of the color separation by the image splitting device. Fluorescence is emitted from each migration lane, which gives a dotted line image on a CCD detector. To detect the fluorescence image color selectively, an image splitting prism coupled with four-color filters is placed in front of the focusing lens. One dotted line image is divided with the optics and the through-pass color filters to produce the four-color images. The transmission wavelengths of the filters are designed so as to pass the emission signals from the set of four fluorophores supplied by PE-ABD. Such a capillary array analyzer with a sheath-flow is available from Hitachi, and from Perkin-Elmer-ABD - Hitachi.

4. DNA Migration in a Gel

From the viewpoint of a rapid analysis, the optimum condition for a DNA analysis is dependent on the target DNA sizes. The typical DNA base reading speed is about 500 bases/2 h; however, a much faster analysis is possible by optimizing the run conditions in accordance with the target DNA sizes. There are two models for DNA fragment migration, the Ogston model and the reptation model. DNAs are separated by their sizes during passing through gel pores.

The way that DNA migrates in a gel is influenced by the size of the fragments. If the size of a DNA is smaller than the pore size, rapid DNA migration and separation analysis is possible by applying a high electric field. **Figure 8** shows that for a DNA

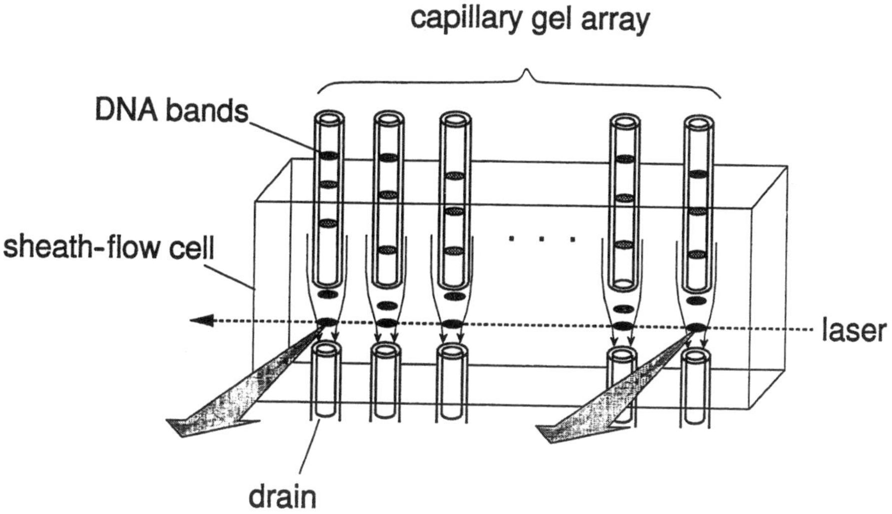

capillary gel array

Fig. 6. A schematic view of a part of the capillary array DNA analyzer system illustrating the "sheath-flow" detection cell. DNA fragments are separated in the gel capillaries and are then eluted from the anodic ends of the capillary into buffer solution in the sheath-flow cell. The separated fragments are carried to the laser-irradiated region by buffer flow and are detected without significant diffusion of the sample. Sheath-flow buffer exits the flow cell through drain capillaries immediately adjacent to the anodic ends of each of the separation capillaries.

fragment smaller than 300 bases in a 6 cm long, 4% T and 5% C crosslinked acrylamide gel can be analyzed with one base resolution in 5 min by using an electric field of 300 V/cm. In such a rapid analysis system, the exposure time of each fluorescence signal is as small as 0.1 s, and a highly sensitive detection system is required. In contrast, such a rapid analysis requires the reduction of the time necessary for prerunning the gel or in preparing the gel, because the time spent on these activities is no longer negligible.

On the other hand, if the DNA fragment is larger than 300 bases, which is larger than the pore size, and the DNA is stretched during migration, which causes a poor resolution for DNA sizing. Therefore, a low electric field is preferable for separating larger DNA fragments and consequently analysis time markedly increases compared to the time necessary for short fragments.

5. Other Capillary Array Gel Analyses and Technology

CAE can be used for DNA band sorting, which is important for gene discovery and gene hunting. There are at least two major gel methods, gel electrophoretic analysis *(31–33)* and "DNA chip" analysis of DNAs *(34,35)* which have been used success-fully for gene expression analysis and for clarifying the functions of genes. Gel elec-trophoresis methods are still important for DNA fragment analysis such as "differential display" and "amplified fragment length polymorphism" (AFLP) *(31–33)* because only electrophoresis methods can analyze unknown DNA fragments by size. DNA frag-

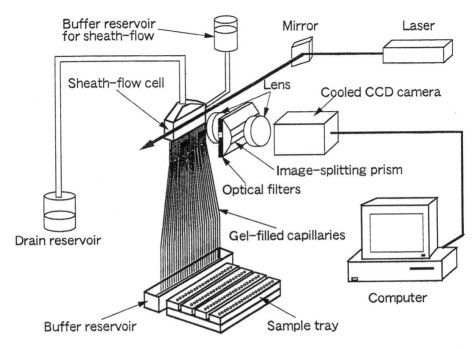

Fig. 7. A schematic view of a capillary array DNA analyzer with a sheath-flow cell. DNA fragments are irradiated in the sheath-flow region to produce a fluorescent, dotted-line image. The fragment images are detected with a cooled CCD camera, coupled with an image splitting prism and color selective filters.

ments expressed in a tissue exposed to one treatment may not be expressed in the same tissue under another treatment regime. The differentially expressed mRNAs can be detected by selective amplification techniques, such as AFLP. An example of differential display by AFLP methodology is shown in **Fig. 9**, in which two or more electropherograms are compared and identify gene fragments that are expressed differentially in the different tissue or treatments.

5.1. DNA Fragment Collection

In conventional gel electrophoresis, the DNA bands are cut out from the gel and extracted to further analyze their DNA sequence. This is both time consuming and labor intensive. However, DNA fragment collection can be done automatically by a capillary analyzer with a sampling tray. Karger's group (*36*) have reported the development of such a fraction collector on a capillary gel electrophoresis instrument.

More recently, we have also developed a DNA fragment sorter, based on an extension of the capillary array sheath-flow cell system (*21*). A schematic view of this instrument is shown in the top panel of **Fig. 10**. The apparatus consists of a capillary array DNA sequencer part, and a DNA fragment sampling part. Additional open capillaries and sampling tubes to collect DNA bands are placed below the DNA separa-

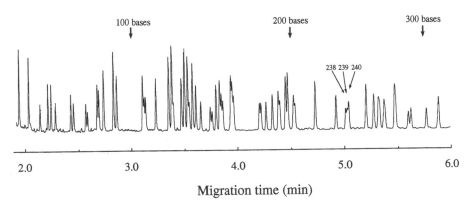

Fig. 8. A part of an electropherogram of DNA fragments obtained in a rapid mode. DNAs up to 300 bases can be separated by one base in 5 min with a short gel-filled capillary.

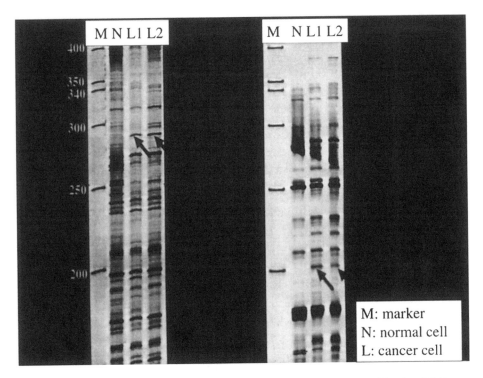

Fig. 9. Electropherogram of tissue-specific cDNA fragments produced by the AFLP procedure *(31–33)*. Several of the DNA fragments (238, 239, 240, indicated by arrows) are differentially expressed in accordance with the treatment imposed on their tissues of origin.

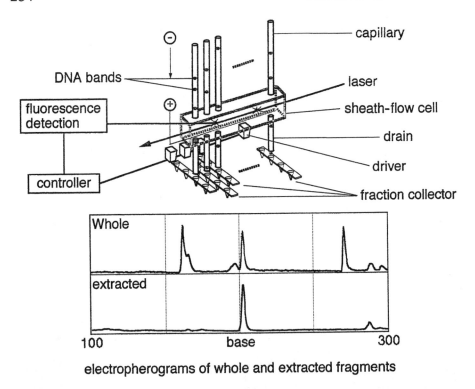

capillary

DNA bands

laser

fluorescence detection

sheath-flow cell

drain

driver

controller

fraction collector

Whole

extracted

100   base   300

electropherograms of whole and extracted fragments

Fig. 10. Illustration of an automated DNA fragment sorter (*upper panel*) with 16 capillaries using sheath-flow detection with sample collecting trays and tubes. The DNAs separated and eluted from the gel-filled capillaries are irradiated with a laser in the sheath-flow cell to emit fluorescence and are detected. The fluorescence signal is used to control the drive of the sampling trays to allow collection of individual DNA fragments. Sheath-flow buffer exits the flow cell through drain capillaries immediately adjacent to the anodic ends of each of the separation capillaries and carries the DNA fractions to the sample trays. Sixteen different trays are controlled independently to target multiple DNA bands from different samples. The *lower panel* demonstrates the purity of individual collected fragments, which are reamplified by PCR after collection and electrophoresized.

tion capillaries, interspaced with a connecting sheath-flow cell. DNA fragment signals are monitored at the sheath-flow cell to control the movement of the sampling tray.

Two DNA fragment groups are labeled with different fluorophores. They are mixed and analyzed by the CAE with a sheath-flow coupled with a crosslinked polyacrylamide gel as the sieving material. Reusable crosslinked polyacrylamide gel capillaries are used because both ends of the capillaries are used, the inlet for sample injection as well as the outlet for DNA band sorting, and there is no space to inject a replaceable gel. The laser irradiation and fluorophore detection are carried out at the sheath-flow cell, then a buffer solution including the desired DNA fragments is drained through open capillaries into collection tubes. This occurs when the signal balance of the two different fluorophores changes to over a threshold value, and then an automated sampling operation takes place. The electrophoresis buffer solution including any DNA bands is captured in a plastic tube on a moving tray. The sampled DNA fragments are in low concentration in the relatively large volume of electrophoresis buffer, however the individual DNA fragments can be recovered by PCR and then identified by sequencing. This capillary gel electrophoresis system is very effective in sorting and recovering individual DNA fragments (**Fig. 10**, bottom panel).

6. Notes

1. To further increase the number of capillaries in an array system, two-dimensional arrays of capillaries are being developed. Mathies's group *(18)* have reported a system having 1000 capillaries in one system, where all the irradiation points of capillaries are placed upon a circle. Both the laser beam emitter and the photodetector are placed at the center of the circle, and both instruments are rotated past the capillary array to irradiate the capillaries and to receive the fluorescence emissions. Such ultra high-throughput CAE systems will be very useful for both DNA diagnostics such as mutation detection and genotyping or for drug discovery experiments.
2. A capillary can be used for at least five times for DNA sequencing if the crosslinked polyacrylamide gel matrix contains urea. When a crosslinked polyacrylamide gel without urea is used, such as if the analysis target is dsDNAs, then the capillary can be used at least for a maximum of 30 times.
3. A high-throughput capillary array system is very useful for a rapid screening of many DNA samples and this demands a fully automated instrument. In order to speed-up capillary operations and to speed instrument turn-around between analysis runs, a replaceable separation gel is superior to the fixed, reusable gel. At present, various different replaceable polymers have been reported to be useful as sieving materials. The properties of these different matrices are reviewed in reference *(37)*.

References

1. Collins, F. S., Patrinos, A., Jordan, E., Chakravarti, A., Gesterand, R., and Walters, L. (1998) New goals for the US human genome project: 1998–2003. *Science* **282,** 682–689.
2. Kambara, H. (1998) Recent progress in fluorescent DNA analyzers and methods. *Current Topics in Anal. Chem.* **1,** 21–36.
3. Hunkapiller, T., Kaiser, R. J., Koop, B. F., and Hood, L. (1991) Large-scale and automated DNA sequence determination. *Science* **254,** 59–67.
4. Wada, A. (1987) Automated high-speed DNA sequencing. *Nature* **325,** 771–772.
5. Drossman, H., Luckey, J. A., Kostichka, A. J., D'Cunha, J., and Smith, L. M. (1990) High-speed separation of DNA sequencing reactions by capillary electrophoresis. *Anal. Chem.* **62,** 900–903.
6. Colon, L. A., Guo, Y., and Fermier, A. (1997) Capillary electrochromatography. *Anal. Chem.* **69,** 461A–467A.
7. Drobyshev, A., Mologina, N., Shik, V., Pobedimskaya, D., Yershov, G., and Mirzabekov, A. (1997) Sequence analysis by hybridization with oligonucleotide microchip: identification of β-thalassemia mutations. *Gene* **188,** 45–52.

8. Drmanac, R., Drmanac, S., Strezoska, Z., Paunesku, T., Labat, I., Zeremski, M., Snoddy, J., Funkhouser, W. K., Koop, B., Hood, L., and Crkvenjakov, R. (1993) DNA sequence determination by hybridization: a strategy for efficient large-scale sequencing. *Science* **260**, 1649–1652.

9. Kostichka, A. J., Markbanks, M. L., Brumley, R. L., Drossmann, H., and Smith, L. M. (1992) High speed automated DNA sequencing in ultrathin slab gels. *BioTechnology* **10**, 78–81.

10. Mathies, R. A. and Huang, X. C. (1992) Capillary array electrophoresis: an approach to high-speed, high-throughput DNA sequencing. *Nature* **359**, 167–169.

11. Huang, X. C., Quesada, M. A., and Mathies, R. A. (1992) DNA sequencing using capillary array electrophoresis. *Anal. Chem.* **64**, 2149–2154.

12. Kambara, H. and Takahashi, S. (1993) Multiple-sheathflow capillary array DNA analyser. *Nature* **361**, 565–566.

13. Takahashi, S., Murakami, K., Anazawa, T., and Kambara, H. (1994) Multiple sheathflow gel capillary array electrophoresis for multicolor fluorescent DNA detection. *Anal. Chem.* **66**, 1021–1026.

14. Anazawa, T., Takahashi, S., and Kambara, H. (1996) A capillary array gel electrophoresis system using multiple laser focusing for DNA sequencing. *Anal. Chem.* **68**, 2699–2704.

15. Anazawa, T., Takahashi, S., and Kambara, H. (1999) A capillary-array electrophoresis system using side-entry on-column laser irradiation combined with glass rod lenses. *Electrophoresis* **20**, 539–546.

16. Ueno, K. and Yeung, E. S. (1994) Simultaneous monitoring of DNA fragments separated by electrophoresis in a multiplexed array of 100 capillaries. *Anal. Chem.* **66**, 1424–1431.

17. Quesada, M. A., and Zhang, S. (1996) Multiple capillary DNA sequencer that uses fiber-optic illumination and detection. *Electrophoresis* **17**, 1841–1851.

18. Kheterpal, I. and Mathies, R. A. (1999) Capillary array electrophoresis—DNA sequencing. *Anal. Chem.* **71**, 31A–37A.

19. Larsen, L. A., Christiansen, M., Vuust, J., and Andersen, P. S. (1999) High-throughput single-strand conformation polymorphism analysis by automated capillary electrophoresis: robust multiplex analysis and pattern-based identification of allelic variants. *Hum. Mutat.* **13**, 318–327.

20. Caetano-Anõlles, G. (1996) Scanning of nucleic acids by in vitro amplification: New developments and amplifications. *Nat. Biotechnol.* **14**, 1668–1673.

21. Irie, T., Oshida, T., Hasegawa, H., Matsuoka, Y., Li, T., Oya, Y., Tanaka, T., Tsujimoto, G., and Kambara, H. (2000) Automated DNA fragment collection by capillary-array gel electrophoresis for hunting differentially expressed genes. *Electrophoresis* **21**, 367–374.

22. Jones, D. A. and Fitxpatrick, F. A. (1999) Genomics and discovery of new drug targets. *Curr. Opin. Chem. Biol.* **3**, 71–76.

23. Collins, F. S. (1999) Genetics: an explosion of knowledge is transforming clinical practice. *Geriatrics* **54**, 41–47.

24. Kambara, H., Nishikawa, T., Katayama, Y., and Yamaguchi, T. (1988) Optimization of parameters in a DNA sequenator using fluorescence detection. *BioTechnol.* **6**, 816–821.

25. Ruiz-Martinez, M. C., Berka, J., Belenkii, A., Foret, F., Miller, A. W., and Karger, B. L. (1993) DNA sequencing by capillary electrophoresis with replaceable linear polyacrylamide and laser-induced fluorescence detection. *Anal. Chem.* **65**, 2851–2858.

26. Kamahori, M. and Kambara, H. (1996) Characteristics of single-stranded DNA separation by capillary gel electrophoresis. *Electrophoresis* **17**, 1476–1484.

27. Kambara, H., Nagai, K., and Kawamoto, K. (1992) Photodestruction of fluorophores and optimum conditions for trace DNA detection by automated DNA sequencer. *Electrophoresis* **13**, 542–546.

28. Ruiz-Martinez, M. C., Salas-Solano, O., Carrilho, E., Kotler, L., and Karger, B. I. (1998) A sample purification method for rugged and high-performance DNA sequencing by capillary electrophoresis using replaceable polymer solutions. A: development of the cleanup protocol. *Anal. Chem.* **70,** 1516–1527.

29. Salas-Solano, O., Ruiz-Martinez, M. C., Carrilho, E., Kotler, L., and Karger, B. L. (1998) A sample purification method for rugged and high-performance DNA sequencing by capillary electrophoresis using replaceable polymer solutions. B. Quantitative determination of the role of sample matrix components on sequencing analysis. *Anal. Chem.* **70,** 1528–1535.

30. Lu, X. and Yeung, E. S. (1995) Optimization of excitation and detection geometry for multiplexed capillary array electrophoresis of DNA fragments. *Appl. Spectrosc.* **49,** 605–609.

31. Vos, P., Hogers, R., Bleeker, M., Reijans, M., van de Lee, T., Homes, M., Frijters, A., et al. (1995) AFLP: a new technique for DNA fingerprinting. *Nucleic Acids Res.* **23,** 4407–4414.

32. Kato, K. (1997) Adaptor-tagged competitive PCR: a novel method for measuring relative gene expression. *Nucleic Acids Res.* **25,** 4694–4696.

33. Matz, M., Usman, N., Shagin, D., Bogdanova, E., and Lukyanov, S. (1997) Ordered differential display: a simple method for systematic comparison of gene expression profiles. *Nucleic Acids Res.* **25,** 2541–2542.

34. Schena, M., Shalon, D., Davis, R. W., and Brown, P. O. (1995) Quantitative monitoring of gene expression patterns with a complementary DNA microarray. *Science* **270,** 467–470.

35. Marshall, A., and Hodgson, J. (1998) DNA chips: an array of possibilities. *Nature BioTechnol.* **16,** 27–31.

36. Müller, O., Foret, F., and Karger, B. L. (1995) Design of a high-precision fraction collector for capillary electrophoresis. *Anal. Chem.* **67,** 2974–2980.

37. Heller, C. (2001) Influence of polymer concentration and polymer composition on capillary electrophoresis of DNA, in *Capillary Electrophoresis of Nucleic Acids,* Vol. 1 (Mitchelson, K. R. and Cheng, J., eds.), Humana Press, Totowa, NJ, pp. 111–123.

24

DNA Sequencing at Elevated Temperature by Capillary Electrophoresis

Peter Lindberg and Johan Roeraade

1. Introduction

The human genome initiative *(1–3)* has spurred the development of analytical instrumentation for high-throughput DNA sequencing. Originally carried out only in poly(acrylamide) slab gels, DNA sequencing is currently to an increasing extent performed in the capillary electrophoresis (CE) format, utilizing narrow bore columns filled with a polymer sieving matrix *(4)*. The advantages of the latter technique are the analysis speed, the possibility of online detection and quantitation, as well as the ease of automation, whereas the advantage of the former technique is the parallel capacity of the gel slab. In order to obtain parallel capacity also in the CE format, capillary array electrophoresis (CAE) systems have been developed in recent years *(5–8)*, for use in high throughput sequencing applications. Another less obvious feature of the slab-gel systems is that the DNA separation is carried out at temperatures above ambient, due to the heat generated by the relatively large currents (compared to CE) that are present in these systems during electrophoresis *(9)*. Elevated temperature has positive effects on the base calling accuracy, the analysis time, and the amount of DNA sequence data that can be obtained from an electrophoretic separation *(10)*. In CE, the generally low currents generated and the efficient heat dissipation from the column prevents any appreciable temperature rise during the analysis, so an external heating source must be employed. A number of CE setups that have been designed to operate at elevated temperatures are described in this chapter, along with some fundamental information on DNA sequencing, on the principles of CE, as well as on the DNA structure and the formation of hairpin loops. The effects of elevated temperatures on the polymer separation matrix and on the migration of DNA fragments are discussed as well.

From: *Methods in Molecular Biology, Vol. 163:*
Capillary Electrophoresis of Nucleic Acids, Vol. 2: Practical Applications of Capillary Electrophoresis
Edited by: K. R. Mitchelson and J. Cheng © Humana Press Inc., Totowa, NJ

2. Methods

2.1. DNA Sequencing According to the Sanger Protocol

1. The Sanger protocol *(11)* is currently the most frequently employed technique for the preparation of DNA-fragment ladders for sequencing experiments. The technique utilizes an enzyme, DNA polymerase, in order to replicate DNA from a template DNA strand. The DNA template can originate from natural materials such as lyzed animal cells, plant cells, or bacteria. Prior to sequencing, the target DNA sequence is isolated and amplified, utilizing the polymerase chain reaction (PCR) technique *(12)*.

2. A primer oligonucleotide is attached to the ssDNA template (**Fig. 1A**), followed by the incorporation of the four deoxynucleotides (dATP, dTTP, dCTP, or dGTP) or one of the four corresponding 2', 3'-dideoxynucleotide analogs (ddATP, ddTTP, ddCTP, or ddGTP) into the growing complementary chain. The chemical structure of a dideoxy-nucleotide is shown in **Fig. 1B**.

3. The occasional incorporation of, e.g., ddATP instead of dATP leads to the termination of a particular chain. The random process eventually results in a ladder of differently sized strands (Sanger fragments) that all terminate with ddATP. Four sets of Sanger fragments that terminate with, respectively, ddATP, ddTTP, ddCTP, or ddGTP, are produced in parallel reactions carried out in individual test tubes. The primer can be tagged with a fluorescent marker *(13)* to facilitate optical detection by laser-induced fluorescence (LIF) *(14)*. When each set of Sanger fragments is analyzed by gel electrophoresis, information on the positions of the bases (A, T, C, or G) in the DNA sequence is obtained *(13)*.

4. In an improved sequencing procedure *(15)*, the various dideoxynucleotides are tagged with fluorescent markers that have different emission spectra. The four dideoxnucleotides can now be added to the same reaction mixture, and the necessary number of analyses that have to be carried out are reduced from four to one.

2.2. Capillary Electrophoresis

1. CE *(16–21)*, is a development from the original slab-gel electrophoresis technique. In CE, the separation is performed in a fused silica capillary column of typically 50–100 μm id filled with a suitable buffer, such as a phosphate or a Tris-borate solution in the mM-range. Due to the small cross-sectional area of the column, currents are low (in the μA-range) and little heat is generated, even when electric field strengths of several hundred V/cm are employed. The use of such elevated electric fields leads to short analysis times (typically 30 min or less).

2. A schematic drawing of a typical CE setup is shown in **Fig. 2**. As can be seen in the figure, the capillary column is connected to a high voltage supply via inlet and outlet buffer vials equipped with platinum electrodes. A detector cell for UV-VIS absorbance or LIF-detection is situated in the far end of the column, near the column outlet.

3. In order to inject a sample, the inlet buffer vial is temporarily substituted for a vial containing the sample solution. A sample plug is introduced into the column by application of a controlled pressure or a defined voltage for a certain period of time. The column is repositioned into the buffer-containing vial and the voltage is reapplied. Under typical CE conditions, i.e., capillary zone electrophoresis (CZE), all sample components, regardless of charge, are carried past the detection zone by a substantial electroosmotic flow *(22)* that originates from charged silanol (SiO) groups on the column inner wall.

2.3. Sieving Polymer Matrices Used in CE

1. Capillary gel electrophoresis (CGE) *(23)* is a special mode of CE, in which the separation column is filled with an *in situ* polymerized crosslinked or linear sieving matrix, such as

A

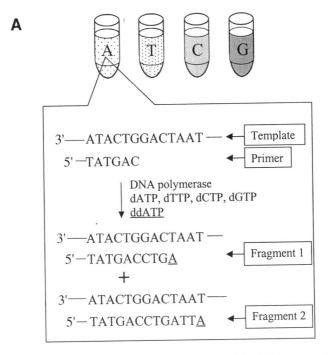

Fig. 1A. A schematic outline of the Sanger protocol for DNA sequencing.

B

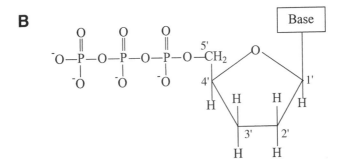

Fig. 1B. The chemical structure of a dideoxynucleotide.

polyacrylamide (PAA). CGE is utilized for the analysis of charged biomolecules, e.g., DNA and SDS-treated proteins, which cannot be separated in free solution *(24)*. *In situ* polymerized polyacrylamide gel columns are very effective for DNA sequencing applications *(25)*.

2. However, when utilized repeatedly for the analysis of more or less impure samples, these columns have a very short lifetime *(26)*. Due to this problem, replaceable, low viscosity polymer solutions *(27)* such as linear polyacrylamide (LPA), e.g., *(28)*, poly(dimethyl acrylamide) (PDMA) *(29)*, or poly(ethylene oxide) (PEO) *(30)* have become more and more popular as sieving matrices for DNA sequencing by CE.

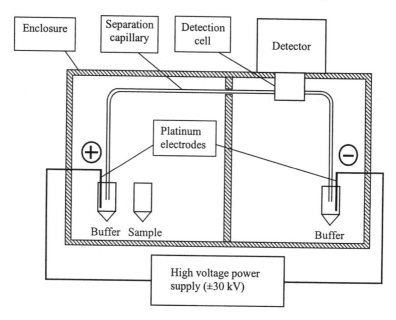

Fig. 2. Schematic of a typical CE setup.

2.4. The Biased Reptation Model of DNA Migration

1. Three principal models exist that describe DNA migration in a polymer matrix, the Ogston model *(31)*, the reptation model *(32)*, and the biased reptation (BRM) model *(33,34)*.
2. The first model applies to DNA fragments that are within the same size range as the average matrix pore, and pictures the migrating DNA as spheres of random coiled polymer strands that are subject to the sieving action of the pores.
3. The second model applies for fragments that are larger (or considerably larger) than the average matrix pore, and visualizes the DNA molecules as more or less elongated rods that migrate head first through available pores in a reptile-like fashion (denominated as reptation).
4. The third model, the BRM-model, describes the electric field effect on DNA fragments, undergoing reptation. According to this model, the charged DNA fragments assume an elongated conformation in the electric field. Size based separation can no longer occur and the fragments start to comigrate in a single band. The BRM-effect on DNA migration becomes most severe in the high-molecular-mass regime of the Sanger-fragment-ladder and is adversely affected by high field strength. Generally, the electric field strengths that can be employed for the large fragments are therefore restricted to 150–200 V/cm.

2.5. Secondary Structures in Single-Stranded DNA Fragments

1. A study of the nucleotide interactions in the dsDNA helix is helpful for understanding how folded secondary structures, i.e., hairpin loops, are formed in ssDNA fragments *(35)*. The nucleotides, which are repeating units of the polynucleotide DNA chain, are composed of a deoxyribose sugar unit, a phosphate unit, and one of the four bases adenine (A), thymidine (T), cytosine (C), and guanine (G).

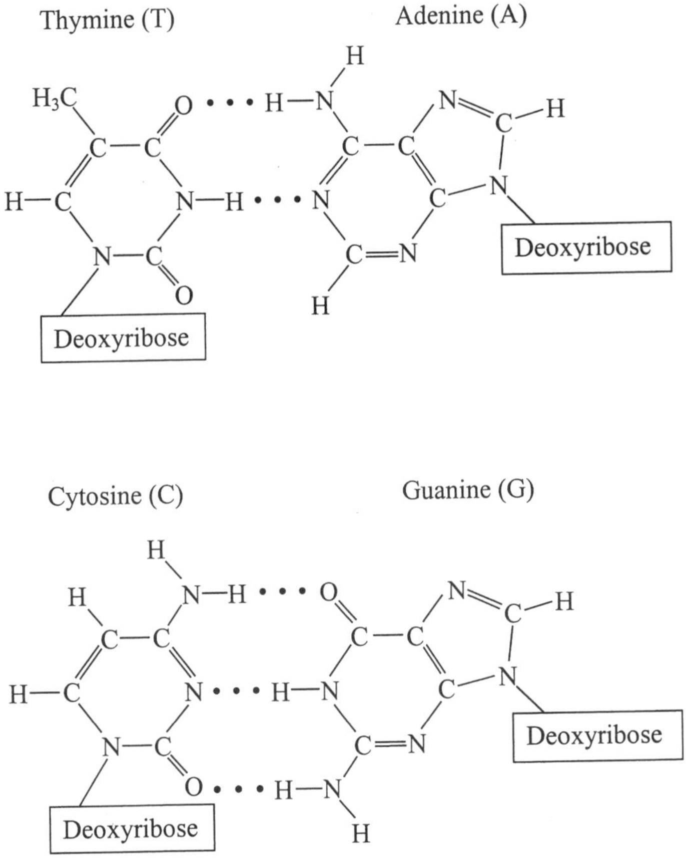
Thymine (T)
Adenine (A)
Deoxyribose
Deoxyribose
Cytosine (C)
Guanine (G)
Deoxyribose
Deoxyribose

Fig. 3. Hydrogen bonds between base pairs.

2. In the DNA helix, two polynucleotide chains are intertwined in a helical pattern, in which the bases occupy the central region whereas the sugar-phosphate backbone is situated on the outside. The bases on opposite chains form base pairs that stabilize the helix. As can be seen in **Fig. 3**, (A) and (T) are connected by two hydrogen bonds, whereas the G-C base pair has a stronger interaction with three hydrogen bonds.
3. An ssDNA fragment in solution is free to assume other conformations than the helical, which facilitates the intramolecular interaction between nucleotides via hydrogen bonding and the formation of secondary structures *(36)*. This is especially true for fragments with base sequences rich in the more strongly interacting bases (G) and (C) *(37)*, as is shown in **Fig. 4**. A DNA fragment that contains a hairpin-loop structure will migrate faster through the gel matrix than an unfolded fragment that has identical base composition. Thus, the dependence of migration time on molecular weight is lost, and the interpretation of the DNA base sequence from the position of the peaks in the electropherogram becomes impossible. In the literature, this situation is referred to as a compression, e.g., *(38)*, or the formation of compressed peaks.

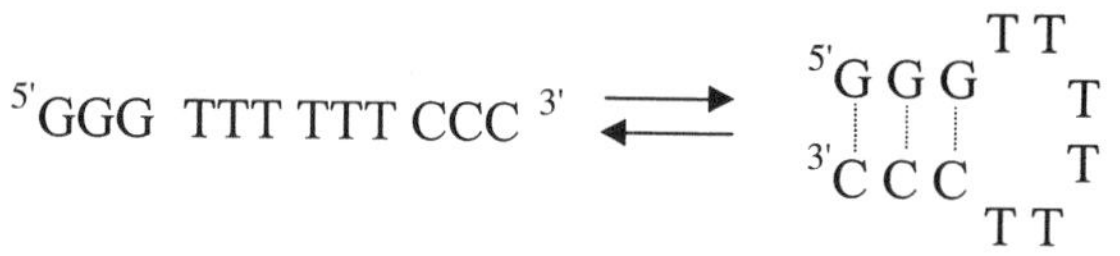

$$^{5'}GGG\ TTT\ TTT\ CCC\ ^{3'} \rightleftharpoons \begin{array}{c} \quad\quad T\ T \\ ^{5'}G\ G\ G\quad T \\ \quad |\ \ |\ \ | \quad\quad T \\ ^{3'}C\ C\ C\quad T \\ \quad\quad T\ T \end{array}$$

Fig. 4. The conformation of a folded and a nonfolded oligonucleotide. Reprinted from **ref. 37**, pp. 307–321 by courtesy of Marcel Dekker, Inc.

2.6. Chemical Denaturation of DNA

In order to suppress compressions, urea and formamide are routinely utilized as hydrogen bond disrupting additives to the gel buffer during the separation of ssDNA fragments *(39)*.

2.6.1. Urea

1. Buffer solutions with a very high urea concentration (7–8 *M*) are often utilized, which is near the urea saturation point. A disadvantage with the use of these solutions is that urea precipitation can occur, and subsequently, the precipitated salt can contaminate sensitive surfaces on analytical equipment, such as optical components. Moreover, when replaceable polymer matrices are employed for the separation, the use of large urea amounts adds to the viscosity of the matrix, leading to a prolonged polymer refill-cycle *(40)*.

2.6.2. Formamide

1. Formamide, which is a stronger denaturant than urea, is frequently added to denaturing buffers to further increase the hydrogen bond disrupting capability *(38–42)*. However, it has been reported that the use of high formamide concentrations in gel buffers can lead to increased separation times and reduced lifetime of polyacrylamide gels *(38,41)*.
2. When G-C rich DNA sequences are analyzed, the use of chemical denaturants may not totally prevent the formation of loop structures in ssDNA fragments *(38)*, and it can be necessary to carry out the separation at a temperature above the ambient.

2.7. Heat-Assisted Denaturation of DNA

1. Three principally different approaches have been utilized to provide a controlled heating of the CE column during the separation of Sanger fragments. In the first approach, the separation column is surrounded with a stream of temperature controlled, hot air *(10,43–47)*. In the second approach, a jacket, in which a temperature-controlled liquid is circulated, encloses the column *(40,48–53)*. In the third approach, the CE column *(28,54–60)*, or the separation channels of chip-based systems *(46,61)*, is in contact with a heated solid surface.

2.7.1. Heated Air

1. **Table 1A** provides a review about the use of heated air devices, and some important experimental parameters from a number of selected references are presented. The first heated air device was described by Lu et al. *(43)*, which utilized a commercial proportional temperature controller (Model CTC-1A, Melabs, CA) to maintain a constant temperature within a Plexiglas box that contained the separation column *(43–45)*.
2. A similar device was constructed by Schmalzing et al. *(46)*, which utilized a thermal air cycler (manufacturer not stated) that was regulated by a temperature controller obtained from Omega Instruments (Stamford, CT).

Table 1A
Review of Methods, in Which a Temperature Controlled Air Stream Is Utilized for Heating the CE Column During the Analysis of Sanger Fragments

Reference	Running conditions		Column parameters		
	t^a (°C)	E^b (V/cm)	id/od (μm)	l^c (cm)	Types of sieving matrices
H. Lu et al. *(43)*	25–50 ± 0.5	300	20/150	35	4% T^d PAA^e *in situ* prepared
J. Zhang et al. *(10)*	60 ± 0.5	150	50/192	39	5% (w/v)^f LPA *in situ* prepared
D. Schmalzing et al. *(46)*	45	200	75/365	36	4% (w/v) LPA replaceable
P. Lindberg et al. *(47)*	50 ± 0.1	200	50/375	40 (30)	10.9% T PDMA *in situ* prepared

^{a}Analysis temperature data.
^{b}Electric field strength employed.
^{c}Effective column length (data on the length of the separation column that is situated within the heated zone is given within parentheses).
^{d}%T = (g acrylamide + g N, N'-methylenebisacrylamide)/100 mL solution.
^{e}LongRanger™ gel (AT biochemicals, Malvern, PA).
^{f}(w/v) = weight of polymer added to a specified volume of buffer solution.

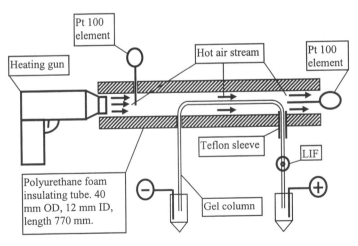

Fig. 5. Setup for CE at an elevated temperature utilizing a heated air stream. Reprinted from **ref. 47**.

3. In **Fig. 5**, a heated air system constructed by Lindberg et al. *(47)*, is depicted. The CE unit consisted of a Plexiglas box with a high-voltage safety interlock and a high-voltage supply, (±) 0–30 kV, constructed from a Spellman CZE 100 (Plainview, NY) unit. The source

Table 1B
Review of Methods, in Which A Liquid-Containing Heating Jacket Is Utilized for Temperature Elevation During the CE Analysis of Sanger Fragments

Reference	Running conditions		Column parameters		
	t^a (°C)	E^b (V/cm)	id/od (μm)	l^c (cm)	Types of sieving matrix
Ruiz-Martinez et al. *(48)*	32	250	75/360	18 (10)	6% (w/v)^d LPA, replaceable
Fung et al. *(40)*	35–65^e	267	75/365	35	2.9% (w/v) PEO, replaceable
Karger^f *(50)*	30	100, 175	75/375	20, 50	4% T^g, 5% C^h PAA, *in situ* prepared

^{a}Analysis temperature data.

^{b}Electric field strength employed.

^{c}Effective column length (data on the length of the separation column that is situated within the heated zone is given within parentheses).

^{d}(w/v) = weight of polymer added to a specified volume of buffer solution.

^{e}2°C/min temperature programmed gradient.

^{f}Utilized the Beckman Coulter P/ACE 5000 CE instrument *(49)*, equipped with a LIF-detection module.

^{g}%T = (g acrylamide + g N,N'-methylenebisacrylamide)/100 mL solution.

^{h}%C = g N,N'-methylenebisacrylamide/(g acrylamide + g N,N'-methylenebisacrylamide).

of warm air was a heating gun, Steinel HL 2002 LE from Germany. The gun had a built-in electronic temperature control so it could be adjusted easily within the required temperature range of 40–60°C. A polyurethane foam insulation tube was employed as a guide, in order to establish a flow of hot air around the separation column. Pt-100 elements positioned directly in front of the heating gun outlet as well as on the far end of the tube were utilized for temperature monitoring with an accuracy of 0.1°C.

2.7.2. Heated Liquid

1. **Table 1B** provides a similar overview to that given in **Table 1A**, but describes CE devices that employ a heated liquid jacket for providing a controlled elevated temperature during the analysis of Sanger fragments. The first to utilize this approach was Ruiz-Martinez et al. *(48)*. In this design, the separation capillary was inserted into a stainless steel capillary column of 380 μm id, which in turn was mounted within a surrounding water jacket. Fung et al. *(40)* also employed a water jacket for temperature control. The jacket that held the capillary had an internal diameter of 1 cm, and water from a thermostat-controlled bath was circulated at a rate of 0.2 L/min. A liquid containing jacket has also been utilized for temperature control in commercial CE equipment, e.g., in the P/ACE series from Beckman Coulter (Fullerton, CA) *(49,50,51)* and in the Biofocus 3000 system from Bio-Rad Laboratories *(52,53)*.

2.7.3. Solid-State Heat Exchangers

1. In the third approach to temperature control, the column is kept in contact with a solid-state heat exchanger element (*see* **Table 1C** for summary). Such a design was first described by Klepárník et al. *(54)*.

Table 1C
Review of Methods, in Which a Solid State Device
Is Utilized for the Controlled Heating of the CE Column
During the Analysis of Sanger Fragments

Reference	Running conditions		Column parameters		
	t^a (°C)	E^b (V/cm)	id/od (µm)	l^c (cm)	Types of sieving matrix
Klepárník et al. (*54*)	25–60	250	75/365	30 (25)	3% (w/v)^d LPA replaceable
Schmalzing et al.^{e} (*46*)	45	200	40×90^f	11.5^g	4% (w/v) LPA replaceable
Liu et al.^{e} (*61*)	35–40	150	50×130^f	7^g	3–4% (w/v) LPA replaceable
Madabhushi^h (*62*)	42	160	50/360	40	6.5% (w/v) PDMA replaceable
Xiong et al.^{h} (*63*)	42	160	50/375	36	6% (w/v) PDMA replaceable

^{a}Analysis temperature data.
^{b}Electric field strength employed.
^{c}Effective column length (data on the length of the separation column that is situated within the heated zone is given within parentheses).
^{d}(w/v) = weight of polymer added to a specified volume of buffer solution.
^{e}A microchip was employed for the separation.
^{f}Depth and width of etched channel structure in a fused silica wafer.
^{g}The entire microchip was heated.
^{h}A Perkin-Elmer-Applied Biosystems CE instrument, the ABI Prism 310 Genetic Analyzer (*60*), was utilized for the separations.

2. A schematic of the column-heating element is shown in **Fig. 6**. A plug of heat conducting epoxy (Norcure 228, Northern Labs, Greenwich, CT) was cast within a copper tube. Utilizing a Teflon tube as a template during the casting process, the epoxy plug was fitted with a centered cylindrical hole with a diameter of 380 µm that could contain a standard fused silica separation capillary. Subsequently, when heated to a temperature above 35°C, the epoxy resin expanded to a tight fit around the separation capillary. The expansion process was reversible, and thus the resin loosened its grip upon cooling, which facilitated an easy replacement of the capillary.

3. The heat was provided by a Capton heating element, KH-108/2 (Omega, Stanford, CT) attached to the copper tube utilizing a high thermal conductivity paste, Omegatherm 2000 (Omega) and Teflon tape. A platinum sensor F31052 (Omega) was attached in a similar manner, and utilized for temperature measurements. A microprocessor controller, CN 76000 (Omega) facilitated temperature settings in the 25–100°C range. The authors suggest that the Capton heating unit may be substituted for a peltier element, in order to provide a more rapid cooling or to more easily create temperature gradients. In the Perkin-Elmer ABI Prism 310 DNA Analyzer (*60–63*), the solid-state heat exchanging principle

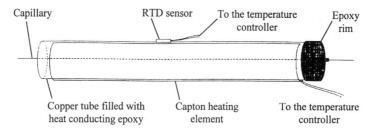

Fig. 6. Setup for CE at an elevated temperature utilizing a solid-state heat-exchanging device. Adapted from **ref. 54**.

is utilized as well, although in this case the separation capillary is positioned between a heated plate and an insulated plate *(60)*.

2.7.4. Heat Exchangers and Microfabricated Devices

1. Schmalzing et al. *(46)* employed the solid-state heating principle for temperature control of a microfabricated chip device. The microchip was positioned on a temperature-controlled alumina heater block. Alternately, Liu et al. *(61)* utilized a hollow aluminum plate that is thermostat controlled by a constant flow of temperature controlled water, in order to heat their chip device with microfabricated CE channels.

3. Results

3.1. Stability and Sieving Performance of the Polymer Matrix

1. Running at elevated temperature during the electrophoretic separation of Sanger fragments can affect the polymer matrix, including the buffer components, in a number of ways. First, chemical degradation of the sieving polymer can occur. Second, the sieving properties of the matrix can change. Third, decomposition of denaturing buffer components such as urea and/or formamide is accelerated. These temperature-induced effects altogether set the upper limit of the temperature range that can be employed to ~60°C *(54)*. Additional data is listed in **Table 1A–C**.

3.1.1. Chemical Degradation of the Sieving Polymer

1. Polymers, e.g., polyacrylamide, which have pendant groups that are prone to hydrolysis, can develop charged sites in the gel matrix and this is a problem that is likely to increase at elevated temperatures *(64)*. Righetti's group *(65)* has suggested a number of *N*-substituted hydrolytically stable acrylamido monomers, which can be utilized for work at elevated temperatures.

3.1.2. Changing Sieving Properties of the Polymer Matrix

1. **Figure 7** shows schematic representations of a DNA molecule migrating through either (A) a crosslinked gel, and (B) a linear polymer matrix *(66)*. CE columns that contain an *in situ* polymerized, crosslinked gel matrix cannot be operated at temperatures significantly above room temperature without a serious degeneration of the matrix *(23)* with a few reported exceptions *(44,47)*.

2. On the other hand, columns that are filled with the more flexible linear polymer matrices are much more tolerant to changes in the operating temperature *(54)*. However, in a linear

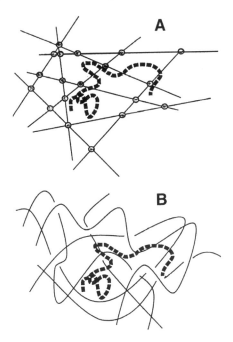

Fig. 7. A schematic, showing the structures of a crosslinked gel and a linear polymer matrix. Reprinted from **ref. 66**.

matrix, the pores are assumed to have a limited lifetime *(67)* that will be reduced even further as the temperature rises and the friction between polymer strands is reduced. Also, under these conditions, large analytes can distort the polymer mesh more easily through the action of constraint release *(34)*.

3. In order to conserve the sieving performance of the separation matrix as the temperature is raised above ambient, it is a sound strategy to increase the molecular mass of the polymers in the matrix and/or to increase the polymer concentration *(68)*.

3.1.3. Decomposition of Chemical Denaturants

1. The chemical denaturants urea and formamide slowly decompose in solutions kept at room temperature, and this effect is accelerated at temperatures above the ambient *(26,54)*. The charged decomposition products of formamide will increase the current during electrophoresis *(26,54)*, which can lead to increased problems with band-broadening due to the formation of a steeper radial temperature gradient in the separation column *(22)*.

2. Fung et al. *(40)* reported that the degradation products of urea in some way interacted with the PEO polymer used in their experiments, resulting in reduced separation efficiency.

3.2. Effects of Elevated Temperature on DNA Migration

1. The use of an elevated temperature during electrophoresis is beneficial for the separation of Sanger fragments in three different ways. First, compressions can be resolved. Second, the analysis time is considerably reduced. Third, the length of DNA sequence that can be read from the electropherogram can increase (*see* **Note 1**).

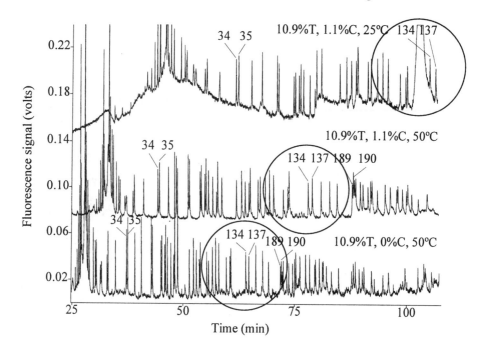

Fig. 8. Examples of electropherograms that show separations of a DNA-sequencing sample obtained from the bacteria *Moraxella*, at room temperature and at an elevated temperature. Adapted from **ref. 47**.

3.2.1. Resolved Compressions

1. Compressed peaks can generally be resolved by utilizing the combined denaturing effect of hydrogen-bond-disrupting buffer additives and an elevated temperature.
2. An example of this effect is shown in **Fig. 8**, depicting separations of T-terminated Sanger fragments from the bacteria *Moraxella* (*47*) performed with 10.9% T crosslinked PDMA gel filled columns. The columns that contained 7 *M* urea, were employed for analysis at 25°C and at 50°C (upper and middle traces, respectively), and the lower trace depicts a reference run at 50°C performed with a 10.9% T linear PDMA-column. Note, how the severe compression of fragments 134 and 137 observed at 25°C (upper trace) has disappeared completely at the higher temperature.
3. Another example is depicted in **Fig. 9** (*54*), showing sequencing data collected from the analysis of a region of M13mp18 DNA. Columns are filled with a replaceable 3% (w/v) LPA matrix containing 3.5 *M* urea and 30% formamide (**Fig. 9A,C**), or containing 7 *M* urea (**Fig. 9B**). The compressions I and II shown in the figure that were present at 25°C (**Fig. 9A**) were equally well resolved at 50°C utilizing either of the denaturant compositions (**Fig. 9B,C**).

3.2.2. Reduced Analysis Time

1. An increase in the migration speed of Sanger fragments with increasing temperature have been observed, both in linear (*45,54*) and crosslinked (*44,47*) polymer matrices, leading to a substantial gain in analysis time.

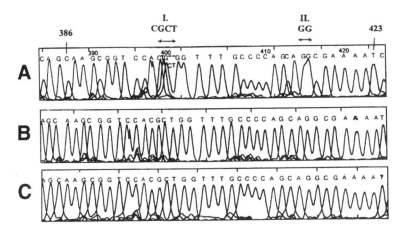

Fig. 9. Sequencing data collected from the analysis of a region of M13mp18 DNA performed utilizing a 3% (w/v) LPA matrix, containing 3.5 *M* urea and 30% formamide (**A** and **C**), or containing 7 *M* urea (**B**). The compressions I and II that were present at 25°C (A) were resolved at 50°C using either of the denaturant compositions (B and C). Reprinted from **ref. *54***.

2. In **Fig. 10** *(45)*, this effect is illustrated with three close-ups of electropherograms depicting separations of Sanger fragments performed in 5% (w/v) LPA at 20°C (upper trace), at 40°C (middle trace), and at 60°C (lower trace). Another example of this effect can be seen in the upper and middle traces of **Fig. 8** *(47)*.

3. In the linear matrices, the increase in migration speed is associated with a reduced polymer solution viscosity due to the decrease in frictional forces between polymer strands discussed in **Subheading 3.1.2.** In the crosslinked matrices, the flexibility of the covalently interconnected mesh cannot be expected to increase with increasing temperature to the same extent. Instead, an enhanced ability of the Sanger fragments to pass through the rigid pores of the mesh due to a temperature induced change in DNA persistence length, is a more likely cause to the observed migration speed increase *(44)*.

3.2.3. DNA Reading Length

1. The DNA reading length is the number of bases of DNA sequence that can be identified in an electrophoretic separation before the onset of biased reptation (*see* **Subheading 2.4.**). The BRM-model predicts that a rise in the temperature employed during electrophoresis will increase the thermal energy (and thermal translation) of the migrating DNA fragments *(33)*. The increase in thermal energy should counteract the electric field orientation of the fragments and lead to an extended reading length.

2. The theoretical predictions have been shown to be correct for separations of Sanger fragments in linear polymer matrices *(45,54)*. However, when crosslinked gels were utilized, the opposite effect, reduced reading lengths at elevated temperatures were observed both by Lu et al. *(44)* and by Lindberg et al. *(47)*. At present, the reason for the discrepancy between theory and practice in the case of crosslinked gels remains unclear. It has been suggested *(44,45)* that the mechanism of DNA migration in such matrices might differ fundamentally from the mechanism of constraint release (*see* **Subheading 3.1.2.**) postulated for linear matrices *(34)*.

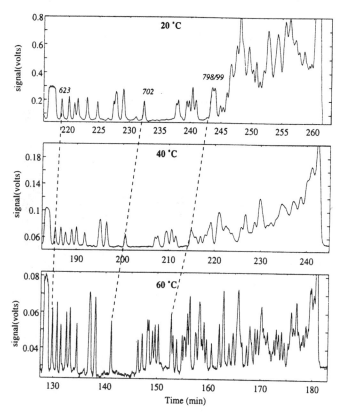

Fig. 10. Examples of electropherograms that show separations of Sanger fragments obtained from the DNA sequence M13mp18, performed in 5% (w/v) linear polyacrylamide at 20°C (*upper trace*), at 40°C (*middle trace*), and at 60°C (*lower trace*). Reprinted from **ref. 45**.

3.3. Analysis of Sanger Fragments by CE at Elevated Temperatures

1. **Table 2A–C** depicts analysis time data and DNA reading length data which originate from the various CE methods for the separation of Sanger fragments at elevated temperatures (*see* **Subheading 2.7.**) listed in **Table 1A–C**. The reported data may not reflect the full-developed potential of the different CE setups described, i.e., other combinations of polymer matrix, analysis temperature, and electric field strength may lead to further improvements. Also, the reading length data from various researchers are difficult to compare due to the lack of a common reading length definition, and due to the use of more and less advanced base-calling software programs in order to extract more information from raw sequencing data *(10,56,59)*. One suggested way is to define the reading length as being the DNA fragment size where the resolution between adjacent peaks in the electropherogram, representing fragments that differ only one base in size, drops below 0.5 *(46,47,63)*.

2. However, the following conclusions can be made from the data presented in the **Table 2A–C**. The state-of-the-art in the separation of Sanger fragments may be represented by a CE method that utilizes the following operating parameters:

Table 2A
Separation Performance of CE Systems, in Which
a Temperature-Controlled Air Stream Is Utilized for Heating the Column

Reference	Analysis time (min)	DNA sequence reading length (bases)
Lu et al. *(43)*	51	517 (700)[a]
Zhang et al. *(10)*	120	640
Schmalzing et al.[b] *(46)*	—	—
Lindberg et al. *(47)*	120	303[c]

[a]The DNA sequence out to 517 bases could be read directly from the electropherogram. Comparison of the electropherogram with the known DNA sequence led to an additional identification of fragments up to 700 bases.
[b]Data not reported.
[c]303 bases of the DNA sequence could be read with a resolution of 0.5.

Table 2B
Separation Performance of CE Systems, in Which
a Liquid-Containing Heating Jacket Is Used
for both Temperature Elevation and for Control

Reference	Analysis time (min)	DNA sequence reading length (bases)
Ruiz-Martinez et al. *(48)*	30	350
Fung et al. *(40)*[a]	—	—
Karger *(50)*	50	500[b]

[a]Data not reported.
[b]500 bases of the DNA sequence could be read with a resolution of 0.5.

a. An elevated temperature of 50–60°C.
b. Moderate electric fields of 150–200 V/cm.
c. High molecular mass linear polymers.
d. A separation column having an effective length of 30–40 cm. An example that falls within the state-of-the-art category is the separation of 640 bases in 120 min performed by Zhang et al. *(10)*, utilizing a column filled with 5% (w/v) *in situ* polymerized LPA. Another example is the separation of 687 bases in 70 min by Klepárník et al. *(54)*, utilizing a 3% (w/v) replaceable LPA. Further examples are the separation of 600 bases in 110 min utilizing a 6.5% (w/v) replaceable linear PDMA matrix by Madabhushi *(62)*, as well as the separation of 660 bases in 110 min by Xiong et al. *(63)* utilizing a linear 6% (w/v) linear PDMA matrix.

3. Very rapid electrophoretic separations of Sanger fragments have been performed at elevated temperatures, utilizing microfabricated chip devices *(46,61)* (*see* **Tables 1C** and **2C**). Recently, Liu et al. *(61)* demonstrated four-color DNA-sequencing up to 600 bases after emergence of the primer within 20 min, on a 7-cm long channel filled with a 4% (w/v) LPA matrix. The sequencing performance of this system is approaching the performances of the above state-of-the-art CE-systems, at a much reduced analysis time (*see* **Note 2**).

Table 2C
Separation Performance of CE Systems, in Which a Solid-State Device
Is Utilized for both Temperature Elevation and for Control

Reference	Analysis time (min)	DNA sequence reading length (bases)
Klepárník et al. *(54)*	70	687 (800)[a]
Schmalzing et al.[b] *(46)*	14	400[c]
Liu et al.[e] *(61)*	20	500 (600)[d]
Madabhushi[e] *(62)*	120	600[f]
Xiong et al.[e] *(63)*	110	660[g]

[a]Single base resolution was obtained up to 687 bases, but the DNA sequence could be read up to 800 bases and longer.
[b]A microchip was employed for the separation.
[c]400 bases of the DNA sequence could be read with a resolution of 0.5.
[d]500 bases of the DNA sequence could be read with a resolution of 0.5. The sequence could be read up to 600 bases, utilizing base calling software.
[e]A Perkin/Applied Biosystems CE instrument, the ABI Prism 310 Genetic Analyzer *(60)*, was utilized for the separations.
[f]600 bases of the DNA sequence could be read with a resolution of 0.59.
[g]660 bases of the DNA sequence could be read with a resolution of 0.5.

4. Notes

1. Performing the electrophoretic separation of Sanger fragments at an elevated temperature has a positive effect on the resolution of compressed peaks *(47,54)*. Also, the reading length can be increased, although the analysis time is reduced *(45,54)*. However, care must be taken not to operate at too high temperatures (above 60°C) for a prolonged period of time, since the polymer matrix and the buffer components will degrade. Furthermore, it is advantageous to utilize a linear sieving polymer of a type that is stable to hydrolysis, e.g., *N*-substituted polyacrylamides *(64,65)*.

2. Presently, CAE instruments for DNA sequencing that dramatically increase the sample processing capacity are becoming commercially available, e.g., the, Perkin-Elmer-Applied Biosystems 3700 DNA Analyzer *(69)*. CAE microplates for DNA sequencing applications are currently under development *(70)*. The next step will most likely be to equip CAE-units for work at temperatures above ambient. We believe that the designs that utilize a temperature controlled air stream *(43–47)* or a temperature-controlled heated plate *(46,54,60)* should be the easiest to adapt for CAE-use, as the capillary array does not need to be arranged within a specially dedicated heating jacket. However, optical components can be affected, such as the alignment between the moving parts of a scanning LIF system *(5)*. Also, if heated air is employed, an air stream can easily stir up dust particles present that could interfere with the light path of the laser beam. Thus, a number of critical factors must be taken into account, which require some creative engineering in the design of future large throughput sequencers.

A rapid development of commercial CAE-instrumentation has occurred after the initial preparation of the manuscript. Also, more technical information has become available from the instrument manufacturers. At present, in July 2000, several manufacturers, e.g., Amersham Pharmacia Biotech, Beckman Coulter, Inc., and Applied Biosystems offer instrumentation that can operate at temperatures above ambient.

References

1. Trainor, G. L. (1990) DNA sequencing, automation, and the human genome. *Anal. Chem.* **62**, 418–426.
2. Dahl, C. A. and Strausberg, R. L. (1996) Human genome project: revolutionizing biology through leveraging technology. *SPIE* **2680**, 190–201.
3. Collins, F., et al. (1998) New goals for the U. S. human genome project: 1998–2003. *Science* **282**, 682–689.
4. Dovichi, N. J. (1997) DNA sequencing by capillary electrophoresis. *Electrophoresis* **18**, 2393–2399.
5. Bashkin, J. S., Bartosiewicz, M., Roach, D., Leong, J., Barker, D., and Johnston, R. (1996) Implementation of a capillary array electrophoresis instrument. *J. Capil. Elect.* **3**, 61–68.
6. Carrilho, E., Miller, A. W., Ruiz-Martinez, M. C., Kotler, L., Kesilman, J., and Karger, B. L. (1997) Factors to be considered for robust high-throughput automated DNA sequencing using a multiple-capillary array instrument. *SPIE* **2985**, 4–18.
7. Tan, H. and Yeung, E. S. (1998) Automation and integration of multiplexed on-line sample preparation with capillary electrophoresis for high-throughput DNA sequencing. *Anal. Chem.* **70**, 4044–4053.
8. Kheterpal, I. and Mathies, R. A. (1999) Capillary array electrophoresis: DNA sequencing. *Anal. Chem.* **71**, 31A–37A.
9. Henry, C. (1996) Knowing the code: Manual and automated gel-based DNA sequencers. *Anal. Chem.* **68**, 493A–497A.
10. Zhang, J., Fang, Y., Hou, J. Y., Ren, H. J., Jiang, R., Roos, P., and Dovichi, N. J. (1995) Use of non-cross-linked polyacrylamide for four-color DNA sequencing by capillary electrophoresis separation of fragments up to 640 bases in length in two hours. *Anal. Chem.* **67**, 4589–4593.
11. Sanger, F., Nicklen, S., and Coulson, A. R. (1977) DNA sequencing with chain-terminating inhibitors. *Proc. Natl. Acad. Sci. USA* **74**, 5463–5467.
12. Mullis, K. B. (1987) Specific synthesis of DNA in vitro: a polymerase-catalyzed chain reaction. *Methods Enzymol.* **155**, 335–350.
13. Ansorge, W., Sproat, B. S., Stegeman, J., and Schwager, C. (1986) A non-radioactive automated method for DNA sequence determination. *J. Biochem. Biophys. Methods* **13**, 315–323.
14. Mac Taylor, C. E. and Ewing, A. G. (1997) Critical review of recent developments in fluorescence detection for capillary electrophoresis. *Electrophoresis* **18**, 2279–2290.
15. Prober, J. M., Trainor, G. L., Dam, R. J., Hobbs, F. W., Robertson, C. W., Zagursky, R. J., et al. (1987) A system for rapid DNA sequencing with fluorescent chain-terminating dideoxynucleotides. *Science* **238**, 336–341.
16. Heiger, D. N. (ed.) (1992) *Capillary Electrophoresis-An Introduction*, 2nd ed. Hewlett-Packard Company, France.
17. Guzman, N. A. (ed.) (1993) *Capillary Electrophoresis Technology*. Marcel Dekker, New York.
18. Baker, D. R. (ed.) (1995) *Capillary Electrophoresis*. John Wiley & Sons, New York.
19. Righetti, P. G. (ed.) (1996) *Capillary electrophoresis in Analytical Biotechnology*. CRC Press, Boca Raton, Florida.
20. St. Claire, R. L. (1996) Capillary Electrophoresis. *Anal. Chem.* **68**, 569R–586R.

21. Beale, S. C. (1998) Capillary electrophoresis. *Anal. Chem.* **70,** 279R–300R.
22. Grossman, P. D. (1992) Factors affecting the performance of capillary electrophoresis separations: Joule heating, electroosmosis, and zone dispersion, in *Capillary Electrophoresis Theory & Practice.* (Grossman, P. D. and Colburn, J. C., eds.), Academic Press, San Diego, CA.
23. Dolnik, V. (1994) Capillary gel electrophoresis. *J. Microcol. Sep.* **6,** 315–330.
24. Grossman, P. D. and Soane, D. S. (1992) Principles of electrophoretic separations in polymer matrices, in *Polymer Applications for Biotechnology, Macromolecular Separation and Identification* (Soane, D. S., ed.), Prentice-Hall, Englewood Cliffs, NJ.
25. Kamahori, M. and Kambara, H. (1996) Characteristics of single-stranded DNA separation by capillary gel electrophoresis *Electrophoresis* **17,** 1476–1484.
26. Swerdlow, H., Dew-Jager, K. E., Brady, K., Grey, R., Dovichi, N. J., and Gesteland, R. (1992) Stability of capillary gels for automated sequencing of DNA. *Electrophoresis* **13,** 475–483.
27. Sunada, W. M. and Blanch, H. W. (1997) Polymeric separation media for capillary electrophoresis of nucleic acids. *Electrophoresis* **18,** 2243–2254.
28. Goetzinger, W., Kotler, L., Carrilho, E., Ruiz-Martinez, M. C., Salas-Solano, O., and Karger, B. L. (1998) Characterization of high molecular mass linear polyacrylamide powder prepared by emulsion polymerization as a replaceable polymer matrix for DNA sequencing by capillary electrophoresis. *Electrophoresis* **19,** 242–248.
29. Menchen, S., Johnson, B., Madabhushi, R., and Winnik, M. (1996) The design of separation media for DNA sequencing in capillaries. *SPIE* **2680,** 294–303.
30. Fung, E. N. and Yeung, E. S. (1995) High-speed DNA sequencing by using mixed poly(ethylene oxide) solutions in uncoated capillary columns. *Anal. Chem.* **67,** 1913–1919.
31. Ogston, A. G. (1958) The spaces in a uniform random suspension of fibres. *Trans Faraday Soc.* **54,** 1754–1757.
32. De Gennes, P. G. (1971) Reptation of a polymer chain in the presence of fixed obstacles. *J. Chem. Phys.* **55,** 572–579.
33. Lumpkin, O. J., Dejardin, P., and Zimm, B. H. (1985) Theory for gel electrophoresis of DNA. *Biopolymers* **24,** 1573–1593.
34. Viovy, J.-L. and Duke, T. (1993) DNA electrophoresis in polymer solutions: Ogston sieving, reptation and constraint release. *Electrophoresis* **14,** 322–329.
35. Stryer, L. (1988) Molecular design of life, in *Biochemistry* (Stryer, L., ed.), W. H. Freeman and Company, New York, pp. 71–90.
36. Bowling, J. M., Bruner, K. L., Cmarik, J. L., and Tibbets, C. (1991) Neighboring nucleotide interactions during DNA sequencing gel electrophoresis. *Nucleic Acids Res.* **19,** 3089–3097.
37. Lindberg, P. and Roeraade, J. (1999) Gel matrices in N-methylformamide for separation of DNA fragments. *J. Liq. Chromatogr.* **22,** 307–231.
38. Konrad, K. D. and Pentoney, S. L., Jr. (1993) Contribution of secondary structure to DNA mobility in capillary gels. *Electrophoresis* **14,** 502–508.
39. Heller, C. (1997) The separation matrix, in *Analysis of Nucleic Acids by Capillary Electrophoresis* (Heller, C., ed.), Friedr. Vieweg & Sohn Verlagsgesellschaft mbH, Braunschweig/Wiesbaden, Germany, pp. 17,18.
40. Fung, E. N., Pang, H., and Yeung, E. S. (1998) Fast DNA separations using poly(ethylene oxide) in non-denaturing medium with temperature programming. *J. Chromatogr. A* **806,** 157–164.
41. Rocheleau, M. J., Grey, R. J., Chen, D. Y., Harke, H. R., and Dovichi, N. J. (1992) Formamide modified polyacrylamide gels for DNA sequencing by capillary electrophoresis. *Electrophoresis* **13,** 484–486.

42. Swerdlow, H., Dew-Jager, K. E., Brady, K., Grey, R., Dovichi, N. J., and Gesteland, R. (1992) Stability of capillary gels for automated sequencing of DNA. *Electrophoresis* **13,** 475–483.

43. Lu, H., Arriaga, E., Chen, D. Y., and Dovichi, N. J. (1994) High-speed and high-accuracy DNA sequencing by capillary gel electrophoresis in a simple, low cost instrument. Two-color peak-height encoded sequencing at 40°C. *J. Chromatogr. A* **680,** 497–501.

44. Lu, H., Arriaga, E., Chen, D. Y., and Dovichi, N. J. (1994) Activation energy of single-stranded DNA moving through cross-linked polyacrylamide gels at 300 V/cm. effect of temperature on sequencing rate in high-electric-field capillary gel electrophoresis. *J. Chromatogr.* **680,** 503–510.

45. Fang, Y., Zhang, J. Z., Hou, J. Y., Lu, H., and Dovichi, N. J. (1996) Activation energy of the separation of DNA sequencing fragments in denaturing noncross-linked polyacrylamide by capillary electrophoresis. *Electrophoresis* **17,** 1436–1442.

46. Schmalzing, D., Adurian, A., Koutny, L., Ziagura, L., Matsudaria, P., and Ehrlich, D. (1998) DNA sequencing on microfabricated electrophoretic devices. *Anal. Chem.* **70,** 2303–2310.

47. Lindberg, P., Righetti, P. G., Gelfi, C., and Roeraade, J. (1997) Electrophoresis of DNA sequencing fragments at elevated temperature in capillaries filled with poly-(*N*-acryloylamino-propanol) gels. *Electrophoresis* **18,** 2909–2914.

48. Ruiz-Martinez, M. C., Berka, J., Belenkii, A., Foret, F., Miller, A. W., and Karger, B. L. (1993) DNA sequencing by capillary electrophoresis with replaceable linear polyacrylamide and laser-induced fluorescence detection. *Anal. Chem.* **65,** 2851–2858.

49. http://www.beckman.com/Beckman/biorsrch/prodinfo/capelec/caphome.asp

50. Karger, A. E. (1996) Separation of DNA sequencing fragments using an automated capillary electrophoresis instrument. *Electrophoresis* **17,** 144–151.

51. Sonada, R., Nishi, H., and Noda, K. (1998) Capillary electrophoresis of oligonucleotides using polymer solutions. *Chromatographia* **48,** 569–575.

52. http://www.bio-rad.com

53. Lowery, J. D., Ugozzoli, L., and Wallace, R. B. (1997) Application of capillary electrophoresis to the measurement of oligonucleotide concentration and purity over a wide dynamic range. *Anal. Biochem.* **254,** 236–239.

54. Klepárník, K., Foret, F., Berka, J., Goetzinger, W., Miller, A. W., and Karger, B. L. (1996) The use of elevated column temperature to extend DNA sequencing read lengths in capillary electrophoresis with replaceable polymer matrices. *Electrophoresis* **17,** 1860–1866.

55. Ruiz-Martinez, M. C., Carrilho, E., Berka, J., Kieleczawa, J., Miller, A. W., Foret, F., Carson, C., and Karger, B. L. (1996) DNA sequencing by capillary electrophoresis using short oligonucleotide primer libraries. *BioTechniques* **20,** 1058–1069.

56. Carrilho, E., Ruiz-Martinez, M. C., Berka, J., Smirnov, I., Goetzinger, W., Miller, A. W., Brady, D., and Karger, B. L. (1996) Rapid DNA sequencing of more than 1000 bases per run by capillary electrophoresis using replaceable linear polyacrylamide solutions. *Anal. Chem.* **68,** 3305–3313.

57. Ruiz-Martinez, M. C., Salas-Solano, O., Carrilho, E., Kotler, L., and Karger, B. L. (1998) A sample purification method for rugged and high-performance DNA sequencing by capillary electrophoresis using replaceable polymer solutions. A. Development of the cleanup protocol. *Anal. Chem.* **70,** 1516–1527.

58. Klepárník, K., Berka, J., Foret, F., Doškar, J., Kailerová, J., Rosypal, S., and Bocek, P. (1998) DNA cycle sequencing of a common restriction fragment of *Staphylococcus aureus* bacteriophages by capillary electrophoresis using replaceable linear polyacryla-mide. *Electrophoresis* **19,** 695–700.

59. Salas-Solano, O., Carrilho, E., Kotler, L., Miller, A. W., Goetzinger, W., Sosic, Z., and Karger, B. L. (1998) Routine DNA sequencing of 1000 bases in less than one hour by capillary electrophoresis with replaceable linear polyacrylamide solutions. *Anal. Chem.* **70,** 3996–4003.

60. http://www.appliedbiosystems.com

61. Liu, S., Shi, Y., Ja, W. W., and Mathies, R. A. (1999) Optimization of high-speed DNA sequencing on microfabricated capillary electrophoresis channels. *Anal. Chem.* **71,** 566–573.

62. Madabhushi, R. S. (1998) Separation of 4-color DNA sequencing extension products in noncovalently coated capillaries using low viscosity polymer solutions. *Electrophoresis* **19,** 224–230.

63. Xiong, Y., Park, S., and Swerdlow, H. (1998) Base stacking: pH-mediated on-column sample concentration for capillary DNA sequencing. *Anal. Chem.* **70,** 3605–3611.

64. Chiari, M., Micheletti, C., Nesi, M., Fazio, M., and Righetti, P. G. (1994) Towards new formulations for polyacrylamide matrices: N-acryloylaminoethoxyethanol, a novel monomer combining high hydrophilicity with extreme hydrolytic stability. *Electrophoresis* **15,** 177–186.

65. Righetti, P. G. and Gelfi, C. (1996) Electrophoresis gel media: the state of the art. *J. Chromatogr. B* **699,** 63–75.

66. Slater, G. W., Kist, T. B. L., Ren, H., and Drouin, G. (1998) Recent developments in DNA electrophoretic separations. *Electrophoresis* **19,** 1525–1541.

67. Bae, Y. C. and Soane, D. (1993) Polymeric separation media for electrophoresis: cross-linked systems or entangled solutions. *J. Chromatogr. A.* **652,** 17–22.

68. Gelfi, C., Perego, M., Libbra, F., and Righetti, P. G. (1996) Comparison of behavior of N-substituted acrylamides and celluloses on double-stranded DNA separations by capillary electrophoresis 25°C and 60°C. *Electrophoresis* **17,** 1342–1347.

69. http://www.appliedbiosystems.com

70. Simpson, P. C., Roach, D., Woolley, A. T., Thorsen, T., Johnston, R., Sensabaugh, G. F., and Mathies, R. A. (1998) High-throughput genetic analysis using microfabricated 96-sample capillary array electrophoresis microplates. *Proc. Natl. Acad. Sci. USA* **95,** 2256–2261.

25

DNA Sequencing in Noncovalently Coated Capillaries Using Low Viscosity Polymer Solutions

Ramakrishna Madabhushi

1. Introduction

Capillary electrophoresis (CE) has become an alternative to slab-gel electrophoresis for DNA separations due to its many advantages such as speed, increased separation efficiency, requires minute amount of sample, and automation of sample loading *(1)*. Currently, high-throughput DNA sequencing is performed exclusively by slab-gel electrophoresis coupled to fluorescence detection systems. However, slab-gel electrophoresis involves labor-intensive steps such as gel pouring, sample loading, and long electrophoretic run times. These disadvantages reduce the overall throughput efficiency of a slab-gel system making it less than ideal. One of the main goals of the Human Genome and other large scale sequencing projects is to increase the throughput rate with a commensurate reduction in the cost-per-base sequenced. A limitation of CE has been that only one capillary could be run and monitored at a time, so that the total experimental throughput is no better than with slower slab-gel system with multiple lanes. However, the recent introduction of the multiple capillary instruments, such as the ABI Prism 3700 DNA Analyzer *(2)* and the MegaBACE *(3)*, may have the potential to meet the high throughput demand.

The two most important components for DNA separations by CE are the separation medium and the capillary inner wall coating. Due to the historic use of polyacrylamide for slab gels, crosslinked polyacrylamide gel-filled capillaries were used initially for the separation of DNA sequencing reaction products by CE. However, gel-filled capillaries are difficult to prepare without the formation of bubbles and have short lifetime because of electroosmosis and capillary inlet fouling with template DNA *(1)*. The use of noncrosslinked sieving media instead of rigid gels has greatly enhanced the utility of CE by eliminating the need to change the capillaries frequently. Instead, due to the flowability of noncrosslinked solution, the medium can be replaced after each electrophoresis run by pumping fresh medium into the capillaries making the process more convenient and economical *(1)*.

From: *Methods in Molecular Biology, Vol. 163:*
Capillary Electrophoresis of Nucleic Acids, Vol. 2: Practical Applications of Capillary Electrophoresis
Edited by: K. R. Mitchelson and J. Cheng © Humana Press Inc., Totowa, NJ

1.1. Capillary Surface Modification

The silica surface is rich in silanol groups and their ionization starts at around pH 3.0 and continues to beyond pH 10.0. Ionization of silanols results in the formation of an electrical double layer near the surface boundary associated with a Zeta potential (ζ). Under the influence of an applied voltage, the positively charged mobile layer migrates toward the cathode, resulting in the phenomenon of electroosmotic flow (EOF). EOF in a bare capillary is generally an order-of-magnitude higher than in coated capillary and prevents DNA migration toward the anode; even low levels of EOF can severely degrade resolution. Additionally, adsorption of analytes onto high-energy surfaces such as silica degrades separation efficiency. Internal coatings of the capillaries can reduce the surface potential as well as reduce the distance between the analyte and the silica surface. This in turn reduces both the EOF and analyte-wall interactions, and makes the separation possible. An equation was derived by Hjertén *(4)* to quantify electroosmotic mobility (μ_{EOF}) for a coated capillary, integrating over the viscosity with respect to potential.

$$\mu_{EOF} = (\varepsilon/4\pi) \int_0^\zeta d\psi(x)/\eta$$

in which, ε is the dielectric constant, $\psi(x)$ is the potential at a distance x from the surface, and η is the viscosity at the interface.

There are several different approaches to exert control of EOF or to eliminate EOF, and each method has some advantages and disadvantages *(5,6)*. However, in all the cases (except the application of a radial electric field), the techniques involve altering the chemical properties of the buffer or the capillary wall to reduce ζ. For DNA separations, capillaries coated with nonionic polymers are the most effective to suppress EOF, by increasing the viscosity at the interface and to reduce DNA-wall interactions. Covalently capping the surface silanols with small molecules may reduce the EOF significantly, however, that approach is not good enough for increased separation efficiency, which indicates that strong DNA-wall interactions continue to occur in the absence of a surface polymer coating *(7)*.

1.1.1. Covalent Coatings

There are several coating methods reported to achieve the surface modification of silica *(6,8)*. However, only notable references in the context of DNA sequencing are discussed in this chapter. Hjertén's method is the most popular covalent coating method, in which the silica surface is first reacted with a bifunctional acrylic monomer, 3-(trimethoxysilyl)propylmethacrylate, followed by graft polymerization with acrylic monomer such as acrylamide *(4)*. This results in polymer coating onto the silica surface. Several groups have adopted this method and have successfully reported DNA sequencing in polyacrylamide coated capillaries *(9)*. These coated capillaries however, have a short lifetime. Cobb and colleagues *(10)* developed coatings via hydrolytically more stable Si-C bonds using a variation to Hjertén's method. In their procedure, the capillary surface is first treated with $SOCl_2$ to generate Si-Cl groups followed by a Grignard reaction with vinyl magnesium bromide to generate polymerizable vinyl groups. Polymerization of acrylic monomers such as acrylamide with surface vinyl moieties results in a polymer coating. Polyacrylamide coatings made by this method were reported to be stable for up to 110 sequencing runs *(11)*, whereas more than 400 runs were performed using poly([acryloylethoxy] ethyl glucopyranose) and poly(bishydroxyethyl acrylamide) coatings *(12)*.

Menchen and colleagues *(13)* demonstrated DNA sequencing using DB®-210 (J&W Scientific, Folsom, CA) fluorocarbon coated capillaries. Whereas Carrilho et al. *(14)* and Salas-Solano et al. *(15)* used poly(vinyl alcohol) (PVA) coated capillaries, which are also commercially available from Beckman Instruments (Fullerton, CA). These PVA coated capillaries gave more than 300 sequencing runs at elevated temperatures, but required thorough washing with deionized water after 5 consecutive runs to maintain the efficiency of the column. In addition to that, these capillaries had to be washed and filled with water for overnight storage *(15)*.

Covalent coating methods require *in situ* synthetic steps which are sometimes cumbersome and are compounded by problems such as capillary fouling with reagents and coating inhomogeniety *(7,16)*. Furthermore, covalently coated capillaries are more expensive than uncoated capillaries. Most importantly, these coatings have finite lifetime and are irreversible in practice, making it less than ideal for high-throughput environment.

1.1.2. Noncovalent Coatings

Silica surface modification by polymer adsorption, a noncovalent method, is an attractive alternative to covalent methods since it is simple, reversible in principle, and involves no additional synthetic steps. This approach eliminates *in situ* polymerization and offers the utilization of tailor-made polymers for surface modification. The polymer solution needs to be simply flushed through the capillaries to coat them. Alternatively, addition of small amounts of these coating polymers to other sieving polymers allows the use of uncoated capillaries. The on-line repeated regeneration of the capillary surface is an attractive feature for high-throughput instruments with potential cost benefits.

Poly(ethylene oxide) (PEO) was utilized to adsorb onto highly acidified capillary surface and 30 sequencing runs were performed in a single capillary by regenerating the surface after every run with an acid wash *(7)*. Poly(dimethyl acrylamide) (PDMA) and polyvinylpyrrolidone (PVP) are also noncovalent coating polymers and were used for DNA sequencing *(9,17,18)*. These polymers suppress EOF and DNA-wall interactions better than PEO due to their strong adsorption characteristics. At least 100 sequencing runs were performed in a PDMA coated capillary before surface regeneration was required *(9)*. A life time of 30 runs was reported for PVP-coated capillaries applied for the separation of the D1S80 allelic ladder *(19)*.

Since noncovalent coating by an uncharged polymer occurs via adsorption, coating stability depends on the temperature, the pH of the medium, and the nature of the polymer as well as the solvent. This of course, dictates the choice of a particular polymer coating for a given set of separation conditions. The lifetime of the coating also depends on the purity of the DNA sample as well as the amount of sample loaded. In this chapter, we will deal primarily with the synthesis and characterization of PDMA for its dual usage as a noncovalent coating material and a sieving matrix in DNA sequencing.

2. Materials
2.1. PDMA Synthesis

All the chemicals are analytical grade reagents from Aldrich (Milwaukee, WI).
1. N,N-dimethyl acrylamide (DMA).
2. Ammonium persulfate (APS).
3. N,N,N',N'-tetramethylethylenediamine (TEMED).
4. Methanol.
5. Acetone.

2.2. DNA Sequencing

1. BigDye® terminator sequencing standard (PE Applied Biosystems, Foster City, CA).
2. 10X TAPS buffer with EDTA (PE Applied Biosystems).
3. Urea (Aldrich, Milwaukee, WI).

3. Methods
3.1. PDMA Synthesis

1. Distill the DMA under vacuum and collect the middle portion of the distillate. DMA should distill at ~30°C under a vacuum of 0.1 mm Hg. Except for DMA, all the other chemicals can be used as received (*see* **Note 1**).
2. Add 65 mL of methanol to 185 mL of deionized water and mix the solution in a screw-capped Erlenmeyer flask.
3. Add 25 g of the distilled DMA (~24 mL) to the methanol-water mixture.
4. Bubble nitrogen gently through the mixture for at least 1 h (*see* **Note 2**).
5. Add 1.25 mL of the APS stock solution (made by dissolving 0.2 g of APS in 1.8 mL of deionized water) to the methanol-water mixture.
6. Add 1.25 mL of TEMED stock solution (made by dissolving 260 μL of TEMED in 1.8 μL of deionized water) to the methanol-water mixture.
7. Place a stirring bar to the mixture before capping the flask and continue polymerization under stirring at room temperature for at least 24 h.
8. Precipitate PDMA by adding approx 600 mL of cold acetone (5–10°C) (*see* **Note 3**).
9. Decant the aqueous layer and air-dry the polymer for about 1 h to remove acetone.
10. Dissolve the polymer in 500 mL of water and freeze-dry it for storage.

3.2. Preparation of the PDMA Separation Medium

1. Add 3.6 g of urea to 6.4 mL of deionized water and 1 mL of 10X TAPS buffer and stir the solution until the urea dissolves.
2. Add 0.5 g of freeze-dried PDMA to the buffer-urea solution and keep it gently stirred overnight at room temperature. This should yield approx 10 mL of 5% (w/v) PDMA solution with 6 *M* urea and 100 m*M* TAPS buffer (*see* **Notes 4** and **5**).

3.3. Sample Preparation, Electrophoresis, and Analysis

1. Add 40 μL of deionized water to the Big Dye DNA sequencing standard. Vortex this solution thoroughly, followed by denaturing at 90°C for 4 min and then place the sample on ice.
2. Take 10 μL of the DNA sample aliquot for electrokinetic injection.
3. Using ABI 310 Genetic Analyzer and the above PDMA separation medium in a 40 cm long (effective length) uncoated capillary (50 mm id) perform electrokinetic injection at 60 V/cm and run at 160 V/cm for 2 h at 40°C (*see* **Note 6**).

4. Reduction of both the urea concentration and the polymer concentration of the medium will increase the speed of analyte separation. The approximate length-of-read (LOR) values for 6%, 4%, and 3% PDMA (120 kDa, with 6 M urea, 100 mM TAPS, and identical electrophoresis conditions) were 635, 575, and 360 bases, respectively; and the run times to the LOR were 85, 56, and 33 min, respectively (*see* **Note 7**).
5. High molecular weight PDMA (420 kDa) can be used at a concentration of 3% (w/v) to achieve even better LOR (680 bases in 53 min) under identical separation conditions. However, such solutions have higher viscosities (~10,000 cP) than expected (*see* **Note 5**). Therefore, it is important that the user to select the concentration and molecular weight of PDMA for a given separation, although keeping the viscosity to a moderate level. This parameter together with urea concentration determines the speed of the run.

3.4. Column Maintenance

1. Finally, the stability of the polymer coating depends on the purity of the DNA sample and the amount of DNA loaded. The capillary surface can be regenerated between runs by washing consecutively with several column volumes of water, ethanol, and 1 N HCl. The regeneration depends on the effectiveness of the washing protocol.

4. Notes

1. DMA should be purified by vacuum distillation to remove impurities such as inhibitor, acrylic acid, and dimethylamine. Once distilled, the monomer should be stored in the refrigerator to prevent polymerization. If acrylic acid is not removed from DMA, it will copolymerize to give an ion-containing polymer. These ions will adversely affect on the adsorption efficiency of the polymer and also increase DNA-wall interactions. The mobility of the separation polymer increases due to the presence of carboxylate groups, which could compromise the resolution of DNA analytes.
2. When purging the polymerization mixture with nitrogen, care should be taken to prevent excessive evaporation of methanol as this could result in high molecular weight polymer. The formation of high molecular weight polymer may not be desirable as the final viscosity of the separation medium increases with increase in molecular weight for a given concentration of polymer.
3. Under cold conditions PDMA precipitates from the acetone, however at room temperature, acetone is a solvent for PDMA. Only a high molecular weight fraction precipitates readily from cold acetone and it may be quickly redissolved if the solvent is warmed. Therefore, it is essential to use cold acetone to increase the yield of precipitated PDMA. Alternatively, the solution may be freeze-dried to recover the PDMA after completion of the polymerization reaction.
4. A simple way to measure the molecular weight of PDMA is by knowing its intrinsic viscosity, [η]. The viscosity measurements were performed with an Ubbelohde viscometer in water at 25°C and the viscosity average molecular weight (M_v) was calculated using Mark-Houwink equation (*9*). Using the above polymerization protocol we determined the M_v of PDMA to be around 120 kDa.
5. The viscosity of the 5% (w/v) PDMA (120 kDa) separation medium measured by Brookfield viscometer Model DV-II (Brookfield Engineering Laboratories, Sloughton, MA) is approx 390 cP. Viscosity measurements were done at a shear rate of 3 rpm at room temperature using small sample adapter and S18 spindle.
6. Low molecular weight PDMA (<100 kDa) coatings are not stable at 40°C and the thermal stability of PDMA coatings can be improved by using high molecular weight PDMA (*9*).

7. To quantitatively analyze resolution of the separation, selectivity per base and peak full width at half-maximum (FWHM) can be plotted against base number *(9)* to get the approximate LOR at a resolution value of ~0.6. Using the separation medium and the electrophoresis conditions described in **Subheadings 3.2.** and **3.3.**, we achieved LOR of 615 bases in 75 min.

Acknowledgments

This chapter was prepared in part under the auspices of the US Department of Energy at the Lawrence Livermore National Laboratory under contract No. W-7405-ENG-48. I would like to thank Avanish Vellanki and Stripe Dibble for their assistance in the experiments. I am thankful to Joe Balch, Pat Fitch, Anthony Carrano, Courtney Davidson, and Glenn Fox for their support and encouragement.

References

1. Heller, C. (1997) The separation matrix, in *Analysis of Nucleic Acids by Capillary Electrophoresis* (Heller, C., ed.), Friedr. Vieweg & Sohn Verlagsgesellschaft, Braunschweig, Germany, pp. 3–23.
2. ABI Prism 3700 DNA Analyzer. (1998) *User's Reference Manual.* Applied Biosystems Division-Perkin Elmer, Foster City, CA.
3. MegaBACE. (1998) *Instrument User's Guide.* Molecular Dynamics-Amersham Pharmacia Biotech, Sunnyvale, CA.
4. Hjerten, S. (1985) High-performance electrophoresis: elimination of electroendosmosis and solute adsorption. *J. Chromatogr. A.* **347,** 191–198.
5. Salomon, K., Burgi, D. S., and Helmer, J. C. (1991) Evaluation of fundamental properties of a silica capillary used for capillary electrophoresis. *J. Chromatogr. A.* **559,** 69–80.
6. Chiari, M. and Gelain, A. (1997) Developments in capillary coating and DNA separation matrices, in *Analysis of Nucleic Acids by Capillary Electrophoresis* (Heller, C., ed.), Friedr. Vieweg & Sohn Verlagsgesellschaft, Braunschweig, Germany, pp. 135–173.
7. Fung, E. N. and Yeung, E. S. (1995) High-speed DNA sequencing by using mixed poly(ethylene oxide) solutions in uncoated capillary columns. *Anal. Chem.* **67,** 1913–1919.
8. Wehr, T. (1993) Recent advances in capillary electrophoresis columns. *LC-GC* **11,** 14–20.
9. Madabhushi, R. S. (1998) Separation of 4-color DNA sequencing extension products in non-covalently coated capillaries using low viscosity polymer solutions. *Electrophoresis* **19,** 224–230.
10. Cobb, K. A., Dolnik, V., and Novotny, M. (1990) Electrophoretic separation of proteins in capillaries with hydrolytically stable surface structure. *Anal. Chem.* **62,** 2478–2483.
11. Dolnik, V., Xu, D., Yadav, A., Bashkin, J., Marsh, M., Tu, O., Mansfield, E., Vainer, M., Madabhushi, R., Barker, D., and Harris, D. (1998) Wall coating for DNA sequencing and fragment analysis by capillary electrophoresis. *J. Microcolumn Seps.* **10,** 175–184.
12. Dolnik, V., Chiari, M., Xu, D., Dell'Orto, N., Melis, A., and Harris, D. (1998) A stable wall coating for DNA sequencing by capillary electrophoresis. *International Symposium on High Performance Capillary Electrophoresis, February 1–5,* Orlando, FL.
13. Menchen, S., Johnson, B., Winnik, M. A., and Xu, B. (1996) Flowable networks as DNA sequencing media in capillary columns. *Electrophoresis* **17,** 1451–1459.
14. Carrilho, E., Ruiz-Martinez, M. C., Berka, J., Smirnov, I., Goetzinger, W., Miller, A. W., Brady, D., and Karger, B. L. (1996) Rapid DNA sequencing of more than 1000 bases per run by capillary electrophoresis using replaceable linear polyacrylamide solutions. *Anal. Chem.* **68,** 3305–3313.

15. Salas-Solano, O., Carrilho, E., Kotler, L., Miller, A. W., Goetzinger, W., Sosic, Z., and Karger, B. L. (1998) Routine DNA sequencing of 1000 bases in less than one hour by capillary electrophoresis with replaceable linear polyacrylamide solutions. *Anal. Chem.* **70,** 3996–4003.

16. Cifuentes, A., Diez-Masa, J. C., Fritz, J., Anselmetti, D., Bruno, A. E. (1998) Polyacrylamide-coated capillaries probed by atomic force microscopy: Correlation between surface topography and electrophoretic performance. *Anal. Chem.* **70,** 3458–3462.

17. Madabhushi, R. S., Menchen, S. M., Efcavitch, J. W., and Grossman, P. D. (1996) Polymers for separation of biomolecules by capillary electrophoresis. *US Patent Number* 5,567,292.

18. Kim, Y. and Yeung, E. S. (2001) Capillary electrophoresis of DNA fragments using poly(ethylene oxide) as a sieving material, in *Capillary Electrophoresis of Nucleic Acids,* Vol. 1 (Mitchelson, K. R. and Cheng, J., eds.), Humana Press, Totowa, NJ, pp. 215–223.

19. Gao, Q.-F. and Yeung, E. S. (1998) A matrix for DNA separation: Genotyping and sequencing using poly(vinylpyrrolidone) solution in uncoated capillaries. *Anal. Chem.* **70,** 1382–1388.

26

Selective Primer Sequencing from a DNA Mixture by Capillary Electrophoresis

Tao Li, Kazunori Okano, and Hideki Kambara

1. Introduction

DNA analysis methods have been greatly developed in the past 10 yr with the progress of the Human Genome Project *(1)*. In addition to rapid and high-throughput DNA sequencers *(2)*, various reagents *(3)*, and sample preparation methods *(4–5)* have been developed as well. In order to suit for large scale DNA sequencing, the following methods have been proposed and widely applied in a sequencing facility: shotgun method *(6)*, nested deletion method *(7)*, primer walking method *(8)*, and modular primer method *(9)*.

In the shotgun method, a sample DNA is digested randomly by sonication, which produces a set of fragments as shown in **Fig. 1**. Each fragment is then subcloned into a plasmid vector for DNA sequencing analysis. The order of the fragments is determined by analyzing the overlaps of fragment sequences. Many fragments have to be sequenced redundantly to determine the whole sequence. The base reading redundancy is about 10, which means that 1 Megabase (Mb) of read is required to accurately determine a DNA sequence of 100 kilobase (kb). In contrast, the nested deletion method produces an ordered set of fragments. Although the preparation of a set of such fragments is very time consuming and labor intensive, the base reading redundancy is low. The fragments are also cloned for DNA sequencing. The whole sequence is determined by confirming the overlapped sequences.

Currently, DNA sequencing by primer walking is easy because the desired oligomers can be easily obtained from various commercial suppliers and a sequencing kit containing labeled terminators has been commercialized *(10)*. Because the sequencing primers have to be synthesized each time, a time delay is always required for designing and obtaining new primers. If a large "primer pool" is prepared prior to the sequencing stages, the primer walking strategy is attractive. Obviously, an oligonucleotide longer than 10-mer is necessary to prime a DNA polymerase chain reaction

From: *Methods in Molecular Biology, Vol. 163:*
Capillary Electrophoresis of Nucleic Acids, Vol. 2: Practical Applications of Capillary Electrophoresis
Edited by: K. R. Mitchelson and J. Cheng © Humana Press Inc., Totowa, NJ

shotgun

nested deletion

primer walking

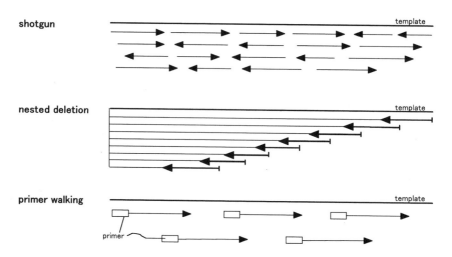

Fig. 1. The principles of three sequencing strategies: shotgun method, nested deletion method, and primer walking method. The shotgun method requires the sequencing of many randomly chosen fragments to reconstruct the entire original sequence. The nested deletion method requires making a set of ordered fragments, which is very labor intensive. The primer walking method requires expensive primer synthesis for each walking step, and takes a lot of time for the synthesis of each primer.

(PCR). This means that almost 1 million primer species have to be prepared in advance when a primer pool is used instead of synthesizing a primer each time. The use of a prepared primer pool is very expensive and seems impractical. An alternative method is the modular primer method, where a rather small number of hexamers are prepared. Two or three hexamers are selected and combined to produce a longer primer for sequencing. The preparation of a sequencing primer in the modular primer method is fast and cost-effective. However, a hexamer pool consisting of about 3000 hexamers still requires lots of synthesis.

With the progress of the automated DNA sequencer *(11)*, DNA sequencing is becoming a popular way for DNA analysis and is used in many laboratories. Cloning *(5)* has been widely accepted as a DNA sample preparation method, however, it is very time consuming and labor intensive. Although PCR amplification is becoming popular, it can only apply to amplify several kilobase of DNA template. Thus, there is still a great need to find and develop a suitable DNA sample preparation method for easy and fast sequencing of large DNA regions and genomes.

1.1. Sequencing Without Subcloning

Here we will introduce an easy method suitable for sequencing a DNA in a size range of several kilobases without subcloning. A set of prepared 16 primers is used instead of thousands of primers from a pool of primers. Let us think about a case that a 5 kb DNA sample is to be sequenced under a condition where the base reading length for one sequencing operation is 500 bases. The best sequencing method will be to

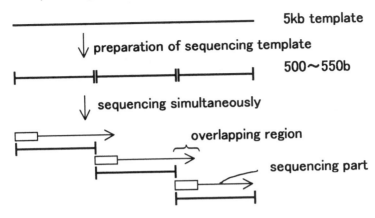

Fig. 2. An ideal sequencing strategy. The least redundancy is expected from the viewpoint of efficiency. When a DNA sequencing sample is about 5 kb, the best method will be to digest the DNA into 10 fragments of 500–550 bp containing small overlaps. Then all the fragments are sequenced simultaneously, followed by overlap sequence analysis to reconstruct the whole sequence.

digest the DNA into 10 fragments of 500–550 bp containing small overlaps, then to sequence all fragments simultaneously, as shown in **Fig. 2**. After the sequencing data is obtained, the fragment order will be determined by the overlaps between the fragments.

Three key points are necessary to realize the above method. First, an appropriate enzyme, which can digest DNA at about 500-bp intervals, should be obtained. However, such an enzyme with a precise interval between cutting sites is not available. Fortunately, several different four-base cutters used individually, can approximately mimic this pattern of cutting. Second, a method to pick out each individual fragment species from a digestion mixture is required. A set of binary primers, where each binary primer consists of a common part (i.e., a sequence to be introduced to the 3' termini of digested fragments as a priming sequence) and a variable part (i.e., the two-base sequence at the 3' terminal, AA, AC, ..., TG, TT), is used to extract individual fragments from a DNA mixture. Each fragment can be sequenced in parallel with a capillary array DNA sequencer. Finally, a method to determine the order of fragments to reconstruct the whole sequence should be developed.

The extraction of a fragment from a mixture, by selective complementary strand extension with a binary primer, is the key technology used here. It is well known that the extension of a complementary strand primer is very sensitive to formation of fully annealed base pairing between the template and the primer at the 3'- terminus *(12)*. If there is a base mismatch between the 3' terminus of the primer and the template DNA, the complementary strand extension of the primer will not occur (*see* **Fig. 3**). Therefore, the complementary strand extension can be used to classify and extract a fragment in a mixture according to the terminal base sequence of corresponding binary primers.

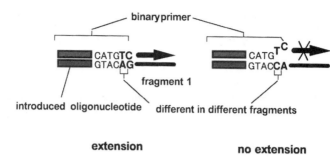

Fig. 3. Principle of selective strand extension of a binary primer. The terminal base sequence selection is used to classify and extract a fragment in a mixture according to its 3'-terminal base sequence. A binary primer consists of a common sequencing (including an introduced oligonucleotide sequencing and four-base recognition site of a particular restriction enzyme) and a variable 2-base sequence at the 3' terminus (16 possible combination of two bases, AA, AC, ..., TG, TT).

1.2. PCR Amplification with Binary Primers

The principle of selective-PCR amplification of a DNA fragment in a mixture with two binary primers is shown in **Fig. 4**. A DNA sample is digested with a four-base recognition endonuclease, followed by ligation of oligomers at the fragment termini. The oligomer sequence and the restriction recognition sequence can be used to form the priming sites for the PCR amplification of the selected fragment. If a two-base sequence is added to the 3'- terminus of the primer, only those fragments perfectly matching to the selective primer can be PCR amplified within this mixture. If the sequence of the oligomers ligated onto the fragment is the same at both termini, various combinations of the binary primers would classify 136 (terminal) "groups" by selective-primer PCR. Thus, if the number of fragments produced by the enzymatic digestion is less than 20, the number of fragments occurring in any group will in most cases be zero, or one. By this means the "PCR amplification with binary primers" can be used to selectively amplify (extract) a single fragment from within a complex mixture of different fragments (*see* **Note 1**).

The application of binary primers for DNA sequencing was first demonstrated for the sequence analysis of a fragment mixture *(13)*. There a DNA sample is digested with a restriction enzyme to produce five double-stranded fragments, which are directly sequenced by binary primers. Recently, the method has been modified for DNA samples containing many more component fragments. In this chapter, a sample preparation method for DNA sequencing is described using binary primers and selective PCR amplification of fragments in a DNA mixture *(14,15)*. Any DNA fragment smaller than 10 kb can be sequenced without subcloning using a small number of prepared primers, instead of a large pool of primers or for the need to continually synthesize new primers required by other methods of random sequencing.

2. Materials

1. A cloned human genomic DNA as sequencing template is supplied by the Human Genomic Center (Institute of Medical Science, University of Tokyo, Japan).

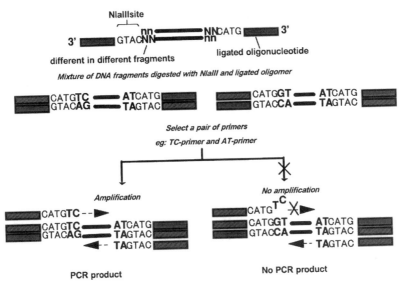

Fig. 4. Principle of selective PCR amplification of a DNA fragment in a mixture. A template DNA is digested by the restriction enzyme *Nla*III followed by ligation of the linker oligonucleotide. The two-base sequence at the 3' terminus adjacent to recognition sequence is used as selective bases in subsequent selective PCR.

2. Terminal deoxynucleotidyl transferase, pUC19 DNA, and T4 DNA ligase are obtained from Takara (Otsu, Japan).
3. *Hha*I is from Toyobo (Osaka, Japan).
4. ThermoSequenase (4 U/mL, product no. US78500), ThermoSequenase dye-primer and dye-terminator sequencing kit, *Taq* DNA polymerase, dNTPs and ddNTPs, and NAP-10 are purchased from Amersham Pharmacia Bioteck (Amersham, UK).
5. The 7-deaza-dGTP is from Boehringer Mannheim GmbH (Germany).
6. The QIAquick spin columns are from QIAGEN GmbH (Hilden, Germany).
7. All other chemicals used are of analytical grade.
8. Oligonucleotides are synthesized by Sawady Technology (Tokyo, Japan) and by Nippon Flower Mills Co. Ltd. (Tokyo, Japan).
9. Oligomers used for ligation reactions are:
   a. For *Nla*III, a pair of 5'-pACTGGCCGTCGTTT-3' and 5'-AAACGACGACGG CCAGTCATGp-3'
   b. For *Alu*I, a pair of 5'-pACTGGCCGTCGTTT-3' and 5'-AAACGACGGCCAGTp-3'.
10. The sequences of binary primer sets used are as follows:
   a. 5'-TCTCCTTTTTTTTTTTTTTTCGCNN-3' (for *Hha*I).
   b. 5'-AACGACGGCCAGTCA TGNN-3' (for *Nla*III).
   c. 5'-AACGACGGCCAGTCT NN-3' (for *Alu*I). Here, NN is a two-base sequence produced by all possible combinations of four nucleotides, which is used to distinguish a fragment in a DNA mixture by selective amplification in a PCR reaction (i.e., selective PCR). Nonlabeled binary primers, as well as Texas Red labeled binary primers are prepared for this purpose.

11. Thermocycler: DNA Engine Tetrad (MJ Research, Inc., Watertown, MA).
12. Fluorescence-image analyzer (FM-Bio 100; Hitachi Software Engineering, Yokohama, Japan.

3. Methods

3.1. Efficient DNA Sequencing Strategy for DNA of About 3 kb or Smaller

A pUC19 DNA (2.7 kb) is sequenced as an example. The size of this DNA template is too long to be sequenced by one sequencing operation. By using a set of binary primers, the whole sequence is easily obtained. The flowchart of this sequencing strategy is shown in **Fig. 5**. First, the DNA is linearized and it is then sequenced from both ends. Then it is digested into small pieces with a restriction enzyme, followed by poly-A tailing to make a priming site (poly A + restriction sequence) at which binary primers can anneal. At first a binary primer is extended to make an "extended complementary strand" of the corresponding fragment in a mixture. Although the binary primer cannot hybridize to the intact DNA, the extended complementary strand" can hybridize to it and further extend the strand along the intact DNA. Thus any "binary primer" can only hybridize to a subset of the modified DNA fragments, and by application of PCR further extend the primer strand along the selected DNA (restriction) fragment template to make one complementary strand of the fragment, which is termed as "extended binary primer." Thus, the extended binary primer can anneal to a complementary locus on the original template DNA and be used as a sequencing primer to sequence the intact DNA using cycle sequencing. Dideoxynucleotides are added to a reaction mixture for sequencing after the extended binary primer is produced. The binary primers are also used to determine the sequence of their respective complementary restriction fragment, whereas the extended binary primers are used to determine the contiguous sequences. The contiguous sequences together with the selected fragment sequences are used to reconstruct the whole sequence of the template DNA.

3.1.1. Template and Sequencing Fragment Preparation

Initially, the template DNA is amplified by PCR reactions. More than 10 pmol of amplified DNA is required to sequence a few kilobase DNA.

1. The PCR product (5 pmol) is digested with an enzyme, such as *Hha*I, followed by synthesis of the poly-A tail at the 3' termini of the fragments.

Digestion of DNA with a four-base cutting restriction enzyme *Hha*I

Content	Concentration	Volume
DNA	76 fmol/μL	66 μL
10X buffer M (including in Kit)		25 μL
*Hha*I	12 U/μL	7.5 μL
H$_2$O		151 μL
Total		250 μL

2. The mixture is incubated at 37°C for 1 h. The small components present in the product are removed by gel filtration with NAP-10. About 1 pmol of the *Hha*I digestion product is poly-A tailed at the 3' termini of the fragments, as follows:

1) DNA sequencing from 3' termini

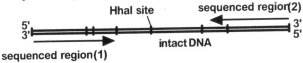

2) enzyme digestion and poly A tailing

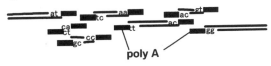

3) sequencing reaction with non-digested DNA

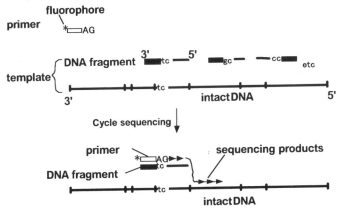

4) assembly of fragment sequences

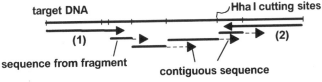

Fig. 5. Flowchart of the strategy for efficient DNA sequencing. *Step 1*, template DNA (about 3 kb) is sequenced from both ends to find cutting sites of a certain restriction enzyme. *Step 2*, template DNA is digested into small pieces with the restriction enzyme followed by poly-A tailing to make a priming site for the binary primers. *Step 3*, a fluorophore labeled binary primer is extended to produce an extended complementary strand of the corresponding fragment in the mixture. The extended binary primer is used as a sequencing primer to sequence the intact DNA. In the presence of the intact DNA, the binary primers are used to determine the fragment sequence, and the extended primers are used to determine its contiguous sequence. *Step 4*, the contiguous sequences of the fragments are used to reconstruct the whole sequence of the template DNA.

Poly A tailing of HhaI digested DNA fragment

Content	Concentration	Volume
DNA fragment	15 fmol/μL	66 μL
10X TdT-buffer		5 μL
CoCl$_2$	20 mM	5 μL
dATP	100 mM	1 μL
TdT	25 U/μL	1 μL
H$_2$O		22 μL
Total		100 μL

3. The mixture is incubated at 37°C for 1 h followed by gel filtration with NAP-10. The amount of the recovered DNA fragments is about 1.6 μg. The recovered DNA fragments are dissolved in 20 μL of H$_2$O (~50 fmol/μL).

3.1.2. Selection of Primers for Sequencing

The selection of primers from a set of 16 binary primers is an important step in preparing for sequencing of a fragment. A binary primer set of the form 5'-*TCTCC TTTTTTTTTTTTTTTCGCNN-3' is used to carry out selective complementary strand extension reactions, where * is a labeling fluorophore, Texas Red. CGC is a part of the enzymatic recognition sequence (CGCG, HhaI), and NN is the possible two-base sequences (16 in total). When a restriction enzyme other than HhaI is used, the sequence CGC has to be replaced by the recognition sequence element corresponding to that enzyme. The sequence of TCTCC is added to the 5' terminus of the poly-T strand to increase the specificity of the primer. Since all the poly-A-tailed DNA fragments in the mixture have the same terminals, it is difficult to distinguish perfectly matched primer-template hybrids from partially matched ones by their hybridization stability, because the only difference is at the two terminal bases. However, fully annealed primers can be distinguished by using complementary strand extension reaction, which is very sensitive to the base mismatch at 3'-terminus of a primer. The reaction occurs only when the two terminal bases of the primer are complementary to the corresponding DNA fragment sequence (see **Fig. 3**). The complementary strand extension of each binary primer is carried out after being hybridized with a digestion mixture. The digestion mixture is divided into 16 fractions after poly-A tailing, each fraction containing 25 fmol of the fragment mixture. The process of binary primer extension reaction is as follows:

1. Adding 0.5 μL (0.5 pmol/ μL) of each of 16-binary primers to 0.2-mL tubes, respectively.

DNA fragments	50 fmol/μL	8 μL
10X ThermoSequenase buffer		4 μL
ThermoSequenase	4 U/μL	4 μL
H$_2$O		24 μL
Total		40 μL

2. Divide the above reaction mixture into 16 tubes of 2.5 μL each and then cover with mineral oil.
3. Incubating at 90°C for 1 min followed by adding 1 μL of dNTPs (1 mM each of dATP, dCTP, 7-deaza-dGTP, and dTTP).

Table 1
The Summary of Fragments Appeared in Fig. 6

fraction	fragment size (approx.)	terminal base species of primer															
		AA	AC	AG	AT	CA	CC	CG	CT	GA	GC	GG	GT	TA	TC	TG	TT
group 1	400			1									1				
	340	1							1								
	330				1				1								
group 2	270		1										1				
	170	(1)		1				1									
group 3	150	(1)															
	130							1				1					
group 4	110												1		1		
	100			1	1	1			1				1				1
	90	1		1													
	< 65			2		2	1					1	1			3	

Thermal cycle 5 times

94°C	30 s
66°C	30 s
72°C	60 s

An annealing temperature of 66°C is critical to allowing DNA strand extension of a fully annealed binary primer although preventing any extension from a primer that is not fully annealed at the 3'-terminus. Cyclic DNA extension reactions performed at an annealing temperature lower than 66°C frequently produce false positive products, especially from binary primers with 3'-termini of primer–GA or primer-GC *(17)*.

4. The extended primer products are analyzed by gel electrophoresis as shown in **Fig. 6**. The number of fragments, their sizes, and the two-base sequences adjacent to the cutting sites of the fragments can be obtained from the electropherograms. There are 32 peaks observed in the electropherogram (summarized in **Table 1**). The total length of all the detected products is about 2.3 kb, which is close to the original DNA size.

5. Size separation of the extension products is necessary prior to sequencing, as several peaks are often produced from a single binary primer. This is seen for example with primer-TC in **Fig. 6**. The extension products are fractionated into four "size groups" by excising individual DNA bands from a 2% agarose gel (20 cm × 15 cm × 0.5 cm run at 2 V/cm) as shown in **Fig. 6**. The groups are chosen such that each group does not contain more than two fragments produced from any one of the 16 primers. Consequently, each fragment can be sequenced independently and directly from the mixture by using the set of 16 binary primers.

3.1.3. DNA Sequencing of the Fragments

1. All DNA sequencing reactions are performed using a ThermoSequenase dye-primer sequencing kit (Amersham, UK), under the condition recommended by the manufacturer. As shown in **Fig. 7**, DNA sequencing is carried out firstly from both ends of the intact

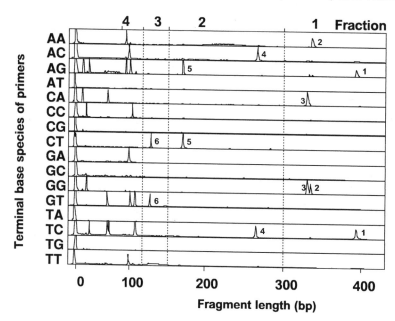

Fig. 6. Electropherograms of the complementary-strand extension products initiated from 16 different primers. The selective two-base NN of primers, 5'*TCTCCTTTTTTTT TTTTTTTCGCNN-3', are indicated. Pairs of fragments appear as the same size, which indicates that they represent the complementary strands to each other from one restriction fragment. The extended products are fractionated into four groups (numbered 1, 2, 3, and 4) followed by gel electrophoresis. The fragments belonging to one group have different terminal base sequences NN, which permit them to be independently sequenced.

DNA, being the regions indicated as (1) and (2). Usually 600–700 bases can be readily sequenced in each operation, and *Hha*I cutting sites which appear in the sequenced regions are identified using computer software in order to identify the fragments for the next sequencing operations.

2. *Hha*I terminated extension fragments numbered 1, 2, 3, and 5, which are shown in **Fig. 6**, are not observed in the two sequenced regions (1) and (2), shown in **Fig 7**. So these fragments are chosen as the next sequencing targets. In addition, fragment 4 is sequenced to find out the fragments adjacent to the sequenced region (1), as shown in **Fig. 8**. The fluorophore-labeled binary primers are added to the reaction mixture which is to be extended to generate "extended binary primers." The extended binary primers act as "unique" sequencing primers by hybridizing to the intact DNA, but not to the polyA-extended restriction fragments, the DNA sequence from both the fragments and the adjacent regions in the intact DNA can be obtained simultaneously from the same mixture, as shown in **Fig. 8**. It can be seen that the fragments 4 and 5 are connected through two small *Hha*I fragments. Using the sequence information of adjacent fragment from the read out of the "extended binary primer," the sequences of the small connecting of fragment are obtained, allowing the entire sequence of the whole fragment to be reconstructed, as shown in **Fig. 9**.

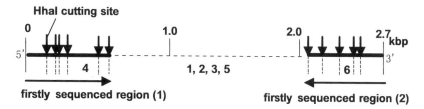

Fig. 7. The cutting sites of *Hha*I appear in the terminal sequenced region of the model template. Fragments 1–6 correspond to those shown in **Fig. 6**.

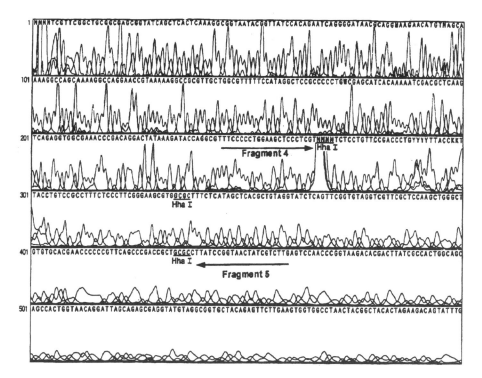

Fig. 8. DNA sequencing base-calling of fragment 4 and its adjacent region obtained with primer-TC. An intense peak appears at the position corresponding to the fragment size (fragment 4 terminal). The binary primer gives the sequence information of the corresponding DNA fragment, and simultaneously the extended binary primer gives the contiguous sequence information beyond the cutting site. By base reading beyond the cutting site, the connection of the fragments is clarified. In this case, fragment 4 is connected with fragment 5 through two small *Hha*I digested fragments.

Overall, this binary primer sequencing strategy requires a large amount of template DNA, because each of the sequencing reaction mixtures should contain the intact original template DNA, in addition to the tailed-restriction fragments.

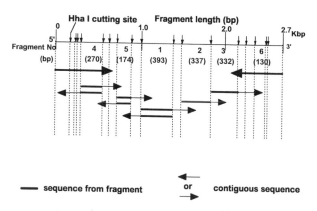

Fig. 9. Reconstruction of the whole sequence by overlaps of fragment and by alignment of adjacent sequences. The fragments 1–6 correspond to those shown in **Fig. 6**. The base reading redundancy is as small as 1.3 per strand. As all the sequencing operations can be carried out simultaneously, this method is very efficient in terms of both time saving and cost saving.

3.2. DNA Sequencing Strategy Using Template Size-Reduction

Another simple sequencing strategy for DNA fragments of several kb is shown in **Fig. 10**. A longer human genomic DNA fragment (~3.8 kb) is used as an example. Firstly, DNA sequencing from both termini is carried out with a long base reading DNA sequencer where 700–1000-base sequence is determined in one sequencing operation. Restriction enzymes sites occurring in the sequenced terminal regions are found with a computer program. Then the template DNA is digested by a restriction enzyme, such as *Nla*III (CATG). After the digestion, a linker oligomer (5'-pTTTGCTGCCGGTCA-3') is ligated to the 3' termini of digested fragments, which introduces a (common) priming sequence to the fragments. Consequently, each binary primer of an appropriate set, based on the selection by two-base sequence at its 3' terminus, can hybridize to one terminus of a selected fragment to produce the complementary strand. The primers are extended and the products are analyzed by gel electrophoresis to identify their sizes and the two terminal bases of the annealing primer.

The *Nla*III sites, observed in the sequenced terminal regions, are selected as the priming sites for a second PCR amplification. The fragments adjacent to these cutting sites are used as template DNAs for the second PCR. Two binary primers (e.g., primer-AC and primer-CT), that can hybridize to the selective fragments respectively, are added to the digestion mixture to produce the corresponding complementary strands. Although the binary primers cannot hybridize to the intact DNA under stringent annealing conditions, the extended primers can hybridize to the intact DNA. In the presence of intact DNA fragment, the extended primers are extended to the termini of the intact DNA, which are the size-reduced template DNAs being used for the second PCR. The size-reduced template DNA is amplified by PCR with a binary primer and one universal primer (5'-TGTAAAACGACGGCCAGT-3' or 5'-CAGGAAA CAGCT ATGAC-3') (*see* **Note 2**). These PCR products are analyzed by agarose gel electro-

1) Terminal sequencing by dye–terminator cyclic sequencing chemistry

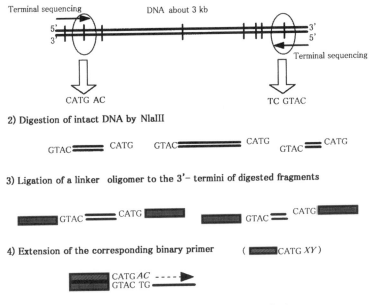

2) Digestion of intact DNA by NlaIII

3) Ligation of a linker oligomer to the 3'– termini of digested fragments

4) Extension of the corresponding binary primer   (CATG *XY*)

5) PCR amplification by extended binary primer and universal primer

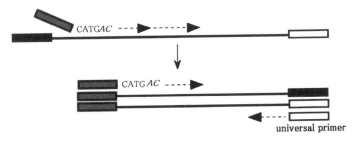

Fig. 10. Principle of template DNA size-reduction sequencing strategy. *Step 1*, DNA sequencing from both termini is carried out with a long base reading DNA sequencer. Restriction enzymes sites in the sequenced regions are identified with computer software. *Step 2*, template DNA is digested with *Nla*III (CATG!). *Step 3*, a linker oligomer (5'-TTTGCTGCC GGTCA-3') is ligated to the 3' termini, which become the priming sites for 16 binary primers. *Step 4*, the extension of the corresponding binary primer is carried out. The binary primer matching to one DNA fragment is extended. *Step 5*, PCR amplification with the extended binary primer and the universal primer are used to produce the size-reduced template DNAs for the next sequencing steps.

phoresis, as shown in **Fig. 11**. A gel extraction step is usually used to separate and purify the template DNAs before sequencing because a binary primer frequently matches to several fragments in the reaction mixture. This produces nested PCR products. Each extracted fragment (having size difference over 200 bp) is amplified again

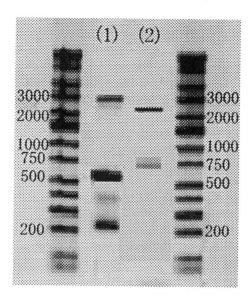

Fig. 11. Electropherogram of PCR products with extended binary primers and universal primers. The DNAs are produced by PCR amplification with the binary primer-AC and reverse primer (lane 1), and with binary primer-CT and forward primer (lane 2), respectively. Multiple fragment bands appear in the gel lanes, which are caused by multiple pairing between the binary primer and the fragments in DNA mixture. These products are recovered by a QIAquick gel extraction kit and used as templates for sequencing.

by PCR for sequencing. This is effective for sequencing a template DNA longer than 2.0 kb. Because sequencing from various positions of the intact DNA is possible by using these products. If there is still nonsequenced regions in the sample, the process to make a size-reduced template DNA is repeated using a different restriction enzyme and an appropriate binary primer set, until the whole sequence is determined. This strategy is similar to the strategy described in **Subheading 3.1.** in which the extended primers act as sequencing primers. However, in the present strategy the extended primers act as primers to amplify the second template DNA (i.e., the reduced template DNA).

3.2.1. Digestion of DNA with a Four-Base Cutting Restriction Enzyme Nla III

The protocols used for digesting DNA with *Nla*III are as follows:

Content	Concentration	Volume
~3.8-kb DNA	41 ng/μL	25 μL
10X Buffer K (Promega)		10 μL
*Nla*III (CATG!)	10 U/μL	4 μL
Distilled H$_2$O		61 μL
Total		100 μL

1. The above mixture is incubated in a 0.5-mL microcentrifuge tube at 37°C for 3 h or more to complete the digestion reaction.
2. The products are heated at 65°C for 15 min to inactivate the restriction enzyme. The restriction products are then recovered by precipitate in ethanol.
3. The sizes of the digestion products are analyzed by electrophoresis in a 2% agarose gel, which is stained with 0.5 μg/mL ethidium bromide.

3.2.2. Ligation of Linker Oligomer to the 3' Termini of Digest Fragments

1. A linker oligomer (5'p-TTTGCTGCCGGTCA-3') stock solution of 100 pmol/μL is ligated to the digested fragments. The ratio of the digestion mixture to the linker oligomer is 1:100.
2. The mixture is incubated at 70°C for 5 min to denature the template DNA. The mixture is allowed to stand for 5–10 min to cool to room temperature. T4 ligase (Toyobo, Japan) in a solution half of the volume of the reaction mixture is added into the reaction tube and the ligation is carried out at 16°C for at least 16 h (overnight).

3.2.3. Strand Extension from a Binary Primer

1. To confirm the sizes and terminal base sequences of digested fragments, a fragment analysis is carried out on single-strand extension products from binary primers labeled with Texas Red. The components of the extension reaction mixture are:

Premixture used for 16 different tube reactions

Digested mixture (10 fmol/μL)	15 μL
Reaction buffer (ThermoSequenase, Amersham kit)	8 μL
ThermoSequenase (4 U/μL) (enzyme:buffer = 1:7)	5 μL
Distilled H_2O	22 μL
Total	50 μL

Reaction mixture in 1 tube:

Above mixture	3.0 μL
dNTPs (0.1 mM each)	1.0 μL
Selected binary primer (1 pmol/μL)	0.4 μL
Total	4.0 μL

2. The thermo-cycle reaction is: commence at 94°C by adding 1 μL of dNTPs (0.1 mM each). Then carry out 5 cycles of 94°C for 30 s, 68°C for 30 s, and 72°C for 1 min. The products are purified by a QIAquick spin column (QIAGEN Inc., Germany).

3.2.4. PCR Amplification for Template Size-Reduction

1. For PCR reactions, a universal primer (5'-TGTAAAACGACGGCCAGT-3' or 5'-CAGGAAACAGCTATGAC-3') is added to the reaction mixture to help the extension of extended binary primers on the intact DNA template.
2. Although a binary primer cannot hybridize to the intact DNA, the extended primer can act as a primer for the intact DNA. It can extend further together with the universal primer to make the second template DNA (i.e., size-reduced template DNA) in the reaction mixture.

3. PCR reaction mixture:

Content	Concentration	Volume
Intact DNA	6 fmol/μL	0.6 μL
PCR primer	10 pmol/μL	0.6 μL
10X *Taq* polymerase reaction buffer		5.0 μL
dNTPs	2.5 m*M* each	4.0 μL
Ex-*Taq* (Takara)		0.2 μL
Distilled H$_2$O		40 μL
Total		50 μL

4. The PCR thermal amplification cycle is the following: 35 cycles of 94°C for 30 s, 68°C for 30 s, and 72°C for 60 s. The PCR reaction commences, with the tubes at 94°C in a DNA Engine Tetrad, upon the addition of 0.125 U of Ex-*Taq* polymerase.
5. The resulting PCR products are subjected to gel electrophoresis in a 0.7% agarose gel saturated by 0.5 μg/mL ethidium bromide, with 40 m*M* Tris-acetate, pH 8.0, 1 m*M* EDTA (*see* **Fig. 11**). The fragments are analyzed with a fluorescence-image analyzer (FM-Bio 100, Hitachi Software Engineering, Yokohama, Japan).
6. A dye-terminator cycle sequencing kit (ABI BigDye) is used with the protocol suggested by the manufacturer. The process is repeated until the whole sequence is determined.

3.3. DNA Sequencing Strategy by Overlapping Fragments Produced with Two Different Restriction Enzymes

This third strategy uses overlap analysis, just like in the shotgun cloning and sequencing method. Multiple sets of binary primers are used to selectively amplify fragments. We demonstrate this sequencing strategy using a 8.7-kb DNA from the human genome. The DNA is independently digested with two different four-base restriction enzymes, such as *Nla*III and *Alu*I, to produce two groups of fragments. The size distribution of the fragments depends on the enzymes. Generally, the restriction sites of four base cutters, such as *Alu*I, *Hae*III, *Hha*I, *Mbo*I, and *Nla*III, appear to be random when used alone or in combination, and are thus suitable for this strategy. The sequence analysis of many small fragments is not desirable if the analysis time is to be minimal. Therefore, determining the size of the digestion products of the different enzymes is an inexpensive procedure, which is useful for subsequent efficient sequencing.

Oligomers are ligated at the termini of the restriction fragments to create priming sites for binary primers. Ethanol precipitation is utilized to remove any very small restriction fragments from the mixture. Complementary strand extension reactions with 16 binary primers are carried out with the digested fragments. These extension products have fluorophore tags and the same sequences as the selective fragments. **Figure 12** shows the analysis of these single-strand extension products. This strategy analyses extension products instead of the digested fragments by gel electrophoresis. Pairs of the same size peaks appear in different binary primer lanes of the electropherogram, with each pair indicated by the same number. These product pairs are complementary strands to each other, extended from each end of the appropriate restriction fragment. Thus, each dsDNA fragment can be amplified with the corresponding pair of binary

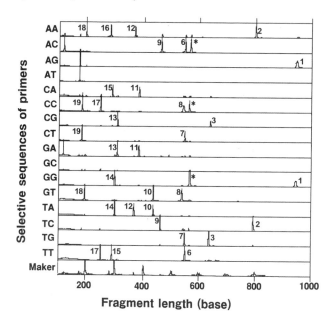

Fig. 12. Electropherograms of the complementary strand extension products from 16 binary primers. The templates are the fragments obtained by digesting an 8.7-kb human genomic DNA fragment with *Nla*III. All the fragments are classified into 16 groups through complementary strand synthesis with binary primers. Pairs of fragments with the same lengths appeared which indicate that they are complementary strands to each other from one fragment, except for the fragments indicated with * that cannot be resolved by the gel electrophoresis.

primers. For example, fragment "1" is selectively amplified by PCR with the binary primer-AG and primer-GG. There are 21 fragments of over 120 bp appearing in the electropherogram (*see* **Fig. 12**), which are to be sequenced. The total coverage sequence of these fragments is 70% of the whole sequence. The sequencing operations are carried out simultaneously for the fragments in the two groups. The overlaps are investigated for reconstructing the whole sequence. The result is shown in **Fig. 13**. There are four contigs, which almost cover the whole sequence of the intact DNA. The first and the fourth contigs are found to be the terminal contigs, by reference to the initial terminal sequence analysis of the intact DNA. Although there are two possible configurations of the remaining contigs [case 1 and case 2], the order of the contigs [2] and [3] can be determined by sequencing the intact DNA with extended complement strand as sequencing primers, which described in the efficient DNA sequencing strategy. The correct order of contigs [2] and [3] was found to be as shown in case [1]. Meanwhile, the gap sequences are easily sequenced using the small fragments, close to the gaps, as sequencing templates. This protocol is described as the template DNA size-reduction sequencing strategy, *see* **Subheading 3.2.** The protocols of this strategy are described in **Subheading 3.3.1.**

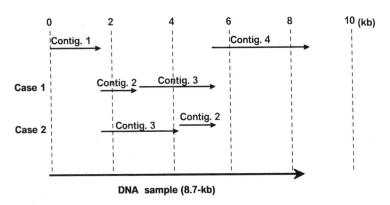

Fig. 13. Connection map of the contigs for an 8.7-kb DNA sample. The contigs can be easily connected using the efficient DNA sequencing strategy. The contig 1 and 4 are contributed to the terminal sequence of the DNA sample. The contig 2 and 3 can be assembled in case [1] or case [2]. It is proved that case [1] is the correct order of the contigs by sequencing the intact DNA with digested fragments as sequencing primers. There are 3 gaps observed in the map, which can be easily sequenced by the reduced template DNA sequencing strategy.

3.3.1. Ligation of Oligomers

1. 0.5 pmol of sample DNA (smaller than 10 kb) is digested with 40 U of *Nla*III (cutting at 3' site of CATG).
2. The restriction fragments (400 fmol) are treated with alkaline phosphatase.
3. The restriction fragments are ligated using 1400 U of T4 DNA ligase with 80 pmol of linker oligomer (5'-pACTGGCCGTCGTTT-3') supported with 32 pmol of helper oligomer (5'-AAACGACGGCCAGTCATGp-3').
4. The phosphate residues are introduced into the 5' and 3' ends of the oligomers to prevent oligomer-oligomer ligation.
5. The ligation products are purified with QIAquick spin column to remove free oligomers.
6. Ethanol precipitation or gel filtration with NAP 10 columns is carried out to remove small fragments from the mixture.

3.3.2. Strand Extension of Binary Primers and Gel Electrophoresis Analysis

1. The ligated fragments are mixed with 4 µL of ThermoSequenase buffer (260 mM Tris-HCl, pH 8.5, 65 mM MgCl$_2$ and 4 µL of ThermoSequenase (4 U/µL), which are divided into 16 fractions.
2. 0.5 µL of one of the 16 different fluorophore labeled binary primer (0.5 pmol/µL) and 1 µL dNTPs (0.2 mM each of dATP, dCTP, 7-seaza-dGTP, and dTTP) are added to each fraction at 90°C.
3. Strand extension is performed using 5 thermal-cycles of: 94°C for 30 s, 70°C for 30 s, and 72°C for 60 s.
4. The extended binary primer products are analyzed by gel electrophoresis to determine the fragment sizes and their terminal two-base sequence adjacent to the cutting sites, as shown in **Fig. 12.**

3.3.3. PCR Amplification of each Fragment

1. Each DNA (restriction) fragment can be PCR amplified with the appropriate pair of corresponding binary primers.

2. For example in **Fig. 12**, fragment 1 is selectively amplified by PCR with the binary primer-AG and primer-GG.
3. The reaction mixture (50 μL) contains 0.125 U of *Taq*-DNA polymerase and 5 pmol of the binary primers.
4. The thermal amplification cycle is carried out for 35 times at: 94°C for 30 s, 66°C for 60 s, and 72°C for 60 s. The annealing temperature is a little higher than that in a conventional PCR. However, it is useful to obtain a large fragment as the dominant PCR product, as two or more fragments can be amplified with the same pair of primers. The PCR products are checked by 2% agarose gel electrophoresis and stained with 0.5 μg/mL ethidium bromide.

4. Notes

1. The strategy of "selective complementary strand extension" is very useful for classification and for the isolation of a specific fragment from a DNA mixture. This is also applied to the analysis of a DNA fragment mixture that includes too many fragment species to be analyzed in an electropherogram at one time. The complementary strands of these fragments are produced with fluorophore tagged binary primers, which classify the fragments according to their terminal base species. Each subgroup then contains a much smaller number of component fragments than those in the mixture, which simplifies individual sequencing procedures. This makes it possible to analyze the mixture by breaking it into component elements. It should be noted that when the fragments are amplified by PCR with binary primers the process is called AFLP (amplified fragment length polymorphism) *(16)*. AFLP analysis is a well-known tool used for gene expression analysis and for linkage mapping analysis.
2. To increase the selectivity of the primers, the fourth nucleotide from the 3' termini, which is in the restriction recognition sequence, is changed to create an artificial mismatch. For example, -CA<u>T</u>GNN is replaced by -CA<u>C</u>GNN (NN, discrimination two-base sequences) in the case of the binary primer set for *Nla*III. By introducing the mismatch at the 3' terminal region, the selectivity of primers for complementary strand extension increases dramatically *(17)*.

References

1. Collins, F. S., Patrinos, A., Jordan, E., Chakravarti, A., Gestrand, R., and Walters, L. (1998) New goals for the US human genome project: 1998-2003. *Science* **282**, 682–689.
2. Kheterpal, I. and Mathies, R A. (1999) Capillary array electrophoresis DNA sequencing. *Anal. Chem.* **71**, 31A–37A.
3. Tabor, S. and Richardson, C. C. (1995) A single residue in DNA polymerase of the *Eschericia coli* DNA polymerase I family is critical for distinguishing between deoxy- and dideoxyribonucleotides. *Proc. Natl. Acad. Sci. USA* **92**, 6339–6343.
4. Innis, M. A., Myambo, K. B., Gelfand. D. H., and Brow, M. A. D. (1988) DNA sequencing with *Thermus aquatics* DNA polymerase and direct sequencing of polymerase chain reaction-amplified DNA. *Proc. Natl. Acad. Sci. USA* **85**, 9436–9440.
5. Deininger, P. L. (1983) Random subcloning of sonicated DNA: application to shotgun DNA sequence analysis. *Anal. Biochem.* **129**, 216–223.
6. Favello, A., Hillier, L., and Wilson, R. K. (1995) Genomic DNA sequencing methods. *Methods Cell Biol.* **48**, 551–569.
7. Hattori, M., Tsukahara, F., Furuhata, Y., Tanahashi, H., Hirose, M., Saito, M., Tsukuni, S., and Sakaki, Y. (1997) A novel method for making nested deletions and its application for sequencing of a 300-kb region of human APP locus. *Nucleic Acids Res.* **25**, 1802–1808.

8. Kaiser, R. J., Makellar, S. L., Vinayak, R. S., Sanders, J. Z., Saavedra, R. A., and Hood, L. E. (1989) Specific-primer-directed DNA sequencing using automated fluorescence detection. *Nucleic Acids Res.* **17,** 6087–6102.

9. Kaczorowski, T. and Szybalski, W. (1998) Genomic DNA sequencing by SPEL-6 primer walking using hexamer ligation. *Gene* **26,** 83–91.

10. Prober, J. M., Trainor, G. L., Dam, R. J., Hobbs, F. W., Robertson, C. W., Zagursky, R. J., Cocuzza, A. J., Jensen, M. A., and Baumeister, K. (1987) A system for rapid DNA sequencing with fluorescent chain-terminating dideoxynucleotides. *Science* **238,** 336–341.

11. Kambara, H. (1998) Recent progress in fluorescent DNA analyzers and methods. *Curr. Topics Anal. Chem.* **1,** 21–36.

12. Crighton, S., Huang, M. M., Cai, H., Arnheim, N., and Goodman, M. F. (1992) Base mispair extension kinetics. Binding of avian myeloblastosis reverse transcriptase to matched and mismatched base pair termini. *J. Biol. Chem.* **267,** 2633–2639.

13. Furuyama, H., Okano, K., and Kambara, H. (1994) DNA sequencing directly from a mixture using terminal-base-selective primers. *DNA Res.* **1,** 231–237.

14. Okano, K. and Kambara, H. (1996) Fragment walking for long DNA sequencing by using a library as small as 16 primers. *Gene* **176,** 231–235.

15. Matsunaga, H., Kohara, Y., Okano, K., and Kambara, H. (1996) Selecting and amplifying one fragment from a DNA fragment mixture by polymerase chain reaction with a pair of selective primers. *Electrophoresis* **17,** 1833–1840.

16. Vos, P., Hogers, R., Bleeker, M., Reijans, M., van de Lee, T., Hornes, M., Frijters, A., Pot, J., Peleman, J., Kuiper, M., and Zabeau, M. (1995) AFLP: a new technique for DNA fingerprinting. *Nucleic Acids Res.* **23,** 4407–4414.

17. Okano, K., Uematsu, C., Matsunaga, H., and Kambara, H. (1998) Characteristics of selective polymerase chain reaction (PCR) using two-base anchored primers and improvement of its specificity. *Electrophoresis* **19,** 3071–3078.

27

Sequencing of RAPD Fragments
Using 3'-Extended Oligonucleotide Primers

John Davis, Ivan Biroš, Ian Findlay, and Keith Mitchelson

1. Introduction

Both directed and nondirected techniques are used in population biology for the identification of the genetic source of variation and the genetic locus of disease or quality traits *(1,2)*. The "nondirected" approaches use random genome scanning methods initially to generate polymorphic map markers, which then can be linked to traits of interest. One nondirected approach is random amplified polymorphic DNA (RAPD) markers *(2–5)*, in which polymorphic polymerase chain reaction (PCR) products are amplified from genomic templates using monomer oligonucleotide primers, typically of 10 nt. Because RAPDs may be applied without prior sequence information, the technique is often used both for genome fingerprinting *(2–4)* and for detailed genetic mapping *(4,5)* in organisms for which other genetic markers have not been developed. The conversion of mapped random markers into sequence characterized loci (STS) necessitates the isolation of the marker DNA fragment and the determination of its DNA sequence.

1.1. Direct Sequencing with Short Oligonucleotides

Short oligomer primers from octamer *(6,7)* to decamer lengths *(8,9)* have been used directly for the direct sequencing of the terminal regions of purified DNA of low complexity, such as cosmid and plasmids inserts. Initially, these sequencing reactions were performed at low temperatures and using T7 DNA polymerase (Sequenase) *(8,9)*, but recently the more efficient cycle sequencing with thermostable polymerase and elevated temperature has also been described *(6,7)*. Other methods for direct sequencing of gel purified DNA fragments with single arbitrary primers, or pairs of primers have also been reported *(2,10,11)*, in which the PCR products possess different terminal sequences that facilitate direct DNA sequence determination. Several similar protocols employing capillary electrophoresis (CE) sequencing have been developed which use short oligonucleotide primers *(12)*, or use primers with two 3'-selective nucle-

From: *Methods in Molecular Biology, Vol. 163:*
Capillary Electrophoresis of Nucleic Acids, Vol. 2: Practical Applications of Capillary Electrophoresis
Edited by: K. R. Mitchelson and J. Cheng © Humana Press Inc., Totowa, NJ

otides *(13)*, and long poly T annealing tracts introduced to the 5'-ends of the selective primers to aid stable annealing to poly A extended template *(14,15)*.

Although RAPD markers are generated from diverse genomic loci, the concurrence between annealing locus sequence and oligonucleotide primer is high. In one study, the maximum number of mismatches observed between primer and primer binding sites was never more than one between positions 1–7 of the primer *(16)*, and the degree of matching between the primer and the primer binding sites increased toward the 3'-end of the primer. This high concordance between primer and binding sequence contributes to the reliability of RAPD marker amplification, and permits use of extended RAPD primers to sequence RAPD fragments directly. This fast and convenient method for DNA sequencing of polymorphic RAPD bands, called "direct oligonucleotide extension sequencing" (DOES) *(17)*, utilizes the prior sequence knowledge of a nonamer RAPD amplification primer, but does not need RAPD fragment cloning or DNA strand separation.

The method is based on the likelihood that the different nucleotides occur immediately internal to the two primer specified terminals in three out of four RAPD fragments amplified from genomic loci. This permits a set of decamer oligonucleotides that extend the nonamer sequence by a single base at the 3'-terminus (A, T, C, or G), to be used as independent sequencing primers for each RAPD strand (*see* **Note 1**). Employing cycle-sequencing dye-terminator chemistry and automated four color sequencing, each strand of the nonamer-primed duplex RAPD fragments may be sequenced directly using the 3'-terminal selective primers which distinguish one strand from the other. The DOES sequencing method can be applied to both conventional PAGE sequencing with an ABI 377 sequencer *(17)* and for capillary sequencing by the MegaBACE 1000 capillary array sequencer (*see* **Note 2**).

2. Materials

1. Deoxyribonucleotide triphosphates are from Pharmacia, and oligonucleotides are from Gibco-Life Technologies Ltd.
2. Agarose gel isolated RAPD fragments are purified using the GeneClean Kit from BIO 101, Inc, according to manufacturers instructions.
3. 96-well, u-bottom PVC plates are from Becton Dickinson Corp.
4. The ABI PRISM BigDye® Terminator Cycle Sequencing Ready Reaction Kit, the ABI PRISM dRhodamine Terminator Reaction Ready Mixture and *Taq* DNA polymerase [FS] and *Taq* DNA polymerase are from Perkin-Elmer Corp.
5. Silica coated paramagnetic beads (Magnacil beads) are from Advanced Biotechnologies Ltd. (XM-600), or Promega Corp. (PT# A220A). Alternatively, carboxyl-coated paramagnetic beads (Amersham) may be used.
6. Bead binding buffer (2 × B³) 2X: 20% polyethylene glycol (PEG 8000), 2 *M* NaCl (Sigma Corp.).
7. MegaBACE 1000® capillary array DNA sequencer (Molecular Dynamics Inc).

3. Methods
3.1. RAPD Amplification

1. RAPD amplification reactions *(3,17)* using nonamer and decamer oligonucleotides are performed in 96-well, U-bottom PVC plates on a MJ Research PTC-100 thermal cycler

with cycling conditions of: 94°C for 1 min; 40°C for 1 min; 72°C for 1 min for 35 cycles. This ensures *Taq* DNA polymerase extension from fully annealed primer-template duplex.

2. The RAPD amplifications use 425 pg/µL genomic DNA and 400 nM of primer, 3.5 mM MgCl$_2$ and 300 mU *Taq* DNA polymerase.

3. Primary-amplification RAPD products are separated in 2% agarose Tris-acetate-EDTA gels. An aliquot of the selected RAPD fragment is collected by gel puncture using a micropipet tip, which is then rinsed into 20 µL of ddH$_2$O. The RAPD fragment is reamplified using RAPD fragment rinse (3.0 µL) and the identical primer and PCR conditions as for the primary amplification reaction.

4. Reamplified RAPD fragments are then agarose gel purified as described above and the entire band excised by scalpel. The DNA is purified using the BIO 101 GeneClean kit and are then dissolved in ddH$_2$O, for direct cycle sequencing (*see* **Note 3**).

3.2. Direct Oligonucleotide 3'-Extension Primer Sequencing (see Note 4)

1. Analysis using the ABI model 377 DNA sequencer: the RAPD fragments (typically 50 ng of DNA) are mixed with 3.2 pmol of decamer 3'-extended primer and are cycle sequenced using the reaction volumes and general protocols recommended by the manufacturer using the ABI PRISM BigDye Terminator Reaction Ready Mixture (Applied Biosystems Inc., protocols: part no. 903526), and *Taq* DNA polymerase [FS]. However, specific thermal regimes are defined for the sequence extension reaction using short 3'-extended primers.

2. Analysis using the MegaBACE sequence analyser: the RAPD fragments (typically 70 ng of DNA) are mixed with 4.0 pmol of decamer 3'-extended primer and are cycle sequenced using the reaction volumes and general protocols recommended by the manufacturer using the DYEnamic Electron Transfer® (ET) terminators (Pharmacia-Amersham) kit and Thermosequenase. However, specific thermal regimes are defined for the sequence extension reaction using the short 3'-extended primers.

3. For analysis using both gel-based analyzer and Array CE sequence analyzer: A combination of factors effects the quality of DOES sequencing reactions, including the annealing temperature, the number of amplification cycles, and to a smaller extent, the RAPD template concentration (*see* **Note 5**). Usually, 0.3–3 pmol of RAPD template is sufficient, and around 30–35 thermal cycles is adequate. Cycle sequencing is a linear PCR amplification and if the template is limiting ≤ 0.5 pmol), increasing the number of amplification cycles results in longer, high quality sequence. An excess of primer with a molar primer/template ratio of ~1–6 is adequate.

4. The 3'-extended sequencing primers are annealed to denatured RAPD template at 40°C and are cycle sequenced using the procedure:- step 1—heat to 96°C for 30 s; step 2—thermal ramp to 40°C for 15 s (*see* **Note 6**); step 3—thermal ramp to 60°C for 4 min; steps 1–3 were repeated for a total of 37 cycles (*see* **Note 7**), and finally step 4—a thermal ramp to 4°C and hold.

5. Unincorporated dye terminators can be removed by ethanol precipitation with 0.46 M sodium acetate, pH 5.2, or by magnetic bead purification (*see* **Subheading 3.3.**) for ABI gel analysis.

6. For ACE analysis, it is preferable to recover the extension products by ethanol precipitation from a solution made at least 0.3 M ammonium acetate, pH 6.0. Ammonium acetate is soluble in ethanolic solutions and repeated extraction of the DNA pellet with ice-cold 80% ethanol efficiently reduces ions for electrokinetic loading.

7. Sequence data was analyzed by the Genetics Computer Group Inc., Wisconsin Package sequence analysis program (version 7.0). MegaBACE sequence is for convenience converted to the ABI sequence format.

3.3. Plate Sequencing Reaction Purification Using Magnetic Beads

1. Wash 5 µL of carboxylated-beads or Magnesil beads 5 times with 50 µL of Millipore dH$_2$O. Concentrate the beads against a magnet with a magnetic beads concentrator (MBC) for each change of water.
2. Add 40 µL of 2X Bead Binding Buffer ($2 \times B^3$) to 5 µL of the beads and then add mixture to the well containing the sequencing reaction.
3. Mix the suspension and incubate at ambient temperature for 10 min (mix it gently by pipeting every 3 min) (*see* **Note 8**).
4. Concentrate the beads with the MBC for 30–60 s.
5. Wash the beads at least 3× with 100 µL of ice-cold 80% ethanol to remove the unincorporated dye terminators. Remove as much ethanol as possible.
6. Allow the beads to dry completely free of ethanol by standing open at room temperature for 30 min, or for 20 min at 37°C.
7. The beads are now desalted and free of dye terminators and ethanol and the sequence reaction products may be recovered for sequencing by ABI gel. Up to 2× additional ethanol extractions may be necessary at **step 5** to further reduce ions if the MegaBACE sequencer is to be used. However, additional washes are accompanied by some loss of short sequencing products.

3.3.1. ABI 377 Sequencer

1. Add 3 µL of formamide loading dye; incubate at 90°C for 2 min.
2. Place the bead suspension immediately onto ice.
3. Load all the samples (including beads) into the gel well.

3.3.2. MegaBACE Sequence Analyzer

1. Add 10 µL of formamide loading-buffer (Amersham) and concentrate the beads on the MCB.
2. Aspirate the sequencing reaction free of bead particles and transfer to a fresh 96-well MegaBACE loading plate (*see* **Note 9**).
3. Load the 96-well plate into the MegaBACE sequencer.
4. The sequencing sample is electrokinetically injected at 2 kV for 20 s (H$_2$O), or 35 s (formamide buffer).
5. The analysis run is performed at 9 kV for 120 min at a capillary array temperature of 44°C.
6. Prior to sequence analysis, the capillary array is hydrodynamically loaded with linear polyacrylamide (LPA) sieving matrix (Molecular Dynamics Corp.) at "high pressure" setting (1000 psi) for 200 s. The LPA matrix is then allowed to relax at atmospheric pressure for 20 min. The capillary array is conditioned by a (pre-) electrophoresis at 10 kV for 5 min.
7. The DYEnamic Electron Transfer (ET) dye terminators *(18)* each have a FAM excitation dye and the secondary emission dyes are respectively, R6G (T), R10 (G), ROX (C), and TAMRA (A) (*see* **Note 10**).

4. Notes

1. DOES sequencing uses the likelihood that the different nucleotides occur immediately internal to the two primer specified terminals of a RAPD fragment (*see* **Fig. 1**). For example, a 9-mer RAPD amplification primer (**Fig. 1A**) and a related-sequence 3'-extended 10-mer DOES sequencing primer (**Fig. 1B**) and their fully complementary genomic annealing site locus (**Fig. 1C**) demonstrate the principle. One of the four possible DOES primers of each set (A, T, C, or G at the 3' terminus) will prime independent direct

A	Nonamer RAPD primer	5'-TCTGACCGG-3'
B	Decamer DOES sequencing primer	5'-TCTGACCGG**T**-3'
C	Genomic DNA annealing locus	3'-AGACTGGCC**A**NNN..-5'

Fig. 1. Illustration of the annealing positions of: (**A**) an "unextended" 9-mer RAPD amplification primer, (**B**) a 10-mer DOES primer and, (**C**) a complementary genomic DNA annealing locus. The discriminatory nucleotide(s) shown in bold permits DNA sequence extension from one terminus of the double-strand RAPD template at which the primer anneals fully. Adapted from *Nucleic Acids Research*, **27**, e28, Copyright (1999), with the permission of Oxford University Press.

sequencing of one strand of a RAPD fragment. Another of the alternate DOES primers from the same set will prime the sequencing of the complementary strand of the RAPD fragment in an independent reaction.

2. Good DNA sequence read of RAPD fragments is obtained using the DOES method with hands-on time of 30 h or less. Although the DOES sequencing method demands relatively pure RAPD fragments, this limitation also applies to sequence from cloned fragments, as consistent sequence data from replicate clones also depends on the purity of the RAPD fragment.

3. Poor or unreadable DOES sequence is often obtained with primary RAPD templates that are gel purified only once. It is important to apply two rounds of RAPD amplification and gel purification to ensure adequate amounts of clean template for direct DNA sequencing. Additional rounds of amplification and gel purification do not further improve DOES sequencing.

4. The DOES sequencing reaction is a modification of conventional sequencing reaction. High quality sequence is obtained over a wide range of RAPD template (0.3–15 pmol), with better quality, longer reads at 0.3–2 pmol template.

5. Longer sequence reads are obtained using primer annealing temperatures of 40–42°C, than at lower or higher temperatures. The correct 3'-terminal nucleotide extension is sufficient for efficient formation of fully annealed duplex at the 5'-terminus of one of the two template strands, allowing sequence reads with DOES primers that 3'-terminate in either T, A, C, or G.

6. The duration of the primer-annealing step also influenced the length of read and longer, high quality reads were obtained with 45-s annealing compared to 15 s. The discrimination of *Taq* DNA polymerase [FS] between correct fully annealed decamer primer and a 3'-terminus mismatched primer that is transiently annealed to opposite strand of the RAPD template is a likely limitation of DOES sequencing.

7. The 25–27 extension cycles used typically for sequencing plasmid template is often not adequate for direct sequencing of linear DNA fragments with decamer primers, and longer reads with higher quality are obtained with 37 cycles.

8. Recovery of sufficient DNA from the beads is necessary for strong sequence reads. Although sufficient DNA is released from the beads after ~1-min incubation, additional mixing and incubation for 5 min can increase the strength of the sequence read appreciably.

9. Consistent sequence reads using the MegaBACE 1000 and DOES sequencing products are more demanding than conventional long primer sequencing of plasmids, and so on. The weaker initiation and extension of products generated by the DOES reaction during the low temperature annealing-extension stages require a high recovery of sequence products during washing steps and an efficient elimination of salts for good ACE reads.

Table 1
Successful Sequence Reads as Percentage of Attempts

		Sequence Cycles			
		27 × cycles		37 × cycles	
Anneal Temp	35°C	40°C	42°C	35°C	40°C
Good Reads	37%	51%	47%	72%	79%

The effects of annealing temperature and PCR cycle number on sequence read. The "% of total" refers to proportion of good readable sequence for 10-mer DOES-sequenced RAPD fragments from *Eucalyptus globulus* with few or no N's (>400 bases read with <5% N's in any contiguous region of 100 bases, and <3 N's in a row) out of total number of attempts. The trend is for increasing length of "good" sequence with higher numbers of linear PCR cycles at 40°C. Modified from *Nucleic Acids Research* **27**, e28, Copyright (1999), with the permission of Oxford University Press.

10. The sequence read using the capillary array sequencer has high concordance to the DOES sequence read from free RAPD strands using the ABI gel sequencer. These DOES sequence reads using conventional ABI gel sequencing *(17)* are adequate for determination of the inner terminal sequence of fragments with runs typically of 300–400 nt, or longer (**Table 1**).

Acknowledgments

The authors would like to acknowledge Forbio Research Pty Ltd for support during development of the gel sequencing protocols.

References

1. Hudson, T. J., Stein, L. D., Gerety, S. S., Ma, J., et al. (1995) An STS-based map of the human genome. *Science* **270**, 1945–1954.
2. Aikhionbare, F. O., Newman, C., Womack, C., Roth, W., Shah, K., and Bond, V. C. (1998) Application of random amplified polymorphic DNA PCR for genomic analysis of HIV-1-infected individuals. *Mutat. Res.* **406**, 25–31.
3. Williams, J. G. K., Kubelik, A. R., Livak, K. J., Rafalski, J. A., and Tingey, S. V. (1990) DNA polymorphisms amplified by arbitrary primers are useful as genetic markers. *Nucleic Acids Res.* **18**, 6531–6535.
4. Reiter, R. S., Williams, J. G. K., Feldman K. A., Rafalski, J. A., Tingey, S. V., and Scolnik, P. A. (1992) Global and local genome mapping in *Arabidopsis thaliana* by using recombinant inbred lines and random amplified polymorphic DNAs. *Proc. Natl. Acad. Sci. USA* **89**, 1477–1481.
5. Xu, J., Yang, D., Domingo, J., Ni, J., and Huang, N. (1998) Screening for overlapping bacterial artificial chromosome clones by PCR analysis with an arbitrary primer. *Proc. Natl. Acad. Sci. USA* **95**, 5661–5666.
6. Jones, L. B. and Hardin, S. H. (1998) Octamer-primed cycle sequencing using dye-terminator chemistry. *Nucleic Acids Res.* **26**, 2824–2826.
7. Hardin, S. H., Jones, L. B., Homayouni, R., and McCollum, J. C. (1996) Octamer-primed cycle sequencing: design of an optimized primer library. *Genome Res.* **6**, 545–550.

8. Slightom, J. L., Bock, J. H., Siemienak, D. R., and Hurst, G. D. (1994) Nucleotide sequencing double-stranded plasmids with primers selected from a nonamer library. *Biotechniques* **17**, 536–544.

9. Studier, F. W. (1989) A strategy for high-volume sequencing of cosmid DNAs: Random and directed priming with a library of oligonucleotides. *Proc. Natl. Acad. Sci. USA* **86**, 6917–6921.

10. Burt, A., Carter, D. A., White, T. J., and Taylor, J. W. (1994) DNA sequencing with arbitrary primer pairs. *Molec. Ecol.* **3**, 523–525.

11. Wang, X. and Feuerstein, G. Z. (1997) Direct sequencing of differential display PCR products. *Methods Mol. Biol.* **85**, 69–76.

12. Ruiz-Martinez, M. C., Carrilho, E., Berka, J., Kieleczawa, J., Miller, A. W., Foret, F., Carson, S., and Karger, B. L. (1996) DNA sequencing by capillary electrophoresis using short oligonucleotide primer libraries. *Biotechniques* **20**, 1058–1069.

13. Li, T., Okano, K., and Kambara, H., (2001) Selective primer sequencing from a DNA mixture by capillary electrophoresis, in *Capillary Electrophoresis of Nucleic Acids*, Vol. 2 (Mitchelson, K. R. and Cheng, J., eds.), Humana Press, Totowa, NJ, pp. 317–336.

14. Furuyama, H., Okano, K., and Kambara, H. (1994) DNA sequencing directly from a mixture using terminal-base-selective primers. *DNA Res.* **1**, 231–237.

15. Okano, K. and Kambara, H. (1996) Fragment walking for long DNA sequencing by using a library as small as 16 primers. *Gene* **176**, 231–235.

16. McGrath, A., Higgins, D. G., and McCarthy, T. V. (1998) Sequence analysis of DNA randomly amplified from the Saccharomyces cerevisiae genome. *Mol Cell Probes* **12**, 397–405.

17. Mitchelson, K. R., Drenthe, J., Duong, H., and Chaparro, J. X. (1999) Direct sequencing of RAPD fragments using 3'-extended oligonucleotide primers and dye-terminator cycle-sequencing. *Nucleic Acids Res. [Methods On Line]* **27**, e28.

18. Bashkin, J. S. (2001) Determining dye-induced DNA mobility shifts for DNA sequencing fragments by capillary electrophoresis, in *Capillary Electrophoresis of Nucleic Acids*, Vol. 1 (Mitchelson, K. R. and Cheng, J., eds.), Humana Press, Totowa, NJ, pp. 95–107.

VI

APPLICATIONS FOR ANALYSIS OF DNA-PROTEIN AND DNA-LIGAND INTERACTIONS

28

Capillary Affinity Gel Electrophoresis

Yoshinobu Baba

1. Introduction

Many human genetic diseases are caused by small alterations in DNA sequence of specific gene(s) *(1,2)*. Many different types of DNA mutations and polymorphisms are found to contribute to the alterations in the DNA sequence of disease-causing genes. These polymorphisms include sequence changes, such as the replacement of one or several nucleotides, the deletion or insertion of a sequence, differences in a variable number of tandem repeat locus (VNTR), and the instability in the numbers of tandem copies of microsatellite repeat elements *(1)*. High-performance technologies need to be developed, for the highly selective recognition of the alterations in the sequence of genes, as new approaches for DNA diagnosis of human diseases are needed for high-throughput analysis.

The recent development of capillary electrophoresis (CE), coupled with the advent of PCR, has facilitated the detection of mutation and polymorphism of human genes *(1,2)*. CE has an extremely high efficiency in the detection of size differences in DNA fragments, and the technology functions with high speed, high resolution, and high accuracy as well as having a high degree of automated operation. One deficiency however, is that the separation of DNA by CE is based on a molecular sieving effect and separation is determined by the size or shape of the DNA fragment, and the technique alone is not applicable to the recognition of specific DNA bases and/or DNA sequence.

Capillary affinity gel electrophoresis (CAGE) *(3–6)* is a methodology based on the combination of the high specificity of bioaffinity ligands and the high resolving power of capillary gel electrophoresis (CGE), and has the potential to overcome that deficiency. CAGE methods have been applied to the separation of DNA molecules based on their sequence. The nucleic acid analog, poly(9-vinyl adenine) (PVAd), which is electrically neutral and stable against both chemical and enzymatic hydrolysis, binds with high affinity to oligonucleotides via complementary hydrogen bonding and is useful as an affinity ligand for the base-specific recognition of a target homooligonucleotide from random pools of synthetic homooligonucleotides *(7–16)*.

From: *Methods in Molecular Biology, Vol. 163:*
Capillary Electrophoresis of Nucleic Acids, Vol. 2: Practical Applications of Capillary Electrophoresis
Edited by: K. R. Mitchelson and J. Cheng © Humana Press Inc., Totowa, NJ

This chapter presents data that suggests that with the use of an appropriate immobilized ligand, which hybridizes to particular deoxynucleotide residues, CAGE can be an effective tool for the rapid detection of point mutations and sequence rearrangements in known native gene regions. Known mutations at "hot spots" are examined and can be detected within 30 min.

2. Materials

1. Polyimide-coated fused silica capillaries with an untreated inner surface (375 μm od and 100 μm id Polymicro Technologies, Phoenix, AZ) are used with an effective length of 30 cm and a total length of 50 cm.
2. Tris-base, boric acid, and urea are all reagent grade.
3. Electrophoresis running buffer: 100 mM Tris-base, pH 8.3, 100 mM borate, 2 M urea is used for the preparation of the gel-filled capillaries, as well as the running buffer in all experiments.
4. Acrylamide, N,N'-ethylenebis(acrylamide) (BIS), N,N,N',N'-tetramethylethylenediamine (TEMED), and ammonium peroxodisulfate are of electrophoretic grade.
5. Oligonucleotides are chemically synthesized using an Applied Biosystems Inc. (Foster City, CA) Model 391 DNA synthesizer.

3. Methods
3.1. Poly(9-Vinyl Axdenine) Synthesis

1. PVAd is synthesized as a water-soluble polynucleotide analog, having a polyvinyl instead of the sugar phosphate backbone found in natural polynucleotides, whereas also possessing some advantages over natural polynucleotides being stable against both chemical and enzymatic hydrolysis, and being electrically neutral in avoiding electroosmotic flow. PVAd is used as an affinity macroligand. Several techniques including imino proton NMR, CE, and UV spectroscopy have established that PVAd forms complexes in vitro with the complementary strand of natural polynucleotides by hydrogen bonding.
2. PVAd is prepared according to Baba et al. *(7)* as follows:
   a. The 9-(2'-chloroethyl)adenine is prepared according to Baba et al. *(7)*.
   b. Add 1.82 g of 1,8-diazabicyclo-[5.4.0]undec-7-ene (DBU) to 1.98 g of 9-(2'-chloroethyl)adenine in 200 mL of dry acetonitrile at room temperature.
   c. Reflux the reaction mixture while stirring for 60 h, then remove the acetonitrile under reduced pressure.
   d. The 9-vinyl adenine product is recrystallized from a benzene/ethanol mixture (9:1 v/v) with a yield of 1.09 g (68%).
   e. The 9-vinyl adenine is polymerized in water in a sealed tube *in vacuo* after repeated degassing. Ammonium peroxodisulfate is used as the radical initiator.
   f. After 48 h of polymerization, the reaction mixture is poured into an excess volume of methanol and the PVAd polymer is collected by filtration and then dried *in vacuo* at 60°C.
   g. The PVAd is size fractionated using ultrafiltration. Long polymers of PVAd with a molecular weight ranging from 30,000 to 50,000 are used for CE.

3.2. Affinity Gel-Filled Capillaries

1. Capillaries filled with affinity gel for CAGE are prepared by *in situ* polymerization from a solution of acrylamide, bisacrylamide and PVAd within the capillary.

2. If a sufficiently large macromolecule is added to the solution of acrylamide monomers, then after the copolymerization reaction, a gel is formed in which the macromolecule is entrapped and thus effectively immobilized. The efficiency of the immobilization depends on the ratio of macromolecular size and gel porosity. PVAd of MW 30,000 to 50,000 is effectively immobilized into the polyacrylamide gel system, since it showed no electrophoretic mobility in slab-gel electrophoresis.

3. Polyimide-coated fused silica capillaries (375 μm od and 100 μm id) are used for preparation of affinity gel-filled capillary. Capillaries filled with polyacrylamide-PVAd conjugated gel are prepared as follows:
   a. The capillary was rinsed with 1 *M* NaOH solution for 15 min and with Milli-Q-purified water for 30 min.
   b. A solution of 0.4% of 3-methacryloxypropyltrimethoxysilane in 6 m*M* acetic acid is continuously passed through the capillary for 10 min.
   c. The capillary is left for 1 h at 40°C and then rinsed with Milli-Q-purified water for 10 min.
   d. A mixed solution of acrylamide and PVAd (8% T, 5% C, and 0–3.0% PVAd) is prepared in a buffer solution (0.1 *M* Tris-borate, pH 8.3 with an appropriate concentration of urea) and carefully degassed in an ultrasonic bath for 5 min.
   e. The percentage of PVAd is calculated by using the equation 100 × PVAd (g)/ [acrylamide (g) + bisacrylamide (g) + PVAd (g)].
   f. Polymerization is initiated with the addition of both 80 mL of 10% *N,N,N',N'*-tetramethylethylendiamine (TEMED) solution and 20 mL of 10% ammonium peroxodisulfate solution into 5 mL of the mixed solution of acrylamide and PVAd. The polymerizing solution is quickly introduced into the capillary for 5 min. Polymerization in the capillary is complete within 5 h at 40°C.

4. A capillary filled with polyacrylamide-PVAd conjugated gel is set on the capillary electrophoretic equipment.

5. Gel-filled capillaries obtained are preelectrophoresed at 10 kV for 30 min before use to remove the nonreacted monomer and the initiators.

6. Samples are electrophoretically injected into the capillary by applying a voltage of 10 kV for 3–10 s. The running voltage is 10 kV (200 V/cm). Oligonucleotides are detected by UV absorption at 260 nm.

4. Results

4.1. Principle of Capillary Affinity Gel Electrophoresis

1. Biospecific recognition during CAGE is based on a change in the electrophoretic mobility of the analyte, caused by some biospecific interaction between the analyte and the affinity ligands.

2. **Figure 1** illustrates the incorporation of affinity ligands into the gel matrix, which can be classified into three groups:
   a. Solubilized ligand in the gel matrix (**Fig. 1A**).
   b. Macroligand entrapped within the gel matrix (**Fig. 1B**).
   c. Immobilized ligand chemically bonded to the gel matrix (**Fig. 1C**).

3. The macroligand or the immobilized ligand used should include bases complementary to the DNA analyte which bind tightly to DNA by specific interactions, in order to realize specific base-recognition using CAGE. Base-paired complexes are formed by the complementary hydrogen bonding and stabilized by the stacking interaction of the bases on the ligand and analyte DNA. Synthetic analogs of nucleic acids will be of most value as affinity ligands for DNA.

A Mobile Ligand   **B Polymerized   C Immobilized
                          Ligand   Ligand**

Fig. 1. Schematic representation of affinity ligands in polyacrylamide gel: **(A)** mobile ligand method; **(B)** polymerized ligand method; **(C)** immobilized ligand method.

4. In contrast, natural oligonucleotides are not useful as the ligand for CAGE, because such anionic molecules will unavoidably result in local electroosmotic flow, which will diminish the high resolving power of CAGE. Further, a mobile monomer nucleotide analog will only interact very weakly with DNA and would not be a useful as an affinity ligand.
5. Theoretical aspects of CAGE have been established *(7,11,12)*. The interaction of DNA (N) and PVAd (L) is expressed as follows.

$$N + L = N \cdot L \tag{1}$$

$$Ka = \frac{[N \cdot L]}{[N][L]} \tag{2}$$

$$t/t_0 = 1 + Ka[L] \tag{3}$$

In which, N • L is the complex, square brackets represent the concentration, and Ka is the apparent association constant. The migration time of N is expressed respectively as t_0 in the absence of affinity ligand, and t in the presence of affinity ligand.
6. Although the value of [L], which is the free ligand concentration in the gel, it depends on the concentration of N. The free ligand concentration [L], is practically equal to the total L concentration in the gel [L]t, when [L]t is much higher than the total N concentration, [N]t.
7. **Eq. 3**, therefore, becomes:

$$t = t_0(1 + Ka[L]t) \tag{4}$$

This equation predicts that an increase in [L]t and Ka will lead to an increase in migration time. Capillary temperature *(11)* and urea concentration *(12)* both strongly affect the migration time of DNA in CAGE.

4.2. Specific Base Recognition of DNA

1. In the CAGE system, a nucleic acid analog, PVAd, is used as an affinity ligand and interacts with the oligodeoxynucleotide analyte, which is injected into capillary as a sample, by forming heteroduplexes via complementary hydrogen bonding within the capillary.

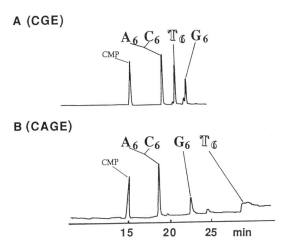

A (CGE)

A_6 C_6 T_6 G_6

CMP

B (CAGE)

A_6 C_6 G_6 T_6

CMP

15   20   25   min

Fig. 2. Separation of a mixture of four homopolymeric hexadeoxynucleotides by (**A**) and CAGE (**B**). The peak at 15 min in both electropherograms is CMP added to the sample as an internal standard. Conditions: capillary, 100 μm id, 375 μm od, 50 cm total length, 30 cm effective length. Running buffer, 100 m*M* Tris-base, pH 8.3, 100 m*M* borate, 2 *M* urea. Gel: 8% T, 5% C, and PVAd concentration: (A) 0%, (B) 0.2%. Electrophoresis conditions: field, 200 V/cm; injection 10 kV for 5 s, and UV detection at 260 nm. Reprinted from **ref.** *14*.

2. First, the electrophoretic behavior of homopolymers of hexadeoxynucleotides is studied by CAGE, as well as by CGE without affinity ligand.
3. **Figure 2** shows electropherograms of four different homopolymers of the hexadeoxy-nucleotides (A_6, C_6, G_6, and T_6). The peak marked CMP corresponds to the internal standard cytidine 5'-monophosphate, which is used because of its rapid migration relative to the migration of the hexadeoxynucleotides.
4. In CGE, these hexamers have different migration times in spite of having the same nucleotide chain length. Homooligomers of the same chain lengths exhibit some differences in mobility in slab-gel electrophoresis, and the order of the mobility is C>A>T>G.
5. The effect of base composition on oligonucleotide mobility is understood in CGE (*7*). The mobility order of the hexamers (A=C>T>G) observed in **Fig. 2A** is in agreement with earlier experiments and with slab-gel electrophoresis.
6. **Fig. 2B** clearly illustrates the unique features of CAGE. Thymidylic acid, which is complementary to the affinity ligand of poly(A) (PVAd) in the gel shows retardation due to its specific interaction with PVAd, whereas the mobility of the other oligodeoxynucleotide homopolymers are unchanged. The result demonstrates that PVAd has a high ability to recognize a base-specific homopolymer and has little or no nonspecific interactions with the other three bases.
7. To optimize the electrophoretic conditions, a mixture of four homopolymers is analyzed by CAGE with varying concentrations of PVAd ranging from 0 to 0.4%. It was found that a concentration of 0.2% PVAd is the most effective for the specific recognition of thymidylic acid from the homopolymer mixture. A lower concentration of PVAd has a low ability to retard thymidylic acid mobility, and higher concentration of PVAd causes severe broadening of the thymidylic acid band due to numerous strong interactions.

A (CGE)   B (CAGE)

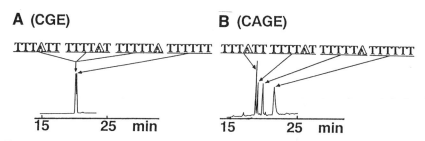

Fig. 3. Detection of mismatch positions on oligodeoxynucleotides (TTTATT, TTTTAT, and TTTTTA), using CAGE. The electrophoresis conditions are as described for **Fig. 2**.

8. The use of DNA affinity ligands is thus tempered by the need to provide only sufficient interaction between ligand and analyte as to retard mobility, without such frequent or strong interaction along the length of the gel as to prevent analyte migration for a significant proportion of the electrophoresis run.

4.3. Detection of Mutations in DNA Using CAGE

1. Since the CAGE system can usefully recognize specific bases in DNA, in this section it is applied to the recognition of specific sequences of DNA. Several oligodeoxynucleotides of identical length that include a base in different positions, which mismatches with the affinity ligand are applied to the CAGE system.
2. **Figure 3** shows that CAGE can completely resolve the mixture of the oligodeoxynucleotides TTTATT, TTTTAT, and TTTTTA, which cannot be separated by CGE. Each of these oligodeoxynucleotides has the same composition (T_6A), but have a different base order (sequence).
3. The apparent association constant, Ka, can be determined by the migration order of these oligodeoxynucleotides, because the migration order reflects the strength of hybridization with the ligand. The estimated Ka values for oligodeoxynucleotides in **Fig. 3** are, respectively:

   $1.08 \times 10^4\ M^{-1}$ for TTTATT.
   $1.39 \times 10^4\ M^{-1}$ for TTTTAT.
   $2.23 \times 10^4\ M^{-1}$ for TTTTTA.
   $3.98 \times 10^4\ M^{-1}$ for TTTTTT.

4.3.1. Detection of Mutations in Genes Using CAGE

1. CAGE shows high selectivity for the recognition of DNA sequence. Thus CAGE may be applied to the detection of known "mutation hot spots," including the point mutation of 248th codon of the *p53* tumor suppressor gene, or as shown in **Fig. 4** the known mutation hot spots of the 12nd and 61st codons of the N-*ras* oncogene.
2. **Figure 4A** shows that conventional CGE fails to separate a mixture 20-nt long oligodeoxynucleotides of the N-*ras* oncogene.
3. In contrast, **Fig. 4B** clearly demonstrates that CAGE succeeds in the separation of the three 20-nt long oligodeoxynucleotides, being the wild-type of N-*ras* oncogene and two of its mutants, which have one or two base replacement compared to wild-type.
4. It is notable that these oligodeoxynucleotides of the 61st codon of the N-*ras* oncogene (*see* **Fig. 4B**) are not homopolymer sequences, but are a native sequence with the wild-type fragment having the composition $T_{10}G_2C_6A_1$. The nucleotide(s) replaced in the two

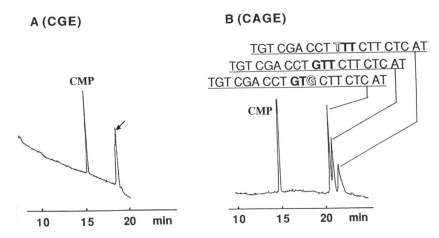

Fig. 4. Detection of mismatch on the 61st codon of the *ras* oncogene using CAGE. The electrophoresis conditions are as described for **Fig. 2.**

mutants separable by CAGE with PVAd ligand are Thymidine, being $T_9G_3C_6A_1$ and $T_8G_4C_6A_1$, respectively. Presumably the discrimination of PVAd is limited to replacement variants in DNA at Thymidine residues as the ligand.

5. These data suggest that with the use of appropriate immobilized ligands, which hybridize to particular deoxynucleotide residues, CAGE can be an effective tool for the rapid detection of point mutations and sequence rearrangements.

References

1. Baba, Y. (1996) Analysis of disease-causing genes and DNA-based drugs by capillary electrophoresis. *J. Chromatogr. B* **687,** 271–302.
2. Righetti, P. G. and Gelfi, C. (1998) Analysis of clinically relevant, diagnostic DNA by capillary zone and double-gradient gel slab electrophoresis. *J. Chromatogr. A* **806,** 97–112.
3. Baba, Y. (1998) Capillary affinity gel electrophoresis: New tool for detection of the mutation on DNA. *Mol. Biotechnol.* **6,** 143–153.
4. Baba, Y. (1999) Capillary affinity gel electrophoresis: New technique for specific recognition of DNA sequence and the mutation detection on DNA. *J. Biochem. Biophys. Methods* **41,** 91–101.
5. Righetti, P. G. and Gelfi, C. (1996) Capillary affinity gel electrophoresis, in *Capillary Electrophoresis in Analytical Biotechnology*, CRC Press, Boca Raton, pp. 461–462.
6. Chu, Y.-H., Avila, L. Z., Gao, J., and Whitesides, G. M. (1995) Affinity capillary electrophoresis. *Acc. Chem. Res.* **28,** 461–468.
7. Baba, Y., Tsuhako, M., Sawa, T., Yashima, E., and Akashi, M. (1992) Specific base-recognition of oligodeoxynucleotides by capillary affinity gel electrophoresis using polyacrylamide-poly(9 vinyladenine) conjugated gel. *Anal. Chem.* **64,** 1920–1925.
8. Akashi, M., Sawa, T., Baba, Y., and Tsuhako, M. (1992) Specific separation of oligodeoxynucleotides by capillary affinity gel electrophoresis using poly(9-vinyladenine)-polyacrylamide conjugated gel. *J. High Resolut. Chromatogr.* **15,** 625–626.

9. Sawa, T., Akashi, M., Baba, Y., and Tsuhako, M. (1992) Recognition of d(TTTATT) and d(TTATTT) by capillary affinity gel electrophoresis using poly(9-vinyladenine)-polyacrylamide conjugated gel. *Nucleic Acids Res. Symp. Ser.* **27,** 51–52.

10. Baba, Y., Tomisaki, R., Tsuhako, M., Sawa, T., Inami, A., Kishida, A., and Akashi, M. (1993) Detection of mismatch positions on the DNA-polyvinyladenine hybrids using capillary affinity gel electrophoresis. *Nucleic Acids Res. Symp. Ser.* **29,** 81–82.

11. Baba, Y., Tsuhako, M., Sawa, T., Yashima, E., and Akashi, M. (1993) Temperature-programmed capillary affinity gel electrophoresis for the sensitive base-specific separation of oligodeoxynucleotides. *J. Chromatogr.* **632,** 137–142.

12. Baba, Y., Tsuhako, M., Sawa, T., Yashima, E., and Akashi, M. (1993) Effect of urea concentration on the base-specific separation of oligodeoxynucleotides in capillary affinity gel electrophoresis. *J. Chromatogr. A* **652,** 93–99.

13. Baba, Y., Inoue, H., Tsuhako, M., Sawa, T., Kishida, A., Yashima, E., and Akashi, M. (1994) Evaluation of selective binding ability of oligodeoxynucleotide to nucleic acid analogues using capillary affinity gel electrophoresis. *Anal. Sci.* **10,** 967–969.

14. Baba, Y., Sawa, T., Kishida, A., Yashima, E., and Akashi, M. (1998) Base-specific separation of oligodeoxynucleotides by capillary affinity gel electrohporesis. *Electrophoresis* **19,** 433–436.

15. Baba, Y., Sawa, T., and Akashi, M. (*personal communication*) Mutation of mismatches on DNA by capillary affinity gel electrophoresis.

16. Muscate, A., Natt, F., Paulus, A., and Ehrat, M. (1998) Capillary affinity gel electrophoresis for combined size- and sequence-dependent separation of oligonucleotides. *Anal. Chem.* **70,** 1419–1424.

29

Capillary DNA-Protein Mobility Shift Assay

Jun Xian

1. Introduction

The interaction of DNA with proteins is a central theme of molecular biology. Although all cells in the organism contain the same genetic "blueprint," this information is utilized in a highly selective and specific manner. Different types of cells activate different subsets of genes. These processes are frequently dependent on the presence of certain DNA-binding transcription factors in different cells.

Transcription factors are the molecular switches that, singly or in combination, interact sequence-specifically with DNA to turn on and off genes involved in the expression of proteins. Highly specific in nature, transcription factors operate at the most fundamental level to regulate any given cell's biochemical functions. The study of the interaction of these transcription factors with DNA is, therefore, a critical aspect of gene regulation.

The most commonly used method to study DNA-protein interaction is the electrophoretic gel mobility shift assay (EMSA). EMSA (also known as the "gel shift" assay) is a process in which a radioactively labeled DNA probe is mixed with a solution containing protein of interest, and after a brief reaction period, loaded on an electrophoretic gel. The DNA-protein complex, if formed, migrates more slowly than does the free DNA probe *(1–3)*. Although used extensively, this method suffers from some significant limitations, such as long assay time, large sample requirements, and problematic quantitation.

Capillary electrophoresis (CE) is a descendant of both electrophoresis and high performance liquid chromatography (HPLC) technology. Combining the inherent resolving power of electrophoresis with on-line, HPLC-like detection CE produces, fast, reproducible separation and sensitive detection using only nanoliter amounts of sample. CE thus well adapted as a replacement for the traditional gel electrophoretic mobility shift.

There are several prior studies in which CE was applied to the study of DNA-protein interactions by using a new mobility shift assay, named capillary electrophoretic

From: *Methods in Molecular Biology, Vol. 163:*
Capillary Electrophoresis of Nucleic Acids, Vol. 2: Practical Applications of Capillary Electrophoresis
Edited by: K. R. Mitchelson and J. Cheng © Humana Press Inc., Totowa, NJ

mobility shift assay (CEMSA) *(4–7)*. Xian et al. *(4)* describe a rapid and quantitative procedure which permits accurate assessment of specific DNA-protein interactions on a scale more than 100-fold below the minimum usually necessary for EMSA that can be carried out on a commercially available CE instrument with a laser-induced fluorescence (LIF) detector. The sensitivity of CEMSA was demonstrated in successful experiments with binding factors from a single sea urchin egg. This procedure can be used for a variety of purposes, quantitative and qualitative, including studies involving the effects of antibodies on DNA-protein complexes or "super-shift" experiments. Stebbins et al. *(5)* applied CEMSA to study the interaction between the *trp* repressor of *E. coli* and *trp* operator by using an in-house assembled electrophoresis and fluorescence detection apparatus. They found that the specificity of the operator for the repressor may be used to selectively separate the repressor from a complex sample that includes nonspecific binding proteins. Using the minimal DNA binding domain of the onco-protein c-Myb (R2R3) and a specific target DNA sequence as a model system, Xue et al. *(6)* ran CEMSA in a coated capillary with linear polymer buffer and a simple UV detector. In their method, DNA and its DNA-protein complex with protein can be directly monitored at 260 nm. Foulds and Etzkorn *(7,8)* presented a fast and simple, precise and general method to run CEMSA in an uncoated capillary with no added gel matrix. They measured the dissociation constant for GCNK58, a DNA-binding region construct of the yeast transcription factor GCN4 and AP1 DNA site by their method.

CEMSA is different from affinity capillary electrophoresis (ACE). The principle of ACE is the analysis of changes in the electrophoretic mobility of a protein complexed with a ligand present in the electrophoresis buffer. When a protein forms a complex with a ligand, the protein-ligand complex may then have a measurable difference in electrophoretic mobility relative to the free protein. Changing the ligand or protein concentration results in the change of electrophoretic mobility of the protein peak. The binding constant (K_b) for the interaction is calculated by analyzing the change in the electrophoretic mobility of the protein, as a function of the concentration of the ligand added into the running buffer *(9)*. When performing CEMSA, protein and DNA (ligand) are mixed in the reaction buffer and then the mixture is injected onto the capillary. Based on the charge/mass ratio, free DNA can be quickly separated from the DNA-protein complex during electrophoresis. Changing the DNA or protein concentration usually does not change their electrophoretic mobilities, but the peak areas for both free DNA and DNA-protein complex. The dissociation constant (K_d) is calculated by measuring the peak areas for both free DNA and its complex.

A critical element in the design of a CEMSA experiment is the selection of DNA-protein reaction conditions, which is based on the understanding of the properties of the target protein. DNA-protein interactions are highly sensitive to the reaction buffer composition, such as pH, salt concentration, and reaction temperature. This chapter will give only a general outline of CEMSA techniques, but will focus on those experimental parameters that require strict control for precise results. This information will hopefully help readers to set up new CEMSA studies, and to make ongoing work more successful.

2. Materials
2.1. Buffers Stored at –20°C

1. Buffer 1: 50 mM Tris-HCl, pH 8.0, 50 mM NaCl, 5 mM EDTA, 7.5 mM β-mercaptoethanol.
2. Urea lysis buffer: 50 mM Tris-HCl, pH 8.0, 8 M urea, 1 M NaCl, 7.5 mM β-mercaptoethanol.
3. Buffer 2: 50 mM Tris-HCl, pH 8.0, 3 M urea, 100 mM KCl, 7.5 mM β-mercaptoethanol.
4. Buffer 3: 10 mM Tris-HCl, 100 mM potassium phosphate, pH 6.3, 3 M urea, 100 mM KCl, 7.5 mM β-mercaptoethanol.
5. Buffer 4: 50 mM Tris-HCl, pH 7.4, 3 M urea, 100 mM KCl, 10% (v/v) glycerol, 7.5 mM β-mercaptoethanol.
6. Buffer 5: 10 mM Tris-HCl, pH 6.3, 300 mM imidazole, 3 M urea, 100 mM KCl, 10% (v/v) glycerol, 7.5 mM β-mercaptoethanol.
7. Buffer A: 10 mM HEPES, pH 7.9, 1.5 mM MgCl$_2$, 10 mM KCl, 0.5 mM DTT, 0.5 mM phenylmethylsulphonyl fluoride (PMSF).
8. Buffer B: 5 mM HEPES, pH 7.9, 26% (v/v) glycerol, 1.5 mM MgCl$_2$, 0.2 mM EDTA, 0.5 mM DTT, 0.5 mM PMSF.
9. Buffer C: 10 mM Tris-HCl, pH 7.4, 1 mM EDTA, 1 mM EGTA, 1 mM spermidine-HCl, 1 mM DTT, 0.36 M sucrose.
10. Buffer D: 10 mM HEPES, pH 7.9, 1 mM EDTA, 1 mM EGTA, 1 mM spermidine-HCl, 1 mM DTT.
11. Buffer E: 20 mM HEPES, pH 7.9, 40 mM KCl, 0.1 mM EDTA, 1 mM DTT, 20% (v/v) glycerol.
12. 5 X Binding buffer: 100 mM HEPES, pH 7.9, 375 mM KCl, 25 mM MgCl$_2$, 2.5 mM DTT.

2.2. Buffers Stored at 4°C

1. 1 X TBE buffer: 89 mM Tris-base, 89 mM boric acid, pH 8.3, 2 mM EDTA.
2. Phosphate-buffered saline: 100 mM NaCl, 4.5 mM KCl, 7 mM Na$_2$HPO$_4$, 3 mM KH$_2$PO$_4$.
3. Probe buffer: 10 mM Tris-HCl, pH 8.0, 10 mM KCl.
4. CE running buffer: 89 mM Tris-base, 89 mM boric acid, pH 8.3, 2 mM EDTA, 0.05–5% linear polyacrylamide (mol wt 750–1000 kDa) (*see* **Note 1**).
5. Ca^{2+}- and Mg^{2+}-free sea water containing 10 mM Tris-HCl, pH 7.4 and 1 mM EDTA.

2.3. Special Reagents and Equipment

1. Linear polyacrylamides (mol wt 700–1000 kDa) are from Polysciences, Inc. (Warrington, PA). These materials are stored at 4°C.
2. Fluorescent dyes for intercalation, such as TOTO-1 or YO-PRO-1 are from Molecular Probe, (Eugene, OR). These dyes should be stored at –20°C.
3. Poly [d(A)•d(T)], poly [d(I)•d(C)], and poly[d(I-C)•d(C-I)] are from Sigma Chemical Co. (St. Louis, MO).
4. Beckman P/ACE 5500 CE instrument with LIF detector (488-nm excitation and 516-nm emission filter) is from Beckman Instruments, (Fullerton, CA).
5. Neutral coated capillary (eCAP neutral) is from Beckman, and linear polyacrylamide coated capillary (BioCap) is from Bio-Rad (Hercules, CA).

3. Methods
3.1. DNA Probe Selection

1. A DNA restriction fragment or synthetic oligonucleotides are two examples of allowable probes for CEMSA. The type and size of probe used depends on the nature of the investigation. If a previously identified protein is to be studied, then an oligonucleotide probe

should be used which is specific for binding interactions with the protein. The use of oligonucleotide probes simplifies the interpretation of the results, as the binding site is typically unique and isolated from other possible (unidentified) sites present within the same regulatory sequence region.

2. DNA restriction fragments are used because they are easy to prepare, and can be used to study the simultaneous binding of several proteins to one probe. The size of the restriction DNA fragment is normally kept below 250 bp to enable clear distinction of the probe from any DNA-protein complexes (*see* **Note 2**).

3.2. Probe Labeling

1. There are two methods for labeling the probe, covalent attachment and intercalation with fluorescent dye. Covalent labeling usually applies to synthetic oligonucleotides (15–30 bp) during the synthesis process, whereas intercalation usually applies to larger restriction DNA fragments (50–200 bp).

2. Intercalation is more versatile and simpler than covalent labeling in situations in which qualitative results are satisfactory. All fluorescent-labeled probes are light sensitive, and therefore requiring an aluminum foil covering during the mobility shift assay.

3.2.1. Protocol 1: Covalent Labeling

1. Oligonucleotides are synthesized and labeled with 6-FAM fluorescent dye (or other fluorescent dye) at the 5'-end using an Expedite 8909 nucleic acid synthesis system from PerSeptive Biosystems (Framingham, MA) and following the manufacturer's recommended protocols.

2. After synthesis, the labeled oligonucleotides are purified by ionic exchange chromatography and then further desalted by reverse phase HPLC.

3. The purified labeled, single-stranded oligonucleotides are then vacuum dried and stored in the dark at –20°C.

4. Equimolar amounts of sense and antisense oligonucleotides are mixed in 0.1 *M* NaCl at 93°C for 5 min and allowed to slowly cool at room temperature to promote strand reannealing (e.g., overnight).

5. Double-stranded DNA is then vacuum dried and stored at –20°C.

6. The purity of the labeled probe can be checked by capillary gel electrophoresis.

7. Prior to use, the probe is dissolved in probe buffer and its concentration is determined by UV absorption measurement.

3.2.2. Protocol 2: Intercalation Labeling

1. A restriction fragment or a double-stranded oligonucleotide in probe buffer is incubated with desired fluorescent dye, such as TOTO-1 or YO-PRO-1, at specific molar ratio, which depends on the length of probe used, in dark room for 30 min before use.

2. The commonly used ratios are 10 bp/1 dye, 20 bp/1 dye, and 30 bp/1 dye (*see* **Note 3**). These types of probes do not need further purification because only double-stranded DNA has been labeled. These probes are not very stable therefore they need to be freshly prepared (*see* **Note 4**).

3.3. Protein Preparation

1. Both recombinant protein and crude protein extracts can be used for the CE mobility-shift binding assay.

2. Purified DNA binding proteins (expressed from bacterial clones) are used to allow the study of the binding interactions of that protein, in the absence of other proteins with which it may interact in vivo. It should be noted that some posttranslational modifications

that may normally occur to eukaryotic proteins may be absent from those proteins when expressed in bacterial systems. However, one of the greatest advantages of using recombinant proteins for DNA-binding and expression studies is that it is possible to manipulate specific regions of the target protein by site directed mutagenesis.

3.3.1. Recombinant Protein Protocol 3 (10)

1. The complete coding sequence of the target protein had been subcloned into pRSET expression vector in the appropriate reading frame. A six-histidine tag is present at the amino terminus of the pRSET fusion proteins, which allows affinity purification by means of nickel-agarose chromatography. Then the construct is transformed into BL21 bacterial host cells.
2. The culture of bacteria expressing the construct was grown at 37°C until $OD_{600} \sim 0.6$. Protein expression is then induced with the addition of 1 mM (final concentration) isopropyl-β-D-thiogalactopyranoside (IPTG) and cells are grown for an additional 4 h at 37°C.
3. The bacteria are harvested by centrifugation and the bacteria pellet is resuspended in 8 μL of Buffer 1, and 800 mL of 10 mg/mL lysozyme is added and the suspension is thoroughly mixed.
4. After incubating at room temperature for 10 min the protein solution is frozen at –80°C until use.
5. Upon thawing at room temperature, 0.7 vol of urea lysis buffer is added and the solution is mixed by vortexing. The mixture is centrifuged for 1 h at 30,000g to remove insoluble matter.
6. A 50% solution of nickel-agarose resin is equilibrated with Buffer 2. The crude protein extract is batch-loaded with 8 mL of the equilibrated resin solution for 1 h at room temperature.
7. The loaded resin is then poured into a Bio-Rad Econo-Prep column, allowed to settle, and is then washed with 250 mL of Buffer 2, then with 250 mL of Buffer 3, and finally with 50 mL of Buffer 4.
8. The purified protein was eluted from the resin with Buffer 5 (about 2 vol of resin) and frozen at –80°C.

3.3.2. Preparation of Protein Extracts

1. Protein extracts may be prepared from whole cells or from isolated nuclei.
2. There are advantages to using both types of protein extract, or even a combination for CEMSA, and then comparing results obtained with both types.
3. The preparation of nuclear extracts results in the isolation of only those DNA-binding proteins with in vivo access to the DNA.
4. The whole-cell extracts enable the entire DNA-binding protein content of the cell to be examined. Some DNA-binding proteins may be present in the cytoplasm rather than the nucleus, and can be identified by the comparison of the binding profiles of nuclear and cytoplasmic extracts.
5. Whole-cell extracts are easier to prepare, requiring less steps in their preparation, which make their use favorable when a tissue sample is limiting because fewer manipulations present fewer stages at which protein will be lost or damaged. It is also necessary to prepare whole-cell extracts when a sample has been frozen. Freezing damages the nuclear membrane, preventing the purification of the intact nuclei, and thus preventing the preparation of nuclear extracts.

3.3.3. Protocol 4 (11)

1. Cell samples and cell fractions should be kept on ice at all times and all centrifugation steps should be carried out at 4°C.
2. Harvest cells (10^7–10^8 cells) and centrifuge at 250g for 10 min.

3. Wash with phosphate-buffered saline and centrifuge at 250*g* for 10 min.
4. Resuspend the cell pellet in 5 vol of Buffer A and then incubate the cells on ice for 10 min. Centrifuge the solution at 250*g* for 10 min to pellet the cells.
5. Resuspend the cell pellet in 3 vol of Buffer A. Add NP-40 to 0.05% and homogenize with 30–50 strokes of a tight-fitting Dounce homogenizer to rupture the cell membrane and release the nuclei.
6. Successful release of cell nuclei may be checked by a phase-contrast microscope (keep a small sample of suspension from before addition of NP-40 and homogenization for comparison).
7. Centrifuge at 250*g* for 10 min to pellet the nuclei.
8. Resuspend the nuclear pellet in 1 mL of Buffer B. Measure total volume and add NaCl to a final concentration of 30 mM. Mix well by inversion.
9. Incubate the nuclei on ice for 30 min to release nuclear proteins and then centrifuge the mixture at 24,000*g* for 20 min at 4°C to remove residual cell debris.
10. Aliquot the extract supernatant and snap-freeze it in dry ice/ethanol. Store the extracts in small aliquots (~0.1 mL) at –80°C.

3.3.4. Protocol 5 (12)

1. *Strongylocentrotus purpurratus* embryos are cultured by standard methods (*12*) to blastula-stage, which are then collected by filtration through a 51-µm Nitex filter. The concentrated embryos are suspended in ice-cold Ca^{2+}- and Mg^{2+}-free sea water containing 10 mM Tris-HCl, pH 7.4 and 1 mM EDTA and washed 1–2 times by low-speed centrifugation.
2. The embryo pellet is resuspended in 10–20 times the pellet volume of Buffer C, frozen in liquid N_2, and stored at –80°C until use.
3. Embryo cells are lysed during thawing by vigorous shaking. Nuclei are washed 2–3 times in Buffer C, and then 2 times in Buffer C to which 0.1% Triton X-100 is added. Nuclei are collected between washes by centrifugation at 3000*g* for 10 min.
4. After a final centrifugation at 3000*g* for 10 min, the nuclei are resuspended in 5–10 times the pellet volume of Buffer D.
5. While mixing the nuclear suspension, one-tenth volume of 4 M ammonium sulfate, pH 7.9 is added dropwise, to prevent local high concentration, to a final concentration of 0.36 M.
6. After incubation for 30–60 min on ice, chromatin is removed by centrifugation at 30,000*g* for 1–1.5 h at 4°C.
7. Protein is precipitated from the supernatant by the addition of 0.3% g/mL ammonium sulfate.
8. After incubation overnight at 4°C, the protein precipitation is collected by centrifugation for 15 min at 10,000*g*, and is then dissolved in 0.5 times the nuclei pellet volume of buffer E. The proteins are dialyzed against Buffer E, overnight at 4°C.
9. Ammonium sulfate insoluble proteins were removed by centrifugation at 30,000*g*, and the protein extracts are stored at –80°C.

3.4. Condition for DNA-Protein Reactions and CE Separation

3.4.1. Reaction Conditions

1. Reaction conditions for the binding of protein to DNA varies from system to system and no single set of binding conditions will be applicable to all interactions. The optimization of binding variables, such as buffer pH, ionic strength, and additives, should be studied for each protein/DNA pair (*see* **Note 5**).

2. In addition, the reaction depends upon the nature of the DNA involved and the specific DNA-protein interactions that consist of two components: sequence dependent components and nonspecific (sequence independent) bindings. All DNA-binding proteins display some nonspecific DNA-binding activity *(14)*. In order to enhance the specific interactions and eliminate nonspecific binding interactions, a nonspecific binding (or competitor) DNA is usually added to the reaction mixture.

3. The binding reaction depends on the amounts of protein, DNA probe, and competitor DNA in the reaction mixture. The addition of an appropriate amount of this competitor DNA in the reaction mixture is also critical for a quantitative measurement of the DNA-protein complex. An optimal amount of the competitor must be determined empirically for each reaction.

4. It is difficult to describe a basic experiment measuring the interaction of a protein with its specific DNA, since the major experimental parameters which control the specificity and stability of such DNA-protein complexes need to be determined for each system.

5. However, several general guidelines can be offered *(3)*.

6. It is usually helpful to carry out the reaction at the highest salt concentration that will allow the detection of specific binding. At high salt concentration, it may be possible to minimize or possibly eliminate nonspecific DNA-protein interactions, without drastically reducing the specific binding of interest (*see* **Note 6**).

7. A simple sequence synthetic polynucleotide (*see* **Note 7**), such as poly [d(I-C)•d(I-C)], poly [d(I)•d(C)], or poly [d(A)•d(T)], is often used as a competitor DNA. The initial range for the competitor DNA would be 0.01–1 µg competitor per fmol of fluorescence DNA probe (*see* **Note 8**). The amount of competitor applied in the reaction also depends on the protein source used. A recombinant protein source will require less competitor oligonucleotide compared to crude protein extracts. Whole-cell extracts and nuclei extracts also contain a multiplicity of proteins that might interact nonspecifically with the target probe.

8. The optimal ratios of protein to specific and nonspecific DNAs depend on the ratios of Kd_S/Kd_{NS}, the dissociation constant ratio of specific binding to nonspecific binding. For a typical 10-µL reaction containing 50 fmol of oligonucleotide probe, 1–2 µL of a target protein is a good starting range.

9. Since many DNA binding proteins are fairly susceptible to protease digestion, even in the presence of protease inhibitors, reactions are typically incubated on ice.

10. Although the binding of protein to DNA is a rapid process, usually 10–30-min incubation is necessary to let the binding reach the equilibrium (*see* **Note 9**).

3.4.2.1. Protocol 6, Reaction Using Recombinant Protein

1. In a 1.7-mL microcentrifuge tube, add 3 µL of Buffer E, 1 µL of FAM-labeled DNA (about 50–60 fmol), 1 µL of recombinant protein (about 300 fmol), 1 µL of 50 ng poly [d(A)•d(T)], 2 µL of 5X binding buffer, and 2 µL of H_2O. Incubate the mixture on ice for 10–30 min, before loading to the ice-cold microvial for injection on to the CE capillary.

3.4.2.2. Reaction Using Nuclear Extract

1. In a 1.7-mL microcentrifuge tube, add 3 µL of Buffer E, 1 µL of FAM-labeled DNA (about 60 fmol), 1 µL of nuclear extract (about 5 mg), 1 µL of 5 mg poly [d(A)•d(T)], 2 µL of 5X binding buffer, and 2 µL of H_2O.

2. Incubate the mixture on ice for 10–30 min, before loading to the ice-cold microvial for injection on to the CE capillary.

3.4.2.3. Reaction of Extract from Sea Urchin Eggs (Whole Cell Extract)

1. Immediately before assaying, the eggs are thawed from –80°C to room temperature for 5 min to break the cell membrane and then the buffer E is added.
2. In a 1.7-mL microcentrifuge tube, add 3 μL of Buffer E, 1 μL of FAM-labeled DNA (about 60 fmol), 2 μL of sea urchin egg extract in Buffer E (from ~1000 eggs), 1 μL of 5 μg poly [d(A)•d(T)], 2 μL of 5X binding buffer, and 1 μL of H_2O.
3. Incubate the mixture on ice for 30 min, before loading to the ice-cold microvial for injection on to the CE capillary.

3.4.3. Practical Example (4)

3.4.3.1. Probe Labeling

1. The wild-type probe (double-stranded) we used in this study contains two adjacent P3A2 sites of differing affinity for SpP3A2 protein and the sense sequence is:
5'-GATCTTTTCGG<u>CTTCTGCGCAC</u>AC<u>CCCACGCGCA</u>TGGGGC-3'
2. Each oligonucleotide (sense or antisense) is labeled with 6-FAM at 5'-end during the synthesis.

3.4.3.2. EMSA Reaction Conditions

1. Reaction mixture (10 μL) is made up of 3 μL of Buffer E, 1 μL of FAM-labeled DNA (about 60 fmol), 1 μL of nuclear extract (about 5 μg), 1 μL of 5 μg poly [d(A)•d(T)], 2 μL of 5X binding buffer, and 2 μL of H_2O. This mixture is incubated on ice for 10–30 min before injection onto the capillary.

3.5. Capillary Electrophoresis Mobility Separation Assay (CEMSA)

1. Because the equilibrium constant specifies the ratio of association and dissociation rates, the kinetic features of a DNA-protein interaction will, in general, be sensitive to any factor that perturbs its equilibrium stability. Thus, in an equilibrium-binding assay, it is essential to verify the attainment of equilibrium.
2. There are two potential sources of errors when CEMSA is employed for the study of DNA-protein interactions. The first involves possible dissociation of the complexes during electrophoresis. This error may not pose a serious problem if the DNA-protein complexes are long-lived relative to the "dead-time" of the experiment (time required for free DNA probe to migrate inside the capillary). Once the zone of free DNA probe moves along the capillary, dissociation of the complex at or near the inlet of the capillary cannot affect the results (*see* **Note 10**).
3. A second source of the possible artifacts is the change in the ionic composition of running buffer used for CE. This may not be a problem in many instances. For example TBE is a low salt buffer, therefore, its presence results in a DNA-protein solution of a low ionic strength. This tends to increase the stability of most DNA-protein interactions and results in a more stable complex.

3.5.1. Capillary Column

1. In order to avoid the adsorption of protein to the capillary during separation, neutral coated columns such as eCAP (Beckman) and BioCap (Bio-Rad) can be employed in this experiment.
2. The internal diameter (id) of capillary can be either 50 or 75 μm and the capillary length is dependent on the DNA/protein pair being analyzed. Usually, 20 cm is long enough for the separation because of the difference of charge/mass ratio between free DNA probe and the DNA-protein complex.

3.5.2. Separation Buffer and Additives

1. The composition of the CE running buffer can strongly affect the lifetimes of DNA-protein complexes during electrophoresis. Running buffers with lower salt concentrations can stabilize interactions that are mediated in part by ionic contacts. However, we also can eliminate some nonspecific interaction by increasing the salt concentrations in the running buffer. The commonly used CE buffer are, 1X TBE or 0.5X TBE. The choice depends on the voltage applied for the electrophoresis.

2. Some polymers, such as linear polyacrylamide, are often added to the CE running buffer. There are several advantages to add some neutral linear polymer to the running buffer (*see* **Note 11**):

3. The polymer helps to prevent the aggregation binding of protein onto the capillary surface and prolongs the lifetime of the capillary.

4. It can enhance the binding of a protein to DNA, even at low-to-moderate concentrations due to a macromolecular crowding effect *(15)*.

5. The polymer acts as a sieving matrix that enhances the separation of free DNA probe from the DNA-protein complex.

6. The amount and molecular weight of polymer added is determined by the strength of complex formation, the size of protein, and the number of specific binding proteins presented in the mixture. A small amount of the polymer helps to eliminate unwanted complexes between DNA and protein *(4)*.

7. Other additives in running buffer, such as small, neutral osmolytes can reduce osmotic stress that limits correct protein folding, and thus also can enhance the stability of specific DNA-protein complexes *(2)*.

3.5.3. Protocol 7

1. The capillary is filled with CE buffer.
2. The sample is introduced by high-pressure injection (10-s injection corresponds to ~10 nL of sample) followed by a second injection of CE buffer for 5 s (*see* **Note 12**).
3. Electrophoresis is run at reversed polarity, i.e., the anode at the detector end, at 18 kV and 18°C.
4. Between each run, the capillary is rinsed with 1X TBE buffer for 2 min and then with CE running buffer for 5 min. Therefore, a complete run lasts 17 min.

3.5.4. Separation Column Temperature

1. The stability of DNA-protein complex is sensitive to the temperature of separation medium. Therefore, it is important to keep constant temperature during the run in order to achieve reproducible results. Since the DNA-protein complex is more stable at lower temperatures, most separation temperatures are set between 18–25°C. It is also true that nonspecific binding is more sensitive to temperature than specific binding. Therefore, we can eliminate some nonspecific complexes by increasing the separation temperature.

3.6. CE Separation

1. Separation is performed in a 50 μm × 37 cm long (30-cm effective length) neutral coated capillary (eCAP neutral, Beckman) with the P/ACE system 2000 (Beckman) upgraded with Gold (version 8.1) software and LIF detector (488-nm excitation and 516-nm emission filter).
2. The capillary is filled with 1X TBE buffer and 0.2% linear polyacrylamide (750–1000 kDa).
3. The sample is introduced by 10-s high-pressure injection followed by a second injection of CE buffer for 5 s.

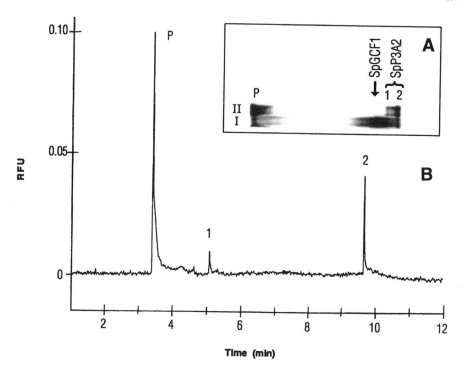

Fig. 1. Identification of DNA-SpP3A2 complexes in nuclear extract, by use of EMSA and CEMSA methods. (**A**) conventional EMSA: lane I, 10-µg nuclear extract of 24-h sea urchin embryos reacted with 40-fmol wild-type probe together with 5 µg poly [d(A)•d(T)], lane II, 300-fmol recombinant SpP3A2 protein reacted with 40-fmol wild-type probe with 0.5 µg poly [d(A)•d(T)]. (**B**) CEMSA of an aliquot of the 5 ng of 24-h embryo nuclear extract reacted with 60-attomol wild-type probe together with 5 ng poly [d(A)•d(T)]. RFU, relative fluorescence unit; P, free probe; 1 and 2, 1:1 and 2:1 DNA-protein complexes of SpP3A2; DNA-protein complex of SpGCF1 is indicated by arrow. Reprinted with permission of the Proceedings of the National Academy of Sciences USA, 2101 Constitution Ave, NW, Washington DC20418. *DNA protein binding assays from a single sea urchin egg: a high sensitivity capillary electrophoresis method.* Jun Xian et al., 1996, Vol. 93. Reproduced by permission of the publisher via Copyright Clearance Center.

4. Electrophoresis is run at reversed polarity, i.e., the anode at the detector end, at 18 kV and 18°C.
5. Between each run, the capillary is rinsed with 1X TBE buffer for 2 min and then in CE running buffer for 5 min.

3.7. Practical Examples

1. The superior resolution available in the CEMSA system is illustrated in **Fig. 1**. These experiments are carried out with the nuclear extract of 24-h sea urchin embryo. P3A2 is present in 24-h embryo nuclei at about 10^4 mol/nucleus (*10*).
2. The extract is reacted with the wild-type probe and the complexes are analyzed by conventional EMSA and by CEMSA. Both methods reveal the same DNA-protein complexes,

but the conventional method took over 6 h, compared to CEMSA which required 12 min, and in this case consumed about 1000X times less sample than did the EMSA.

3. Conventional EMSA does not resolve the P3A2 complexes clearly because of the presence in 24-h embryo extract of another DNA-binding factor, SpGCF1. SpGCF1 is relatively prevalent and interacts weakly at the CCCC site on the wild-type probe *(16)*. SpGCF1 complexes account for the broad set of bands that extend below the P3A2 complexes in the EMSA shown in lane I of **Fig. 1A** (compare the complexes formed with recombinant P3A2 in lane II).

4. In the absence of the EMSA "caging effect," these complexes do not survive in the CEMSA experiment and are not observed with the nuclear extract shown in **Fig. 1B**. Furthermore, in the CEMSA experiment, the bimolecular and the trimolecular complexes are widely separated, whereas in this particular EMSA system they are much more difficult to separate.

5. The CEMSA peaks obtained in the nuclear extract are also efficiently competed by excess unlabeled wild-type probe, and the inhibiting monoclonal antibody again eliminates both complex peaks when added to the nuclear extract.

4. Notes

1. All buffers used in this protocol should be filtered and sterilized with 0.2-μm nylon filter and CE running buffer needs to be degassed before use.

2. The fluorescent label should be designed at least 5 bp away from the binding site to avoid the quenching effect following protein binding. If a large amount of DNA has been applied (this is especially true for weak DNA-protein interactions), or a larger restriction DNA fragment has been used we might directly monitor the DNA by UV detection at 260 nm.

3. For a double-stranded DNA with size of 120 bp, 10 bp/1 dye, 20 bp/1 dye, or 30 bp/1 dye ratio has the correlation of 1 DNA molecule/12 dye molecules, 1 DNA molecule/6 dye molecules, or 1 DNA molecule/4 dye molecules, respectively.

4. Because dye intercalation into DNA is relatively weak compared to specific DNA-protein interactions and DNA probe is always in excess amount in reaction, the specific DNA-protein interaction will compete away the dye intercalate at the binding site. Therefore, the intercalating dye does not interfere with the specific protein binding. The huge excess amounts of competitor DNA present in the reaction mixture prevent the addition of the fluorescent dye to the reaction or to the CE running buffer.

5. The selection of reaction buffer, pH and ionic strength is based on the properties of the protein. One of the commonly used buffers is HEPES. Other buffering systems, such as Tris-HCl, MOPS, and phosphates are also used as reaction buffers.

6. Regardless of the extraction method used or the starting tissue or cell source, the protein extract is usually dissolved in a high salt buffer. It is important to carefully control the amount of extract added or, if a titration is done with extract, to add the same amount of salt to each sample. If the salt concentration is allowed to vary, estimates of DNA-protein binding will not be reproducible.

7. The selection of nonspecific competitors might depend on the sequence of the probe and the source of the target protein. It is often found that one competitor is better than the others in certain DNA/protein pairs.

8. The amount of competitor DNA for each system can be easily determined by CEMSA. If binding to the target DNA is not detected, decrease the amount of competitor. If free target DNA is detected, increase the competitor amount. Finally, small change in the amount of competitor will not greatly affect the peak ratios for both free DNA probe and for its complex with protein.

9. The reaction time needed for each system is easily tested by CEMSA. We can apply an aliquot of sample onto CE after equilibration interval, t, and another aliquot later after equilibration time much greater than t. The absence of detectable change in the peak areas of complex and free DNA with prolonged equilibration is an indication that the interval t may be sufficient for the attainment of equilibrium under those reaction conditions.

10. For weak bindings or fast off (k_{off}) kinetics, the complex is a front running peak. Calculation should be careful when measuring the peak areas.

11. Other polymers can be used for this purpose, such as methyl cellulose (MC), poly(ethylene glycol) (PEG), and so on. If the capillary column is coated with particular other polymer, that polymer will be the first choice as the additive into the running buffer. The size of the polymer also affects the CE separation and should be considered.

12. Pressure injection (or hydrodynamic injection) usually will not change the composition of the reaction mixture. We should avoid the electrokinetic injection in this application, because it changes the ratio of free probe to its protein complex.

Acknowledgments

The experiments for this chapter were performed at the California Institute of Technology while working with Eric H. Davidson. I am grateful for his support and guidance that made this research successful. I would also like to acknowledge Charles Romano and Jennifer Paturzo for their useful discussions and assistance in preparing this chapter.

References

1. Fried, M. G. and Crothers, D. M. (1981) Equilibria and kinetics of *lac* repressor-operator interactions by polyacrylamide gel electrophoresis. *Nucleic Acids Res.* **9,** 6505–6525.

2. Garner, M. M. and Revzin, A. (1981) A gel electrophoresis method for quantifying the binding of proteins to specific DNA regions: application to components of the *Escherichia coli* lactose operon system. *Nucleic Acids Res.* **9,** 3047–3060.

3. Fried, M. G. and Garner, M. M. (1998) The electrophoretic mobility shift assay (EMSA), in *Nucleic Acid Electrophoresis*, (Tietz, D., ed.). Springer Press, Berlin, Heidelberg, and New York, pp. 239–271.

4. Xian, J., Harrington, M. G., and Davidson, E. H. (1996) DNA-protein binding assays from a single sea urchin eggs: a highly sensitive capillary electrophoresis method. *Proc. Natl. Acad. Sci. USA* **93,** 86–90.

5. Stebbins, M. A., Hoyt, A. M., Sepaniak, M. J., and Hurlburt, B. K. (1996) Design and optimization of a capillary electrophoretic mobility shift assay involving trp repressor-DNA complexes. *J. Chromatogr. B* **683,** 77–84.

6. Xue, B., Gabrielsen, O. S., and Myrset, A. H. (1997) Capillary electrophoretic mobility shift assay (CEMSA) of a DNA-protein complex. *J. Capillary Electrophor.* **4,** 225–231.

7. Foulds, G. J. and Etzkorn, F. A. (2001) Protein-DNA binding affinities by capillary electrophoresis, in *Capillary Electrophoresis of Nucleic Acids*, Vol. 2 (Mitchelson, K. R. and Cheng, J., eds.), Humana Press, Totowa, NJ, pp. 369–378.

8. Foulds, G. J. and Etzkorn, F. A. (1998) A capillary electrophoresis mobility shift assay for DNA-protein binding affinities free in solution. *Nucleic Acids Res.* **26,** 4304–4305.

9. Chu, Y.-H., Avila, L. Z., Biebuyck, H. A., and Whitesides, G. M. (1992) Use of affinity capillary electrophoresis to measure binding constants of ligands to proteins. *J. Med. Chem.* **35,** 2915–2917.

10. Zeller, R. W., Britten, R. J., and Davidson, E. H. (1995) Developmental utilization of Spp3A1 and Spp3A2: two proteins which recognize the same DNA target site in several sea urchin gene regulatory regions. *Dev. Biol.* **170**, 75–82.

11. Dent, C. L. and Latchman, D. S. (1993) The DNA mobility shift assay, in *Transcription Factors* (Latchman, D. S., ed.), Oxford University Press, Oxford and New York, pp. 1–26.

12. Calzone, F. J., Thézé, N., Thiebaud, P., Hill, R. L., Britten, R. J., and Davidson, E. H. (1988) Developmental appearance of factors that bind specifically to *cis*-regulatory sequences of a gene expressed in the sea urchin embryo. *Genes Dev.* **2**, 1074–1088.

13. Calzone, F. J., Höög, C., Teplow, D. B., Cutting, A. E., Zeller, R. W., Britten, R. J., and Davidson, E. H. (1991) Gene regulatory factors of the sea urchin embryo 1. Purification by affinity chromatography and cloning of P3A2, a novel DNA-binding protein. *Development* **112**, 335–350.

14. von Hippel, P. H. (1979) On the molecular bases of the specificity of interaction of transcriptional proteins with genome DNA in *Biological Regulation and Development*. (Goldberger, R. F., ed.), Plenum Publishing, New York, pp. 279–347.

15. Zimmerman, S. B. and Minton, A. (1993) Macromolecular crowding: biochemical, biophysical, and physiological consequences. *Annu. Rev. Biophys. Biomol. Struct.* **22**, 27–65.

16. Zeller, R. W., Coffman, J. A., Harrington, M. G., Britten, R. J., and Davidson, E. H. (1995) SpGCF1, a sea-urchin embryo DNA-binding protein, exists as 5 nested variants encoded by a single mRNA. *Dev. Biol.* **169**, 713–727.

30

Protein-DNA Binding Affinities by Capillary Electrophoresis

Glenn J. Foulds and Felicia A. Etzkorn

1. Introduction

Described in this chapter is a capillary electrophoresis (CE) method for the accurate determination of protein-DNA binding constants *(1)*. The advantages of the method include the fast separation and quantitation of protein-DNA complex and free DNA, in uncoated capillaries without sieving matrices, coupled with LIF detection of fluorescently-labeled DNA at nM to pM concentrations.

Mass-to-charge ratio is the basis for separation of analytes by CE. Migration times depend on the applied electric field, their electrophoretic mobility, and the electroosmotic flow (EOF). Our method takes particular advantage of the electroosmotic flow generated by the negatively charged silanol groups on the capillary at pH >6.0 *(2)*. EOF is a plug-like flow of the run buffer, which enables all species, charged and neutral, to migrate in the same direction along a fused silica capillary *(3)*.

1.1. Capillary Electrophoresis Mobility Shift Assay

Our CE mobility shift assay (CEMSA) *(1)* uses the reverse polarity of standard EMSA *(4)*, and the order of elution is: free protein, protein-DNA complex, then DNA. All three species can be observed simultaneously with UV detection. However, laser induced fluorescence (LIF) detection of fluorescently-labeled DNA is far more sensitive (μM to pM) due to higher emission intensity and low fluorescent background, enabling quantitation of biologically relevant DNA concentrations.

CE has found many applications in the analysis of biomolecules *(5)*. A few previous examples of application to protein-DNA interactions exist *(6–11)*. Affinity CE has been demonstrated as a useful and sensitive method for measuring both the binding stoichiometries and equilibrium and kinetic constants of ligands binding to receptors *(12)*. Protein-DNA interactions have traditionally been studied by gel EMSA, however CEMSA has the advantages of automatic sampling, high resolution, fast assay times, real-time detection, and small sample size.

From: *Methods in Molecular Biology, Vol. 163:*
Capillary Electrophoresis of Nucleic Acids, Vol. 2: Practical Applications of Capillary Electrophoresis
Edited by: K. R. Mitchelson and J. Cheng © Humana Press Inc., Totowa, NJ

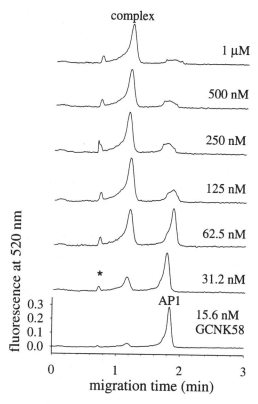

Fig. 1. Electropherograms of 2 s injections of mixtures of AP1 at 1 n*M* in 133 µg/mL poly(dA-dT) · (dA-dT), 50 m*M* KCl, 100 µg/mL BSA, 5% glycerol, 50 m*M* MOPS/Et$_3$N, pH 7.5 and GCNK58 at 15.6 n*M*, 31.2 n*M*, 62.5 n*M*, 125 n*M*, 250 n*M*, 500 n*M*, and 1 µ*M*. Reprinted from *Nucleic Acids Res.* **26,** 4304,4305, Copyright (1998), with permission from Oxford University Press.

Protein-DNA interactions are important in the investigation of the mechanisms of gene expression. At the molecular level, complex assemblies of transcription factors and coactivators bind to specific DNA sequences to regulate transcription. In order to elucidate the mechanisms involved it is important to have information about the relative binding strengths of interactions between these biomolecules.

The CEMSA method described below was developed with the DNA-binding protein GCNK58 *(13)* and a complementary 20-bp oligonucleotide incorporating the DNA binding site AP1 (*see* **Figs. 1** and **2**). In order to demonstrate its general applicability, the method was tested successfully with DNA-binding constructs of the RXR *(14)*, E47, and MyoD proteins *(15a)*. A further application is being developed in this lab for the detection and investigation of endocrine disrupters, which are environmental pollutants implicated in fertility disorders, low sperm counts, cancer, and wildlife birth defects *(16)*. The endocrine disrupter assay is designed to probe ligand-dependent bind-

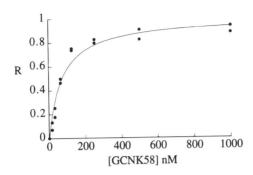

Fig. 2. Saturation (R) of AP1 (1 n*M*) vs concentration of GCNK58. K_d is calculated using Eq. 1 to be 35 ± 4 n*M*. Reprinted from *Nucleic Acids Res.* **26,** 4304,4305, Copyright (1998), with permission from Oxford University Press.

ing of steroid receptor-DNA complexes. CEMSA may provide a fast, effective method for the evaluation of DNA-binding peptidomimetics as well *(17)*.

We envisage that CEMSA can be applied to a variety of protein-DNA systems. Factors to consider in adapting the method to a particular protein-DNA system are described in the **Subheading 4**. The method is performed in uncoated capillaries, with no sieving or caging matrix. This is possible because the on-rate is substantially slower for protein-DNA interactions (10^7 $M^{-1}s^{-1}$) *(18,19)* than for small molecule-protein binding (diffusion-controlled), allowing the complex to remain associated during the separation. For example, a binding affinity of 10 n*M* and diffusion controlled on-rates would give an off-rate of approx 1 s^{-1}. In this example, the complex would dissociate during our 2-min separation (*see* **Note 1** and **2**). It has been observed that sequestration *(20)* and caging *(21)* effects stabilize some protein-DNA complexes during gel electrophoresis *(22)*. Fast separation is possible without sieving gels due to the differences in charge between free DNA and complex.

The main concerns in CEMSA are protein–wall interactions *(23)* and dissociation of the complex (*see* **Note 3**). We suggest that the protein is protected from wall interactions by its association with negatively charged DNA. Proteins with higher molecular weight and pI tend to have increased wall interactions and therefore strategies for decreasing protein adsorption should be utilized (*see* **Fig. 3** and **Notes 4–8**) *(23,24)*. Dissociation of the complex can be observed in the electropherograms under extreme conditions (*see* **Fig. 4** and **Note 6**). The systems we investigated, which showed no significant dissociation, have equilibrium dissociation constants in the n*M* range.

2. Materials

1. CE instrument with LIF detector. Our CEMSA was performed on a Beckman P/ACE® 2100 series instrument equipped with the Beckman Laser Module 488 and LIF detector, excitation λ 488 nm and detection λ 520 nm (*see* **Note 9**). A sample cooling tray allowed the sample mixtures to be maintained at 4°C. The instrument was interfaced with an IBM PC running Beckman System Gold software (version 8.1).

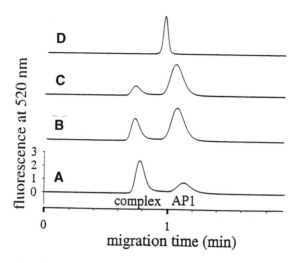

Fig. 3. Protein adsorption onto the capillary walls is investigated by a 60-s injection of 4 μM GCNK58 followed by a 1-min wash with run buffer and 2-min electrophoresis at 30 kV. **(A–C)** subsequent consecutive 1-s injections of 100 nM AP1 DNA with no washes between runs. **(D)** capillary was washed for 2 min with 0.1 M NaOH and 2-min run buffer prior to AP1 injection.

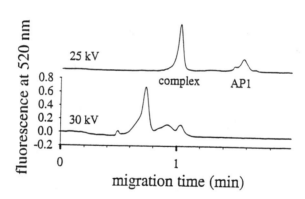

Fig. 4. Dependence of DNA-protein complex integrity on electric field strength. Consecutive runs with 1-s injections of 50 nM AP1 + 200 nM GCNK58 in 100 μg/mL BSA, 5% glycerol, 10 mM MOPS/Et$_3$N, pH 7.5 at 30 kV and 25 kV (27-cm capillary).

2. Fused silica capillary of 75 μm inner diameter (id) and 375 μm outer diameter (od) (Polymicro Technologies), with an outer polyimide coating for strength, of 27 cm total length with the detector at 20 cm on the Beckman P/ACE 2100. The polyimide coating can be removed at the detection window with fuming sulfuric acid at 120°C, instead of burning in a bunsen flame.

3. 10 mM MOPS/Et$_3$N run buffer is prepared by dissolving 1.047 g (5.0 mmol) of 3-(N-morpholino)propanesulfonic acid (MOPS, Sigma) into approx 400 mL deionized distilled water and adjusting the pH to 7.5 with freshly distilled triethylamine (Et$_3$N). The

solution is made up to 500 mL, filtered through a 0.45 μm nylon membrane filter and stored at 4°C in the dark for up to 3 mo (*see* **Note 3**).

4. 0.1 *M* NaOH and deionized water, filtered through a 0.45 μm nylon membrane, for rinsing the capillary.
5. Fluorescently labeled and complementary unlabeled DNA strands for annealing. The fluorescent label should be compatible with the LIF detector (*see* **Subheading 2.**, **item 1**). Target DNA for our protein was synthesized (Biomolecular Research Facility, University of Virginia) following standard phosphoramidite protocol. The fluorescent dye 5'-carboxyfluorescein phosphoramidite was attached to the 5' end of one strand only.
6. After column purification the single-stranded oligonucleotides were vacuum dried, redissolved in 10 m*M* Tris-HCl, 1 mM EDTA buffer (pH 7.4), and stored at –20°C. The stock oligonucleotide concentrations were determined from their UV absorbances at 260 nm (1 A.U. = 30 μg/mL). Fluorescently labeled DNA is light sensitive and was stored in amber Eppendorf tubes.
7. A stock of DNA-binding protein. 145 μ*M* GCNK58 was stored in 50 m*M* ammonium acetate (NH$_4$OAc) (pH 7.0), 2 m*M* DTT at –60°C.

3. Methods
3.1. Sample Preparation (see Note 1)

1. Sense and antisense AP1 DNA, 3 n*M* of 5'-GTACGCACACCTGCTGCCTGA-3' and 2 n*M* of 5'-X-TCAGGCAGCAGGTGTGCGTAC-3' where X is fluorescein, were freshly annealed daily in 50 m*M* MOPS/Et$_3$N (pH 7.5), 100 mM KCl, 266 μg/mL poly(dA-dT) · (dA-dT) at 90°C for 5 min, slowly cooled to room temperature over 3 h, and stored at 4°C.
2. Serial dilutions (2 μ*M*–31.25 n*M*) of freshly thawed GCNK58 stock protein were prepared in a buffer containing 200 μg/mL BSA, 10 % glycerol, and 50 m*M* MOPS/Et$_3$N, pH 7.5.
3. For each mixture, 10 μL of the GCNK58 solution is mixed separately with 10 μL of freshly annealed 2 n*M* AP1 solution. Then this mixture is preincubated in the dark at 4°C for 2–3 h. The Eppendorf tubes used for each mixture are snipped in half with a tool made for this purpose and then placed in sample vial holders, just prior to the injection of the mixture onto the CE.

3.2. Electrophoresis

1. At the beginning of the day, the capillary is rinsed with 0.1 *M* NaOH (30 min), distilled water (20 min) and run buffer (20 min), in order to ensure regeneration of the capillary wall. During this time, allow the laser module to warm up and sample cooling tray to come to temperature. Shorter washes of 0.1 *M* NaOH and run buffer are subsequently used between runs.
2. Utilizing System Gold software, a method is set up which allows automatic operation of the instrument. The capillary is rinsed for 2 min with 0.1 *M* NaOH and 2 min with run buffer. The sample is injected onto the capillary under pressure for 2 s.
3. Electrophoresis is performed at 25 kV (*see* **Note 10**) for 3 min in 10 m*M* MOPS/triethylamine run buffer at pH 7.5 (*see* **Notes 1, 3,** and **4**) at 21°C (*see* **Fig. 1**). Electrophoresis is at reversed polarity to conventional EMSA, with the cathode at the detector end.

3.3. Determination of the Dissociation Binding Constant

1. **Figure 1** shows a set of electropherograms obtained at varying concentrations of GCNK58 and 1 n*M* AP1. Utilizing the Beckman System Gold software, the complex and free DNA peak areas are integrated. The ratio of bound to total DNA, defined as:

$$R = \frac{[protein_2 \cdot DNA]}{[protein_2 \cdot DNA] + [DNA]}$$

is calculated for each electropherogram based on the peak integrations. The term *[protein2 · DNA]* represents the concentration of the protein dimer/DNA complex.

2. R is plotted against concentration of protein, shown in **Fig. 2**. Kaleidagraph (version 3.0.5) software was used to determine the equilibrium dissociation constant K_d, by a nonlinear least squares fit of the data to the equation:

$$R = \frac{0.5[protein]}{K_d + 0.5[protein]} \qquad (1)$$

in which, R is determined as above and [*protein*], the monomer protein concentration, is measured. For GCNK58 homodimer-AP1 complex formation, K_d was found to be 35 ± 4 n*M*. **Eq. 1** is derived from the expression for the equilibrium dissociation constant (*see* **Note 11–13**).

4. Notes

1. The sample buffer components can have a large effect on the binding affinities of protein-DNA complexes. The sample buffer for our GCNK58-AP1 CEMSA was chosen in order to imitate closely the sample buffer used in the literature EMSA study. This enabled useful comparisons to be made between the binding affinities obtained by the two methods. However, different sample mixtures from those described may be utilized, depending upon the chosen system. High ionic strength sample buffer, relative to the run buffer, can lead to band broadening. In our separations, the resolution of complex from free DNA is large enough to tolerate some peak broadening. Ideally, to tolerate a higher ionic strength sample, the run buffer ionic strength should be as high as possible.

2. Protein-DNA complexes with relatively weaker binding affinities and faster off rates will be more susceptible to dissociation during electrophoresis. The alternative for weak binding affinities or fast off-rates is capillary zone electrophoresis or affinity capillary electrophoresis, in which the retention time of one analyte changes as a function of concentration of the other analyte dissolved in the buffer *(12)*.

3. The disadvantages of using a high concentration run buffer include higher currents and Joule heating. Zwitterionic buffers can increase conductivity without high currents. Joule heating can be alleviated by the use of narrower 50-μm id capillaries. The larger surface area to volume ratio more effectively dissipates the heat generated by high currents. The Beckman P/ACE 2100 can also cool the capillary down to 15°C to help remove Joule heat.

4. In order to achieve reproducible runs and results, it is important to regenerate the fused-silica wall at the beginning of the day and between runs with a suitable washing protocol. After a number of days the silica wall may degenerate, as observed by peaks shifting to longer migration times (a change in EOF) and loss of resolution and peak shape. If this is not rectified with long washes for 30 min with 0.1 *M* NaOH and deionized water, then the capillary should be replaced.

5. Protein adsorption to the wall of the fused silica capillary is a common problem in CE *(23)*. The effect of any protein-wall interactions on the GCNK58-AP1 equilibrium and the amount of complex observed is investigated. Removal of adsorbed ions and regeneration of the charged capillary wall was usually achieved by rinsing for 2 min with 0.1 *M* NaOH and 2 min with run buffer between runs. Six consecutive injections of the same

GCNK58-AP1 sample mixture were run without washing the capillary between electrophoresis. Although some resolution was lost and changes in EOF were observed, the amount of complex relative to total DNA (23.7 ± 0.8 %) did not change systematically during these runs without interim washing. This result indicates that binding affinity is not affected by adsorption of protein onto the capillary wall. We believe that the protein bound by the negatively charged DNA is prevented from adsorbing to the wall.

6. A large injection of free protein (60 s of 4 μM GCNK58, followed by rinsing with run buffer) did lead to wall-adsorbed protein, which was subsequently observed by injections of AP1 DNA followed by electrophoresis. Protein desorbed from the wall was observed as complex during consecutive runs A–C without interim washing (*see* **Fig. 3**). A fourth consecutive electrophoresis run D, with prerun washing of 2 min 0.1 M NaOH and 2 min run buffer, showed no protein desorption to form complex, only free DNA. This indicates that our washing cycle was effective in cleaning and regenerating the capillary wall.

7. In adapting this method to a different protein-DNA system, protein adsorption needs to be investigated in order to ensure reproducible results. Protein adsorption can be correlated to a protein's pI and molecular weight *(24)*. Electrostatic binding of proteins to the walls has been observed when operating at a pH below the pI of the analyte. Altering the run buffer pH can help to reduce wall interactions, but this approach is limited to the physiological range required for protein-DNA interactions. Based on sequence, the pI of GCNK58 is expected to be higher than the pH 7.5 of the run buffer, however, protein adsorption was shown not to affect the ratio of complex to DNA obtained in this system. Proteins having a molecular weight above 50 kDa are reported to have a stronger tendency to adsorb to both coated and uncoated capillaries *(24)*.

8. Strategies to minimize protein-wall interactions include the use of zwitterionic buffers (*see* **Note 2**), buffer additives, and wall coatings. Various coatings have been employed in CE to prevent analyte-wall interactions *(25)*. Many examples of covalently bound coatings have been used, however a more practical approach is to use run buffer additives which act as a dynamic coating, shielding the wall from the analyte. Methyl cellulose and other derivatives have been used in uncoated capillaries to reduce wall interactions but still retain some EOF. The use of a dynamic coating also enables simple washes to rinse and regenerate the capillary wall. For strongly adsorbing proteins, a thicker covalently bound coating may be necessary. A number of capillaries with a range of hydrophobic and hydrophilic covalent coatings are commercially available *(25)*. The effect of a coating on the EOF must be considered. Complete elimination of EOF will necessitate that the voltage polarity be reversed (negative to positive) so that the negatively charged DNA, the detected species, will migrate toward the detector.

9. The sensitivity of LIF detection compares favorably with that of other detection methods and makes the method applicable to the analysis of small biological samples. For example, the Amersham catalog reports that ^{32}P-labeled nucleotides can be detected at 0.01 pg, whereas fluorescein-labeled nucleotides can be detected at 0.05 pg by gel electrophoresis. On the Beckman P/ACE 2100, a 2-s (25 nL) injection of 1 nM fluorescein-labeled AP1 contains 0.03 pg. Longer sample injections of 10 s or more can be performed in order to increase the signal-to-noise ratio. The minimum detection limit of the method was investigated by running dilutions of fluorescently-labeled AP1 from 1 μM to 50 pM. A 10-s injection of 50 pM AP1 gave a signal-to-noise ratio greater than 2. Sample loading techniques such as longer injection times and stacking could be applied to give better signal-to-noise ratios at levels below 50 pM. A ratio of electrophoresis buffer to sample buffer ionic strength (*see* **Note 1**) of approx 10 allows sample stacking. Stacking is the

concentration of a large volume injection of sample into a sharp band, owing to the differences in field strength between the sample plug and run buffer. More of the sample can be loaded in this way to increase sensitivity.

10. The field strength voltage was optimized to keep the complex intact and maintain fast migration times. A series of injections of the same GCNK58-AP1 sample mixture were run in a 27-cm capillary at various applied voltages from 30 kV to 5 kV. The ratio of complex peak area to total DNA (R, **Eq. 1**) was found to be independent of the applied voltage at 25 kV and below. At 30 kV, the R dropped from 0.85 to approx 0.70 with broad, intermediate peaks that indicated complex dissociation (*see* **Fig. 4**). In sample mixtures containing smaller relative amounts of complex, this degradation was less significant. Therefore, 25 kV was used as the standard applied voltage.

11. **Eq. 1** was derived by substituting:

$$\frac{[DNA]}{[protein_2 \cdot DNA]} = \frac{(1-R)}{R}$$

into the equation for K_d:

$$K_d = \frac{[protein_2][DNA]}{[protein_2 \cdot DNA]}$$

12. Cooperativity has been observed as S-shaped binding curves in plots of R vs concentration of protein. Such data needs to be fit to the related **Eq. 2**, describing the cooperative binding of dimeric proteins to DNA. The method of preparation of our protein DNA mixtures affected whether cooperativity was observed or not. GCNK58 dilutions prepared in the absence of 100 µg/mL bovine serum albumin (BSA) and 5% glycerol gave strongly S-shaped curves. This suggested an instability of the GCNK58 folded homodimer at low concentrations in the absence of the stabilizing additives BSA and glycerol. In the absence of AP1, folding and dimerization have been observed previously for GCNK58 by ^{1}H NMR and circular dichroism through the change from helical homodimer at 155 mM to unfolded monomer at concentrations of less than 2 µM *(26)*. The cooperative binding of dimeric proteins to DNA have been previously observed *(15,27,28)*. Cooperativity is not normally accounted for in the determination of DNA-binding affinities due to the inherent difficulty of separating the protein dimerization constant from the protein-DNA dissociation constant.

13. **Eq. 2** incorporates protein dimerization to account for S-shaped cooperativity of binding to DNA. It was derived from **Eq. 1** and the equilibrium constant for protein dimer dissociation:

$$R = \frac{K_{dimer}^{-1}[protein]^2}{K_d + K_{dimer}^{-1}[protein]^2} \tag{2}$$

where $K_{dimer} = \dfrac{[protein]^2}{[protein_2]}$

Other equations have also been used for the determination of binding affinities *(15)*.

Acknowledgments

This work was supported by New Zealand Science and Technology Postdoctoral Fellowship UOV701 to GJF and by NIH Grant GM52516-01 to FAE.

References

1. Foulds, G. J. and Etzkorn, F. A. (1998) A capillary electrophoresis mobility shift assay for protein-DNA binding affinities free in solution. *Nucleic Acids Res.* **26**, 4304–4305.
2. Grossman, P. D. and Colburn, J. C. (1992) *Capillary electrophoresis: Theory and Practice*, Academic Press, San Diego.
3. Rice, G. L. and Whitehead, R. (1965) Electrokinetic flow in a narrow cylindrical capillary. *J. Phys. Chem.* **69**, 4017–4024.
4. Garner, M. M., and Revzin, A. (1981) A gel electrophoresis method for quantifying the binding of proteins to specific DNA regions: application to components of the *Escherichia coli* lactose operon regulatory system. *Nucleic Acids Res.* **9**, 3047–3060.
5. Karger, B. L., Foret, F., and Berka, J. (1996). Capillary electrophoresis with polymer matrices: DNA and protein separation and analysis. *Methods Enzymol.* **271**, 293–319.
6. Heegaard, N. H. H. and Robey, F. A. (1993) Use of capillary zone electrophoresis for the analysis of DNA-binding to a peptide derived from amyloid P component. *J. Liq. Chromatogr.* **16**, 1923–1939.
7. Janini, G. M., Fisher, R. J., Henderson, L. E., and Issaq, H. J. (1995) Application of capillary zone electrophoresis for the analysis of proteins, protein-small molecules, and protein-DNA interactions. *J. Liq. Chromatogr.* **18**, 3617–3628.
8. Stebbins, M. A., Hoyt, A. M., Sepaniak, M. J., and Hurlburt, B. K. (1996) Design and optimization of a capillary electrophoretic mobility shift assay involving Trp repressor-DNA complexes. *J. Chromatogr. B* **683**, 77–84.
9. Xian, J., Harrington, M. G., and Davidson, E. H. (1996) DNA-protein binding assays from a single sea urchin egg: a high-sensitivity capillary electrophoresis method. *Proc. Natl. Acad. Sci. USA* **93**, 86–90.
10. Xian, J. (2001) Capillary DNA-protein mobility shift assay, in *Capillary Electrophoresis of Nucleic Acids*, Vol. 2 (Mitchelson, K. R., and Cheng, J., eds.), Humana Press, Totowa, NJ, pp. 355–367.
11. Xue, B., Gabrielsen, O., and Myrset, A. (1997) Capil. electrophoretic mobility shift assay (CEMSA) of a protein-DNA complex. *J. Capillary Electrophor.* **4**, 225–231.
12. Chu, Y. H., Avila, L. Z., Gao, J., and Whitesides, G. M. (1995) Affinity capillary electrophoresis. *Acc. Chem. Res.* **28**, 461–468.
13. Ellenberger, T. E., Brandl, C. J., Struhl, K., and Harrison, S. C. (1992) The GCN4 basic region leucine zipper binds DNA as a dimer of uninterrupted α-helices: Crystal structure of the protein-DNA complex. *Cell* **71**, 1223–1237.
14. Rastinejad, F., Perlmann, T., Evans, R. M., and Sigler, P. B. (1995) Structural determinants of nuclear receptor assembly on DNA direct repeats. *Nature* **375**, 203–211.
15. Wendt, H., Thomas, R. M., and Ellenberger, T. (1998) DNA-mediated folding and assembly of MyoD-E47 hetero-dimers. *J. Biol. Chem.* **273**, 5735–5743.
15a. Foulds, G. J. Etzkorn, F. A. (1999) DNA-binding affinities of MyoD and E47 homo- and hetero-dimers by capillary electrophoresis mobility shift assay. *J. Chromatogr. A* **862**, 231–236.
16. Colborn, T., Dumanoski, D., and Myers, J. P. (1996) *Our Stolen Future: How We Are Threatening our Fertility, Intelligence, and Survival—A Scientific Detective Story*, Dutton, New York.
17. Travins, J. M. and Etzkorn, F. A. (1997) Design and enantioselective synthesis of a peptidomimetic of the turn in the Helix-Turn-Helix DNA-binding protein motif. *J. Org. Chem.* **62**, 8387–8393.
18. Crute, B. E., Lewis, A. F., Wu, Z., Bushweller, J. H., and Speck, N. A. (1996) Biochemical and biophysical properties of the core-binding factor $\alpha 2$ (AML1) DNA-binding domain. *J. Biol. Chem.* **271**, 251–260.

19. Hart, D. J., Speight, R. E., Cooper, M. A., Sutherland, J. D., and Blackburn, J. M. (1999) The salt dependence of DNA recognition by NF-kB p50: a detailed kinetic analysis of the effects on affinity and specificity. *Nucleic Acids Res.* **27**, 1063–1069.

20. Fried, M. G. and Liu, G. (1994) Molecular sequestration stabilizes CAP-DNA complexes during polyacrylamide gel electrophoresis. *Nucleic Acids Res.* **22**, 5054–5059.

21. Fried, M. G. and Crothers, D. M. (1981) Equilibria and kinetics of lac repressor-operator interactions by polyacrylamide gel electrophoresis. *Nucleic Acids Res.* **9**, 6505–6525.

22. Lane, D., Prentki, P., and Chandler, M. (1992) Use of gel retardation to analyze protein-nucleic acid interactions. *Microbiol. Rev.* **56**, 509–528.

23. Lee, K.-J. and Heo, G. S. (1991) Free solution capillary electrophoresis of proteins using untreated fused-silica capillaries. *J. Chromatogr.* **559**, 317–324.

24. Cordova, E., Gao, J., and Whitesides, G. M. (1997) Noncovalent polycationic coatings for capillaries in capillary electrophoresis of proteins. *Anal. Chem.* **69**, 1370–1379.

25. Wehr, T., Rodriguez-Diaz, R., and Zhu, M. (1999) Capillary electrophoresis of proteins, in *Chromatographic Science*, Vol. 80. Marcel Dekker, New York.

26. Weiss, M. A., Ellenberger, T., Wobbe, C. R., Lee, J. P., Harrison, S. C., and Struhl, K. (1990) Folding transition in the DNA-binding domain of GCN4 on specific binding to DNA. *Nature* **347**, 575–578.

27. Kim, B. and Little, J. W. (1992) Dimerization of a specific DNA-binding protein on the DNA. *Science* **255**, 203–206.

28. Park, C., Campbell, J. L., and Goddard, W. A., III. (1996) Can the monomer of the leucine zipper proteins recognize the dimer binding site without dimerization? *J. Am. Chem. Soc.* **118**, 4235–4239.

31

Ligand Binding to Oligonucleotides

Imad I. Hamdan, Graham G. Skellern, and Roger D. Waigh

1. Introduction

We have been developing a method for the study of small-molecule interactions with DNA, typically with ligands of less than 1000 Dalton. Such interactions are of interest in biochemistry, where cell signaling may involve DNA at various stages, but the main driving force in our studies has been the potential for drug development, particularly in cancer, parasitic diseases, and inflammation.

Drugs binding to DNA fall primarily into two categories of ligand: those ligands which bind in the minor groove of the DNA double helix, and those which slide between the base pairs by "intercalation." These primary binding processes may be followed by covalent bonding. There are examples of small molecules where some of the binding extends to the major groove, but in general, the major groove is too wide to provide a stable binding site for ligands, other than for proteins. Typically, ligands binding in the minor groove are long, relatively thin molecules (**Fig. 1**), which have the capacity to become isohelical with the DNA, to achieve a snug fit in the base of the groove, along the path of the groove. Intercalators are primarily flat (**Fig. 2**), to allow them to slide between the base pairs, although some ligands, such as actinomycin D also have large cyclic peptide units which also confer some sequence selectivity to the site of binding.

In virtually every example of which we are aware, irrespective of the binding mode, the ligand possesses a positive charge that may provide the first attractive force, for interaction with phosphates of the DNA backbone. Thereafter, the binding energy is provided by a combination of weak interactions, the most powerful, in many cases, being hydrophobic. The potential neutralization of DNA negative charge by the positive charge on the ligand provided the theoretical basis for the assumption that capillary electrophoresis (CE) would provide a means for separating DNA from DNA-ligand complexes, since the charge/mass ratio would change on binding.

From: *Methods in Molecular Biology, Vol. 163:*
Capillary Electrophoresis of Nucleic Acids, Vol. 2: Practical Applications of Capillary Electrophoresis
Edited by: K. R. Mitchelson and J. Cheng © Humana Press Inc., Totowa, NJ

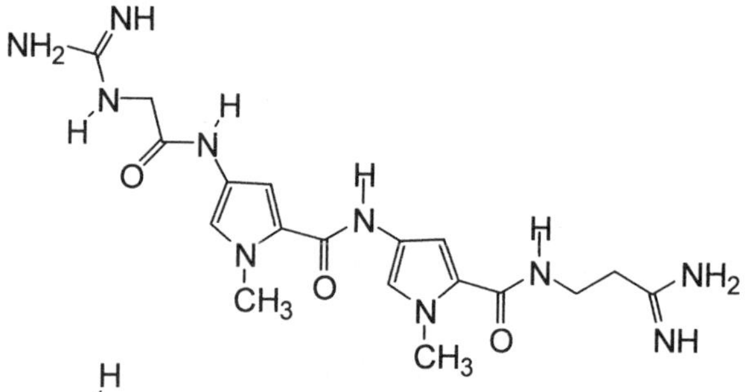

netropsin

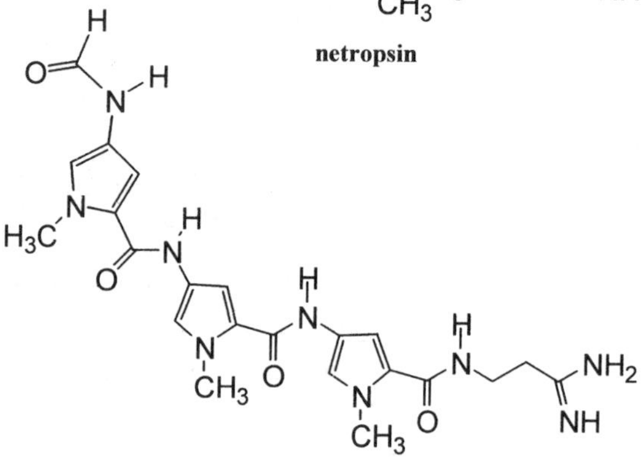

distamycin

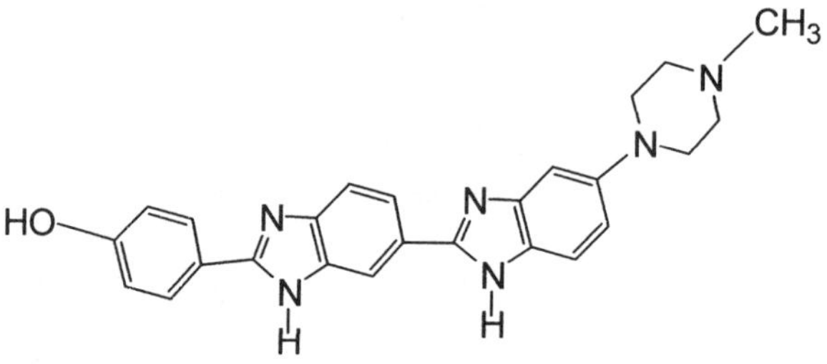

Hoechst 33258

Fig. 1. Chemical structures of some minor groove binding ligands.

1.1. Separation of Oligonucleotides in Free Solution CE

We had expected that in order to carry out competition experiments with ligands it would be necessary to use gel- or polymer-filled capillaries *(1,2)* to separate oligonucleotides with small differences in mass. The intention, from the beginning, was to develop a method to determine sequence-binding preference: This is the basis for all future attempts to reduce the toxicity of DNA-binding drugs. Whereas the separation

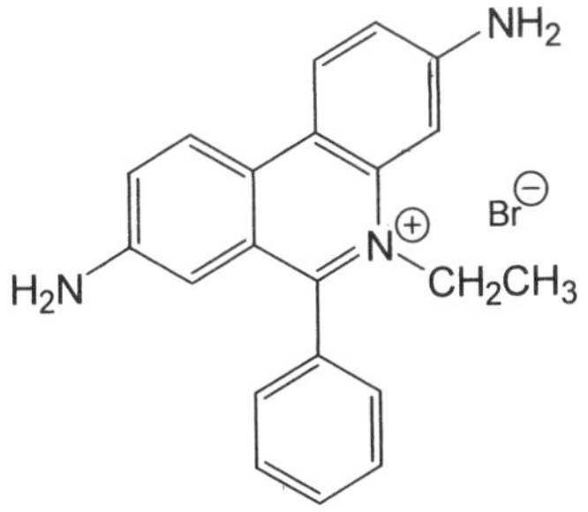

ethidium bromide

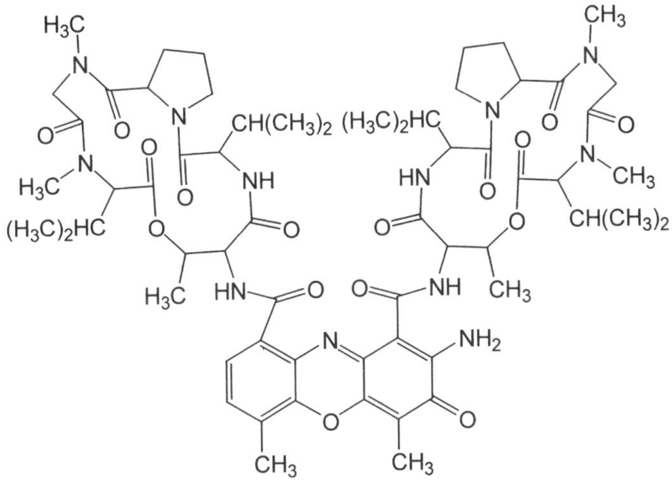

actinomycin D

Fig. 2. Chemical structures of two intercalators.

of DNA fragments, based on length, is well established in gels, the finding that sequence differences would lead to separation in free solution capillary electrophoresis (FSCE) *(3,4)* was unexpected. The mass of an AT base pair is only one unit different from that of a GC pair and the charge is the same. The separation that we have observed must therefore be based on shape differences, typically involving base pair roll, slide, and twist *(5)*. There is insufficient space here for a discussion of the effect of sequence on helix shape, and we are at present unable to make predictions of CE migration time based on sequence. We must content ourselves with the observation that separation can often be achieved in practice and that, even where the simple separation of oligonucleotides of the same length is not possible, there are straightforward means to bring about the desired result (*see* **Notes 1** and **2**). The use of FSCE has a major cost advantage, compared to the use of gel-filled capillaries. Since we began our

Table 1
Volume (mL) of Boric Acid Solution (0.5 *M*) Required
to Be Mixed with Disodium Tetraborate (DSTB) Solution (100 mL)
to Produce Solutions of Specified pH[a]

DSTB (M)	pH 7.5	pH 8.0	pH 8.5	pH 9.0
0.02	71 (0.22)	37 (0.15)	17 (0.09)	2 (0.03)
0.04	84 (0.25)	61 (0.22)	24 (0.13)	7 (0.07)
0.06	131 (0.31)	83 (0.26)	33 (0.17)	13 (0.11)
0.08	180 (0.35)	121 (0.31)	50 (0.22)	20 (0.15)

[a]The final borate concentrations *(M)* are given in brackets.

work, a method has been developed for the determination of binding constants between calf thymus DNA and a series of synthetic tetrapeptides, using FSCE *(6)*.

1.2. Structures of Ligands

We have worked with examples of all the major types of DNA binding ligand, including distamycin, netropsin, Hoechst 33258, which are all minor groove binders, ethidium and actinomycin D, which are base-stack intercalators. We have also studied a variety of "in house" molecules, from our own and from the research programs of other investigators. The chemical structures of the known compounds are given in **Figs. 1** and **2**.

1.3. Oligonucleotide Sequences

The first sequences that we found to be separable *(3)* were $(AT)_{12}$ and $(GC)_{12}$ (all sequences are here written as starting at the 5' end). These two 24-mers are self comple-mentary and both give rise to multiple forms of double helix, presumed to be hairpins as well as full-length duplexes. To avoid such ambiguity, we first used the analogous 12-mers AAATTATATTAT and GGGCCGCGCCGC, which separated very well, as duplexes *(4)*. A major step forward was the finding that the 12 mers CGCAAATTACGC and CGCTATTATCGC would also separate, in the same simple system of 0.22 *M* total borate concentration (TBC) at pH 7.5 *(see* **Table 1** for buffer details). The latter two 12-mers differ only in the base sequence and not in composition. We have since found (unpublished observations) that octamers of the general formula CGXXXXCG, where XXXX contains two As and two Ts, will separate into two groups. The group with shorter migration times (~3.2 min) comprised CGAATTCG, CGTAATCG, and CGATATCG, whereas CGTTAACG and CGTATACG migrated ~0.2 min more slowly than the former oligonucleotides.

1.4. The Desirability of Competition Experiments

The ultimate goal of our work, and that of many others in the area of chemotherapy, is to identify molecules that interfere with DNA processing, particularly with defec-tive DNA processing. The final and most important experiment is a competition for the ligand between diseased and normal cells, i.e., between the desired binding site and other sites on DNA where binding results in unwanted biological effects. If we are to find molecules that show the ability to discriminate finely between similar sites, competition experiments are most likely to replicate the in vivo situation for binding site discrimination. In addition, such test protocols have the characteristic of always possessing internal measures of experimental reliability: The oligonucleotide that does not complex with the ligand acts as an internal standard. If further insight is desired into the binding process, titration experiments using a single oligonucleotide with increasing proportions of ligand can show, for example, cooperativity of binding and can give an estimate of the number of binding sites *(4)*.

2. Materials

2.1. Column Specifications

1. Untreated fused silica capillaries are available from Composite Metal Ltd, UK. Our capillaries are an id 50 µm and an od of 375 µm, with a total length of 40 cm and with an effective length of 32 cm. These capillaries are coated in transparent material, and signal may, in principle, be detected at any point along the capillary.

2.2. Buffers and Additives

1. Disodium tetraborate (Sigma Chemical, St. Louis, MO) solution (0.08 M) is prepared by dissolving sodium tetraborate decahydrate (30.5 g) in distilled water (800 mL) while heating to 50°C. After cooling to room temperature, the volume is adjusted to 1000 mL with distilled water.
2. This solution is diluted as appropriate to give solutions of lower molarity, if required. Boric acid (Sigma Chemical, St. Louis, MO) solution (0.5 M) is prepared in a similar way, using 30.9 g of boric acid in a final volume of 1000 mL.
3. As described in **Table 1**, the two solutions are mixed to provide solutions of varying pH. All final solutions are filtered through 0.2-µm pore filters (Whatman International Ltd, England) (*see* **Notes 3–5**).
4. It is not usually necessary to use an internal standard in the competition experiments, since the noncomplexing oligonucleotide acts as an *ad hoc* standard. If it is felt necessary to achieve some additional measure of standardization, methyl orange (BDH Ltd, Poole, England) (0.5 µM) may be used. Generally, it is better to keep the conditions as simple as possible, to avoid any possible effects of the added standard on binding affinity of the ligand under study.

2.3. Nature of DNA-Binding Ligands

1. Actinomycin D, ethidium bromide, distamycin, Hoechst 33258, and netropsin are all available from Sigma Chemical and used without further purification.
2. It should be noted that any DNA-binding compound may potentially be either carcinogenic or mutagenic and should be treated accordingly (*see* **Subheading 3.** for suggested precautions).

2.4. Sources and Storage of Oligonucleotides

1. Oligonucleotides are obtained from Cruachem Ltd, Glasgow, UK, or from Oswel DNA Service, Southampton, UK. All are claimed to be 90–95% pure and are used without further purification.
2. They are stored either as received (vacuum dried), or are dissolved in distilled water, and kept at –20°C until required. We have not had problems with loss of material when stored in this way; the majority of samples are still useable up to at least 6 mo after preparation.

3. Methods

3.1. Capillary Electrophoresis

1. The apparatus used throughout is a TSP-CE1000 CE separation system (ThermoSeparation Products) (*see* **Note 6**). The equipment may be regarded as fairly standard for this purpose, with the capability for recording UV/visible spectra of peaks and for variable wavelength detection in the range 200–350 nm. Automated sample handling is standard.

2. New capillaries are conditioned by washing with sodium hydroxide (0.1 *M*) for 10 min followed by distilled water (10 min) and finally with running buffer (5 min). Comparisons of electroosmotic flow are made by loading a sample of acetone (reagent grade) directly onto the capillary and measuring the migration time: in our system the value is normally in the range 1.2–2.2 min, depending on the composition of the running buffer (*see* **Note 5**).

3. All experiments are carried out at 25 kV, using a capillary temperature of $20 \pm 0.1°C$. All samples are loaded hydrodynamically for 5 s. Before loading each sample, the capillary is washed with 0.1 *M* NaOH for 1 min before flushing with running buffer for 2 min (*see* **Notes 7–12**).

3.2. Sample Preparation

1. Gloves (preferably lightweight, disposable ones) should be worn at all stages when handling DNA-binding compounds. All DNA-binding compounds must be weighed out in an enclosed fine balance, located in a fume hood. Solutions should also be made up in the fume hood and any dilutions carried out before transferring the analytical solution to the equipment. After use, residual solutions and contaminated glassware should be disposed of in the locally approved manner.

2. Pairs of complementary oligonucleotides are obtained in single-strand form (we term as A and B). The concentration of single-stranded oligonucleotide may vary with the scale of synthesis and is specified by the supplier. Since the concentration of strand A and strand B may be different (as is often the case), we routinely confirm the concentration using a Genequant II spectrophotometer. Concentrations are usually in the range 80–170 μ*M*. Calculations are made of the appropriate volumes of the solutions containing A and B that need to be mixed for each experiment.

3. Usually, the NaCl solution is added to the A/B mixture to 0.02 *M* to increase the stability of the DNA duplex; this solution is then ready for binding studies. A further simple calculation is required to give 25 μL of solution containing 10–15 μ*M* duplex with 0.02 *M* NaCl, for injection onto the capillary. In an alternative procedure, the mixture A/B may be diluted with buffer to give the solution of duplex required for injection on the column.

4. We normally use 0.22 *M* total borate buffer at pH 7.0 (*see* **Table 1**). The solutions containing the mixtures of two single-strands are left at room temperature for at least one hour after mixing, prior to injection, to allow equilibration as duplex.

5. For DNA binding studies, the prepared oligonucleotide samples also require addition of various increasing concentrations of the drug ligand. The molar ratio of the drug to the duplex may range from 0:1, 1:1, 2:1 and so on up to 10:1. A relatively high concentration stock solution of the drug is prepared, and concentration confirmed using UV absorption. The stock solution of drug is diluted in a series: 1:1, 1:2, 1:3, 1:4, and so on, such that 5 μL of each of the dilutions contains an appropriate amount of drug to mix with a fixed volume of the duplex DNA solution. The series of drug-DNA mixtures in the required ratio are incubated for 1 h at room temperature prior to analysis by FSCE (*see* **Notes 13–16**).

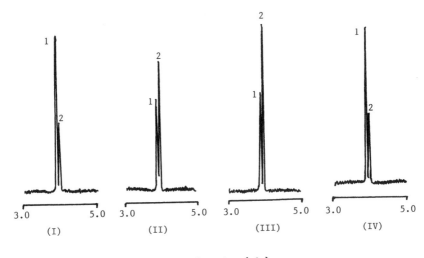

Migration time (min)

Fig. 3. Confirmation of duplex formation with the sequence GGGCCGCGCCGC. The single-strands are not separated and give rise to peak 2, whereas peak 1 is the duplex. Electropherogram (I) is from an equimolar mixture of the two strands in distilled water, in which duplex formation is incomplete even after equilibration. E-gram (II) is spiked with 75 pmol of strand A. Electropherogram (III) is spiked with 150 pmol of strand A and electropherogram (IV) is from the solution used for (III), spiked with 75 pmol of strand B. The migration time of peak 2 is consistent with those of the two single strands.

3.3. Confirmation of Duplex Formation

1. The melting temperature of many short oligonucleotides is below room temperature in distilled water, even though the presence of sodium ions in the running buffer is expected to stabilize the double helix. It is a wise precaution to confirm the formation of duplex by obtaining electropheric evidence of duplex formation (*see* **Note 13**).
2. One method of confirmation is given in **Fig. 3**, where the A/B mixture was spiked separately with an excess of each of the single-strands, which act as mass markers. It should be noted that, almost invariably, the single-strands separate from the duplex, and that as shown in **Fig. 4**, the electrophoretic separation can be enhanced further by increasing the buffer pH above 7.0.

3.4. Effect of pH on the Separation of Duplexes

1. We prefer to work wherever feasible, at pH values close to physiological pH (*see* **Notes 3–5**). **Figure 5** indicates that it is possible to improve separation in many cases by increasing the pH, however the peaks tend to become broader as the pH is increased.

3.5. Choice of Detection Wavelength

1. The standard detection wavelength for DNA solutions is 260 nm, irrespective of whether the bases giving rise to the absorption are AT or GC. At this wavelength, ssDNA as well

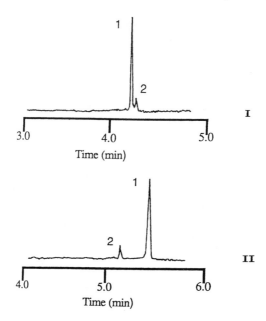

Fig. 4. The effect of buffer pH on the separation of single strands (peak 2) of the sequence GGGCCGCGCCGC from the duplex. Electropherogram (I), 0.22 M total borate concentration (TBC), pH 7.5; (II), TBC 0.3M, pH 8.5. Note that the order of migration is different at the two different pH values. Solutions are equilibrated in buffer and show broader peaks, with less single-stranded form, compared to the electropherograms shown in **Fig. 3**.

as dsDNA will be detected. In a few cases, where annealing is incomplete, electropherograms may be difficult to interpret in the presence of a ligand.

2. However, the bound ligand (particularly a minor groove binder) confers a long-wavelength absorption to the complex (**Fig. 6**). This signature can be used advantageously and only the complex(es) which may be formed are detected, simplifying the electropherogram (**Fig. 7**). This detection mode is also particularly useful where the migration time of the complex is the same as that of the unbound oligonucleotide.

3. An example of the use of a longer wavelength for detection is given in **Fig. 8**. The electropherogram obtained by detecting DNA absorption at 260 nm displays several extraneous peaks arising from ssDNA. However, detection at 315 nm shows only the two complexes, in this case showing that there is no great difference between the octamer CGAAAACG and the dodecamer CGCGAAAACGCG in affinity for netropsin.

4. Notes

1. There are methods for coping with a lack of resolution between putative complexes and free oligonucleotides. Electropherograms (*see* **Subheading 3.**) show that the success of our approach depends on the resolution of the oligonucleotide peaks in the first instance. If the uncomplexed oligonucleotides are not separated, the essential information on binding preference may still be obtained as long as the peaks for the complexes are separated from each other and from that for the pure oligonucleotide.

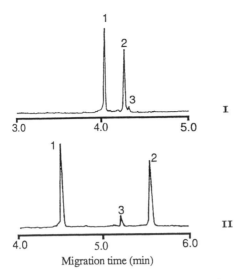

Fig. 5. The effect of pH on the separation of the duplexes GGGCCGCGCCGC (peak 2) and AAATTATATTAT (peak 1). Panels are: (I), 0.22 *M* TBC, pH 7.5; (II), 0.22*M* TBC, pH 8.5. Peak 3 represents traces of ssDNA.

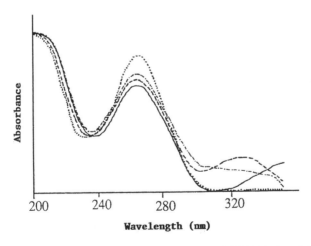

Fig. 6. Normalized UV spectra of free 12-mer AAATTATATTAT (·····) and the same 12-mer complexed with netropsin (-·-·-·-), distamycin (-----), and Hoechst 33258 (——). All are obtained from electrophoresis in buffer at 0.22 *M* TBC, pH 7.5 (*see* **Notes 3** and **4**).

2. If neither the nucleotides nor the complexes are separated, it is necessary to be a little more inventive: the simplest approach that we have found so far is to increase the overall length of one of the oligonucleotides, without altering the putative binding site. Generally, a dodecamer CGCGXXXXCGCG, for example, will migrate more slowly than the

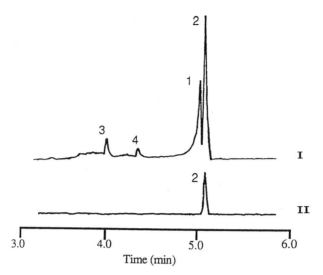

Fig. 7. Electropherograms of the duplex CGCAAATTACGC complexed with netropsin, molar ratio 0.78, in 0.3 *M* TBC, pH 8.0. Electropherogram traces are: (I) UV absorbance at 260 nm; (II) ligand absorbance at 320 nm. Peak 1 is free duplex, peak 2 is the complex, peak 3 and peak 4 are traces of ssDNA.

octamer CGXXXXCG with the same central sequence. So far, we have not found the flanking sequence to have a large effect on the binding affinity, although there is a slight preference for the longer duplex in some cases. With the added flexibility of the increased length of flanking sequence, we have so far been able to carry out competition experiments for any given pair of four-base-pair binding sites.

3. From inception, we wished to select buffer solutions to be optimally effective in the pH range 7–9, and with a preference for the physiological value of about 7.4, in order to maintain the ionization of DNA in its native state. The obvious candidates are phosphate, which has good buffering up to about pH 8, and borate with good buffering at higher pH values.

4. In the event, for reasons which are not understood, good electropherograms were obtained with borate, and very poor results were obtained with phosphate. The latter buffer did not seem compatible with maintenance of adequate voltages, nor with acceptable peak shape. As a result, we have used borate throughout, even at pH 7.5, which is close to the limit of its useful buffering capacity. We have not observed, nor are we aware of any problems arising from the theoretical lack of buffering effect.

5. Variations in run buffer pH have a marked effect on both peak separation and peak shape. Higher pH values often produce better separation of molecules, but at the risk of peak broadening. We prefer to use experimental conditions similar to physiological pH, if necessary at the expense of resolution.

6. Since this chapter was completed, this supplier has ceased selling CE equipment. Any of the alternative suppliers will provide equipment capable of the same function.

7. The origin of the electroosmotic force is the ionization of the silanol groups on the silica surface, which causes the flow of liquid in the capillary under the applied voltage. The resulting surface negative charge results in an affinity for the majority of DNA-binding

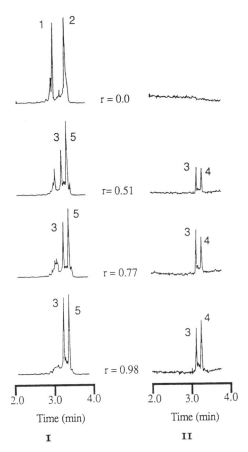

Fig. 8. Electropherograms of an equimolar mixture of CGAAAACG and CGCGAAAACGCG duplexes, complexed with increasing amounts of netropsin. Molar ratios (r) are for netropsin to either nucleotide. Electropherograms are: (I) UV detection of DNA at 260 nm; (II) ligand detection at 315 nm. Peak 1 is free octamer, peak 2 is free dodecamer, 3 is octamer complex, 4 is dodecamer complex, and 5 is unresolved free and bound forms of dodecamer duplex.

ligands that are positively charged, some of which bind very strongly to the capillary wall and may render the column unusable.

8. Many ligands that bind in the minor groove to DNA have a low dissociation constant (typically 10^{-6} or smaller), and/or have a slow dissociation rate. Such ligands usually remain associated with DNA for the duration of the analysis, and there is no problem with binding to the capillary wall.

9. However, for many ligands that intercalate into DNA, such as ethidium, the dissociation constant is larger. Then the dissociation of ligand does occur as the complex passes along the capillary, and eventually there may be an accumulation of silica-bound ligand as successive experiments are carried out. This has a direct effect on migration time, by changing the nature of the capillary wall. The wall-bound ligand also offers to interact

with passing DNA, with the possibility that the DNA will be held back through transient DNA-ligand interactions. The effect is to increase migration time and to cause peak-broadening. It is good practice to inject a drug-free oligonucleotide solution onto the capillary after each sample loaded with the drug, or at least after a few drug-loaded samples have been run, to confirm that the capillary is behaving consistently.

10. The practical solution, if problems are encountered, is to clean the capillary between runs and to avoid undue exposure to ligand; in particular, it is not wise to attempt electrophoresis of the ligand alone. Attempts to construct calibration curves for unbound ligand are generally not successful and may require replacement of the column with a fresh length of capillary.

11. It might be useful to experiment with coated capillaries to limit the interaction of capillary wall with positively charged ligands.

12. Typically, we have found that the minor groove binders form stable complexes under the conditions of the experiment, which survive to reach the detector after passage down the capillary and have distinctive UV spectra. There is a tendency for the DNA intercalating class of ligands to dissociate during passage, resulting in detection of peaks at 260 nm that show no ligand present. For the intercalator-ligands, binding is detected through changes in migration time, and from peak broadening. There is no obvious reason why intercalator-ligands with higher affinity for DNA should not form complexes which would survive to reach the detector.

13. The most uniform peak shapes are observed when samples are prepared in distilled water. Unfortunately, this is sometimes inconsistent with duplex formation, which may or may not occur at all, or may only occur when the sample is exposed to the running buffer on the column. Our preference has been to prepare samples in buffer, or to add sodium chloride to encourage duplex formation. The elevated ionic strength often results in a distortion of peak shape. In particular, it is possible to obtain an electropherogram of two oligonucleotides, where one has regular, Gaussian peak shape and the other is broad or distorted with a trailing edge. The effect is consistent from one run to the next and can be repeated.

14. It has become fairly routine procedure in the FSCE of oligonucleotides to add EDTA to the running buffer *(7)*. EDTA can competitively bind traces of heavy metals that could form coordination complexes with the DNA, and help to give uniform DNA peak shape. In our experience, the effect is variable: sometimes peak shape is considerably improved, but in many cases there is little effect. Although EDTA may not affect ligand binding, our present view is that it is best to keep the system as simple as possible. Subject to the results of future experiments, we may add EDTA routinely, but at present we are leaving it out.

15. Although FSCE will happily cope with lengths of DNA up to several tens of base pairs, competition experiments that test the ability of a ligand to distinguish between various combinations of four or more bases in a sequence may require a large number of oligonucleotides of different sequence. A dodecamer duplex will cost about £48, at UK price of ~£2 sterling for each base. This is a cost implication which has caused us to examine the use of shorter sequences, in particular the use of 5'-CGXXXCG-3', where the central AT region (X = either A or T) is the binding domain. Such duplexes have "melting temperatures" (Tm) in the range 11–13.5°C in distilled water, which increases to 13–17.5°C in 0.22 *M* total borate, pH 7.5 buffer. CE experiments run at 20°C impose a danger that the oligonucleotides will disassociate into single-stranded forms during the run. In practice, this has not been a problem. It is possible to show very clearly that the ssDNA has a different migration time from the duplex in almost every case studied so far, and that the form obtaining during the experiment is the duplex. Electrophoresis of oligonucleotides

in capillaries, or the use of borate buffer, may contribute to the thermal stability of the duplex in a way that is not properly understood.

16. It is reassuring that the results observed, by CE of DNA-ligand complexes, agree well with those obtained by gel electrophoresis footprinting *(8)*. Complex formation in itself appears to stabilize the oligonucleotide duplex and raises the Tm of DNA well above 20°C. Of all the sequences used in our most recent work, the only octamer of the general sequence CGXXXXCG, which would not anneal is CGTTTACG. This sequence also failed to anneal as the dodecamer CGCGTTTACGCG.

Acknowledgment

We thank the Dr. Hadwen Trust for Humane Research, for assistance toward purchase of equipment, Al-Hikma Pharmaceuticals and the Jordanian Government for support of I.I.H.

References

1. Pariat, Y. F., Berka, J., Heiger, D. N., Schmitt, T., Vilenchik, M., Cohen, A. S., Foret, F., and Karger, B. L. (1993) Separation of DNA fragments by capillary electrophoresis using replaceable linear polyacrylamide matrices. *J. Chromatogr. A* **652,** 57–66.

2. Khan, K., Van Schepdael, A., and Hoogmartens, J. (1996) Capillary electrophoresis of oligonucleotides using a replaceable sieving buffer with low viscosity-grade hydroxyethyl cellulose. *J. Chromatogr. A* **742,** 267–274.

3. Hamdan, I. I., Skellern, G. G., and Waigh, R. D. (1998) Separation of pd(GC)$_{12}$ from pd(AT)$_{12}$ by free solution capillary electrophoresis. *J. Chromatogr. A* **806,** 165–168.

4. Hamdan, I. I., Skellern, G. G., and Waigh, R. D. (1998) Use of capillary electrophoresis in the study of ligand-DNA interactions. *Nucleic Acids Res.* **26,** 3053–3058.

5. Dickerson, R. E. (1999) Helix structure and molecular recognition by B-DNA, in *Nucleic Acid Structure* (Neidle, S., ed.), Oxford University Press, Oxford, UK, pp. 145–197.

6. Li, C. and Martin, L. M., (1998) A robust method for determining DNA binding constants using capillary zone electrophoresis. *Anal. Biochem.* **263,** 72–78.

7. Hows, M. E. P., Alfazema, L. N., and Perrett, D. (1997) Capillary electrophoresis buffers: approaches to improving their performance. *LC. GC International* **10,** 656–668.

8. Abu Daya, A., Brown, P. M., and Fox, K. R. (1995) DNA sequence preferences of several AT-selective minor groove binding ligands. *Nucleic Acids Res.* **23,** 3385–3392.

9. Hamdan, I. I., Skellern, G. G., and Waigh, R. D., unpublished observations.

Index